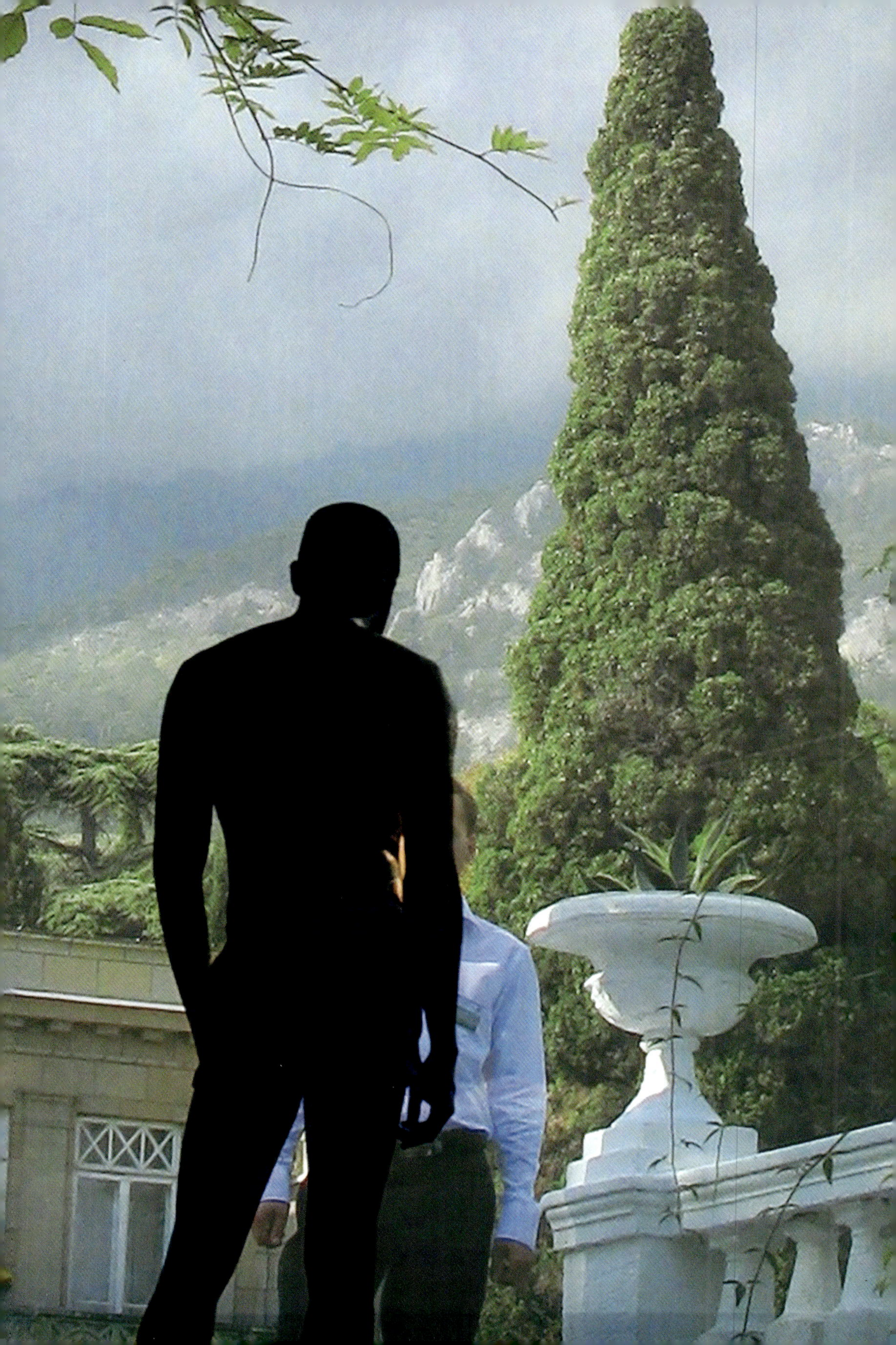

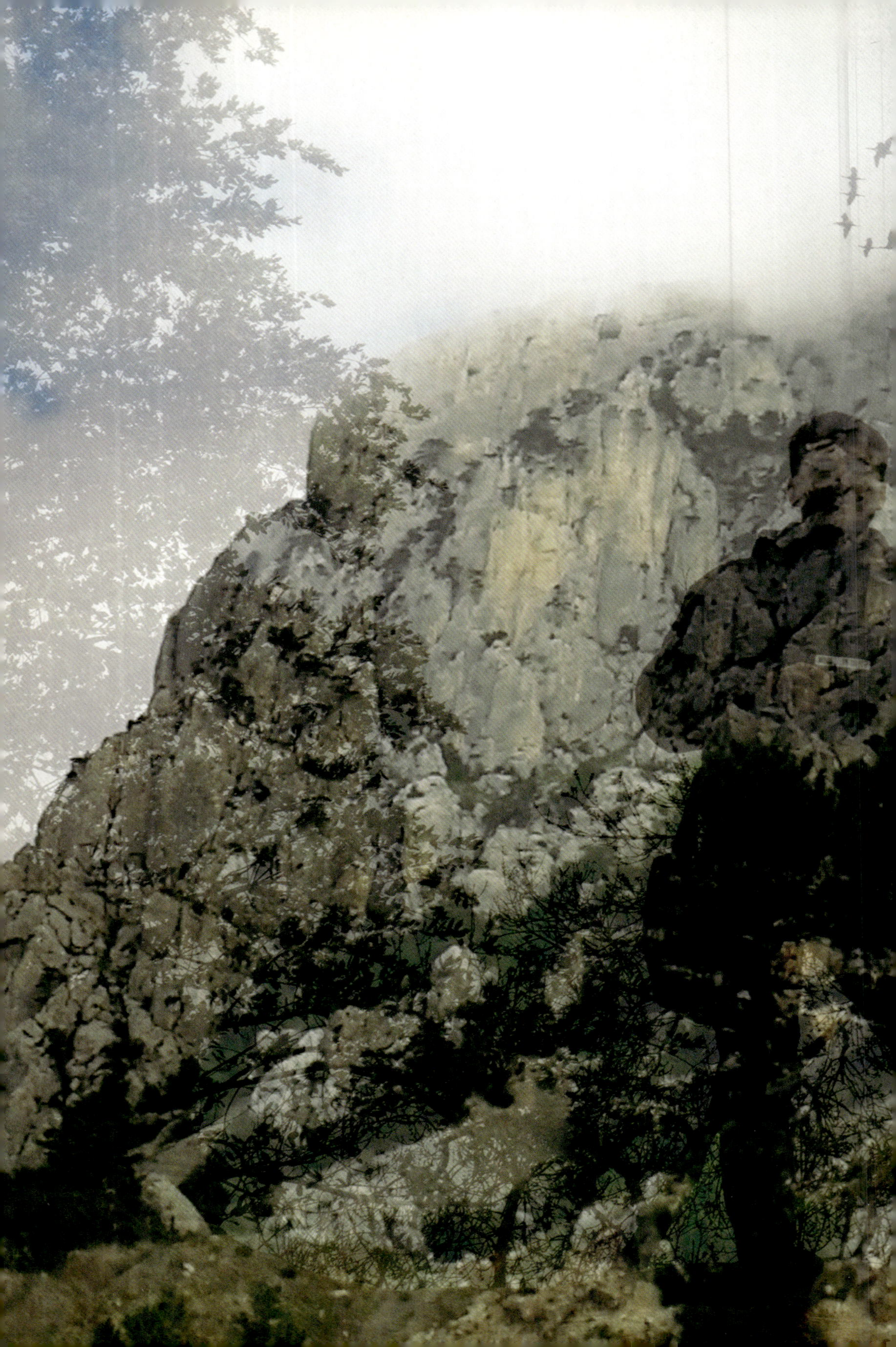

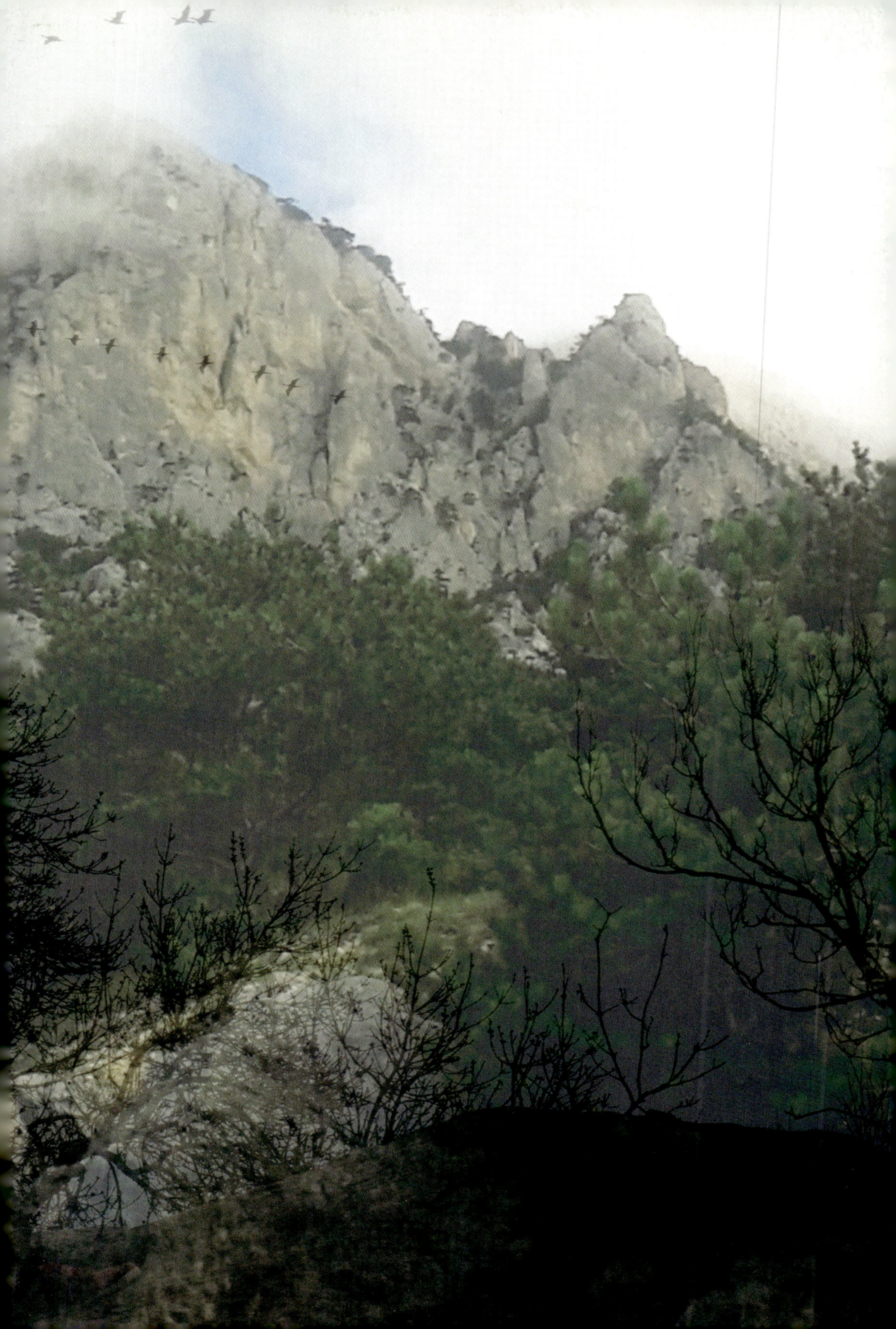

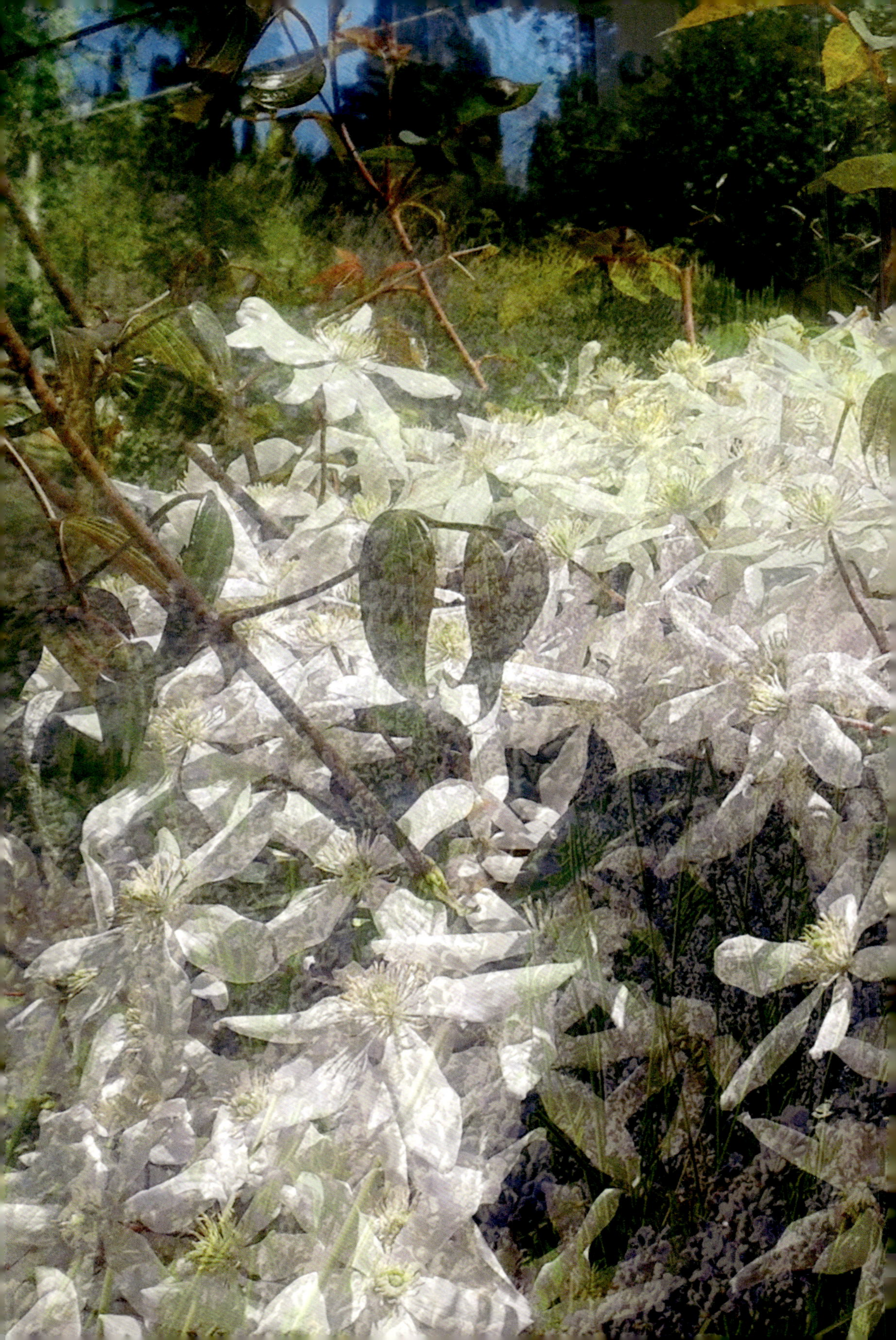

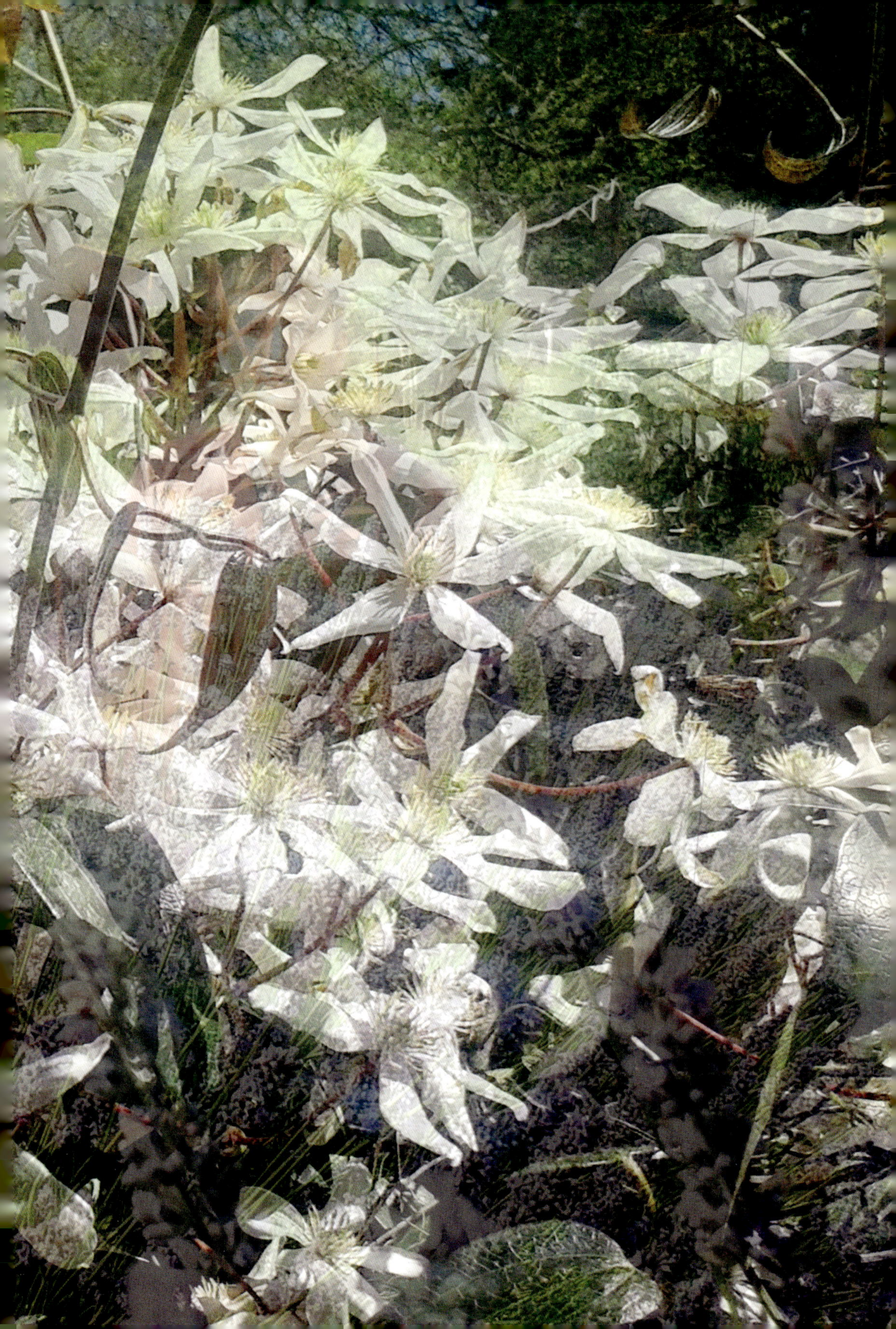

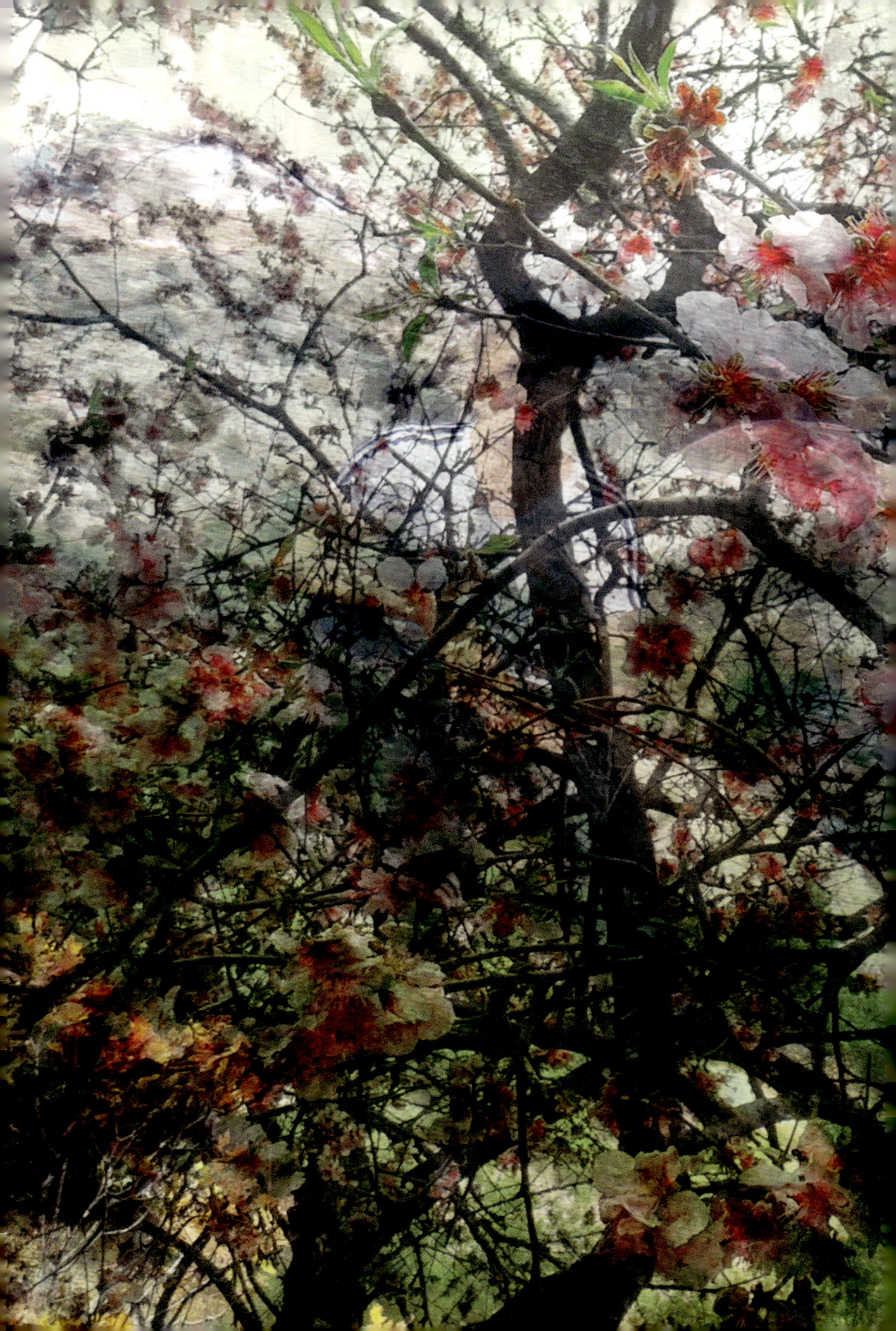

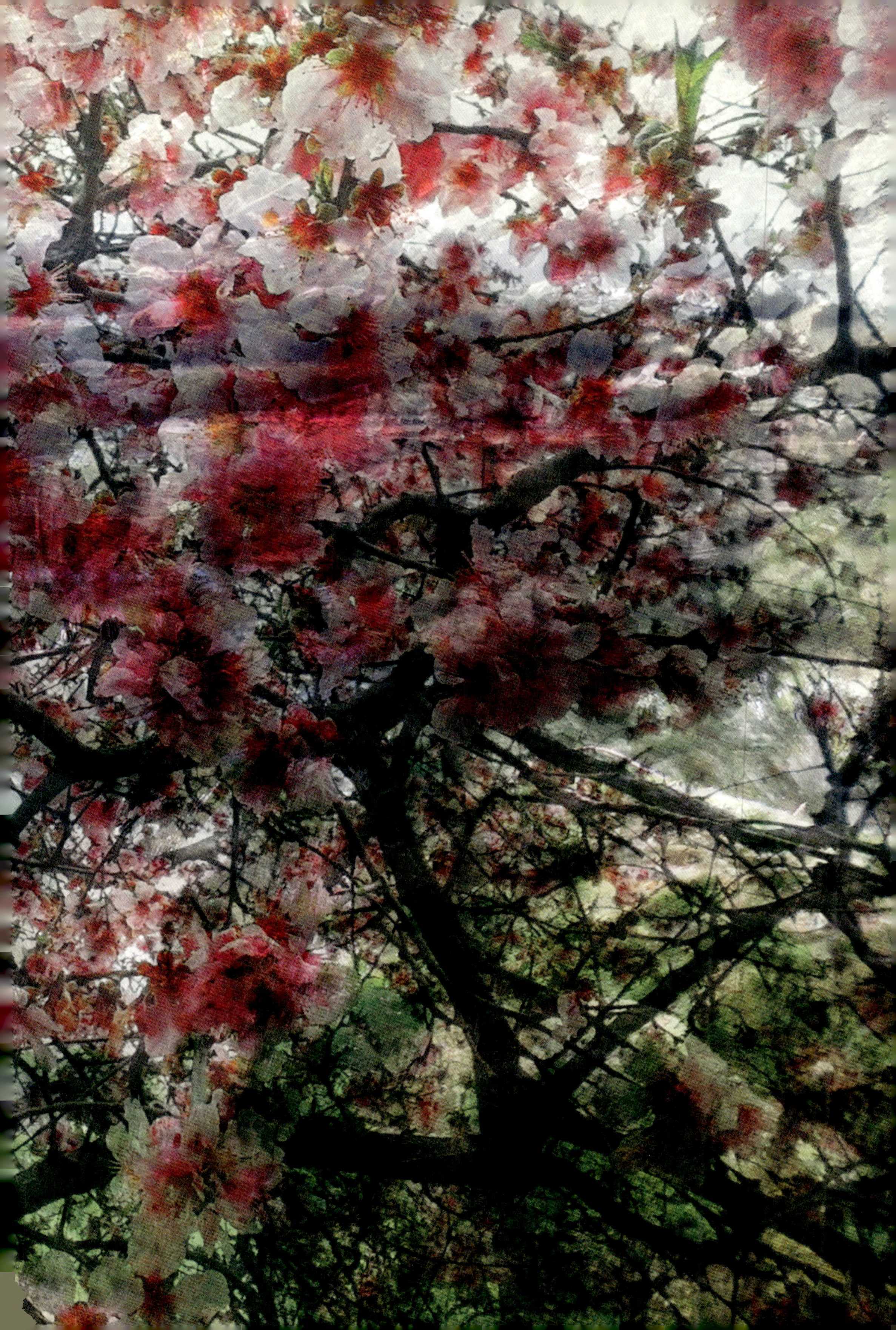

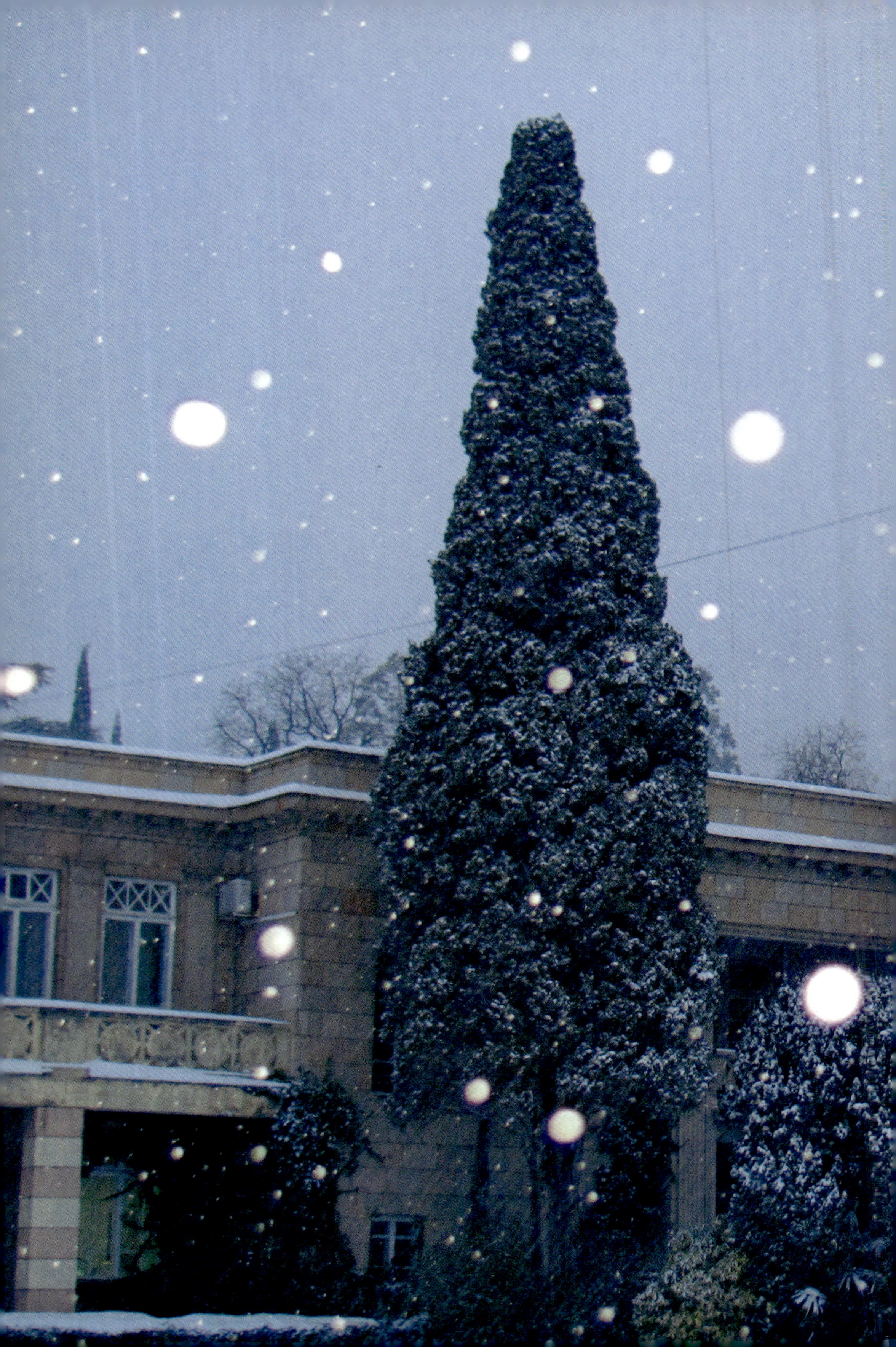

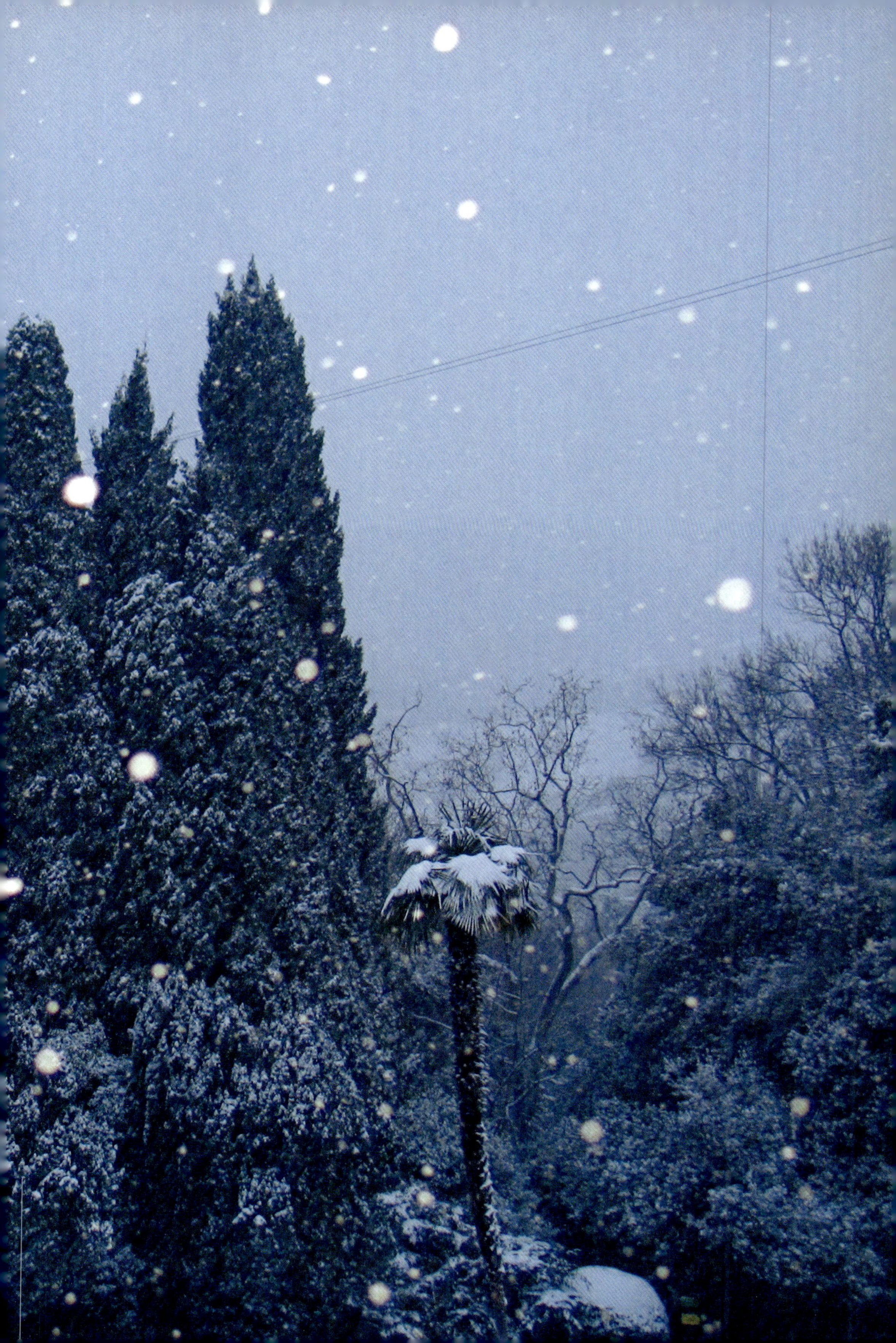

Somewhere, on an unlabeled hard drive or a misplaced memory card, there is a video. It's grainy, with an ochre hue, shot on a pocket camera back in 2007, when the digital dark was still thin. Two drag queens stand apart at the back of a mirror-lined stage, in a nightclub called Sinners, in Saint Petersburg, Russia. They are captured from above, and cropped in close, so we see only the performers, backed against their own reflections: a blonde in a light dress, a brunette in black.

The song begins and at first they lip sync in place, taking turns. But with the chorus, the two queens dance briskly away from their mirrored doubles, towards the invisible audience and out of the camera's frame. Their ankles, in tall heels, are the last to leave the scene.

Moments later, as the chorus ends, they skip backwards into the video, their bodies rejoining their re-appearing reflections in the glass behind them.

And thus it repeats, the ritual of running away from oneself and returning, of dashing into oblivion, which is death but also the world, and retreating to oneself, a solitude; trapped against a mirrored wall. A dumb-show of oscillating fates, set to a Spice Girls tune.

IN

MOST

TIDES

AN

ISLAND

IN MOST TIDES AN ISLAND

Nicholas Muellner

SPBH EDITIONS

In Most Tides An Island by Nicholas Muellner

First edition published in March 2017 by SPBH Editions
SPBH Editions is the publishing house of Self Publish, Be Happy
selfpublishbehappy.com

Publisher: Bruno Ceschel
Designer: Andrew Sloat
Fonts: Adobe Caslon, HTF Didot, and Berthold Akzidenz Grotesk
Printer: KOPA, Lithuania

Table of Contents

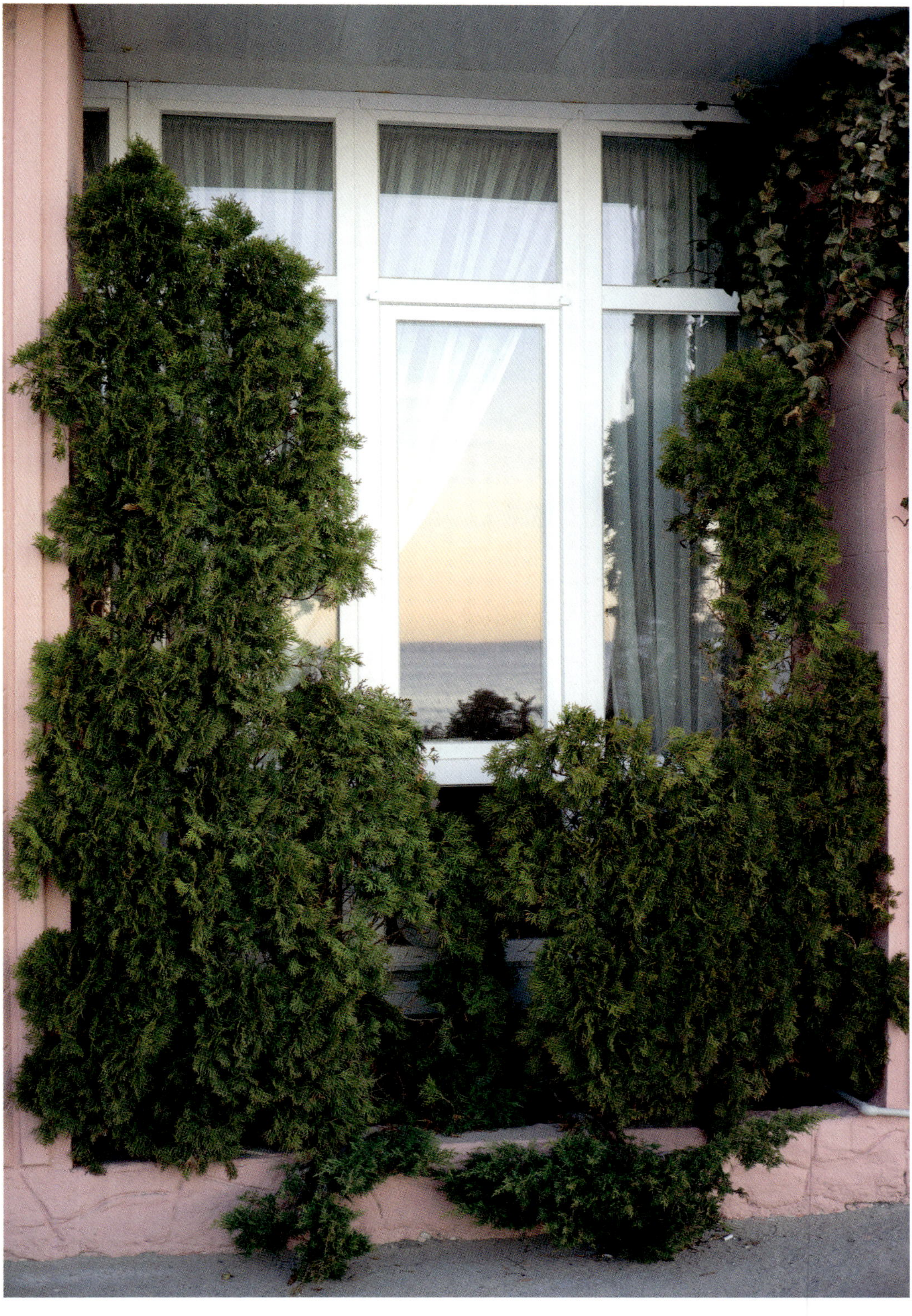

Chapter
I

The words solitude and loneliness share a certain lunar roundness, as if the moon tugs their sounds like the sea. Loneliness, a riptide, steals one away along rocky coastline. We don't know how we'll ever get back. But solitude is a voluptuous dark wave, the color of opals in a curtained room, pulling us into ourselves, and out to sea. Solitude draws us ever deeper, drifting alone atop warm currents, self and ocean indistinguishable, without impediment or foothold. Embraced by its swells, the white balustrades of the human promenade become deliciously obscure.

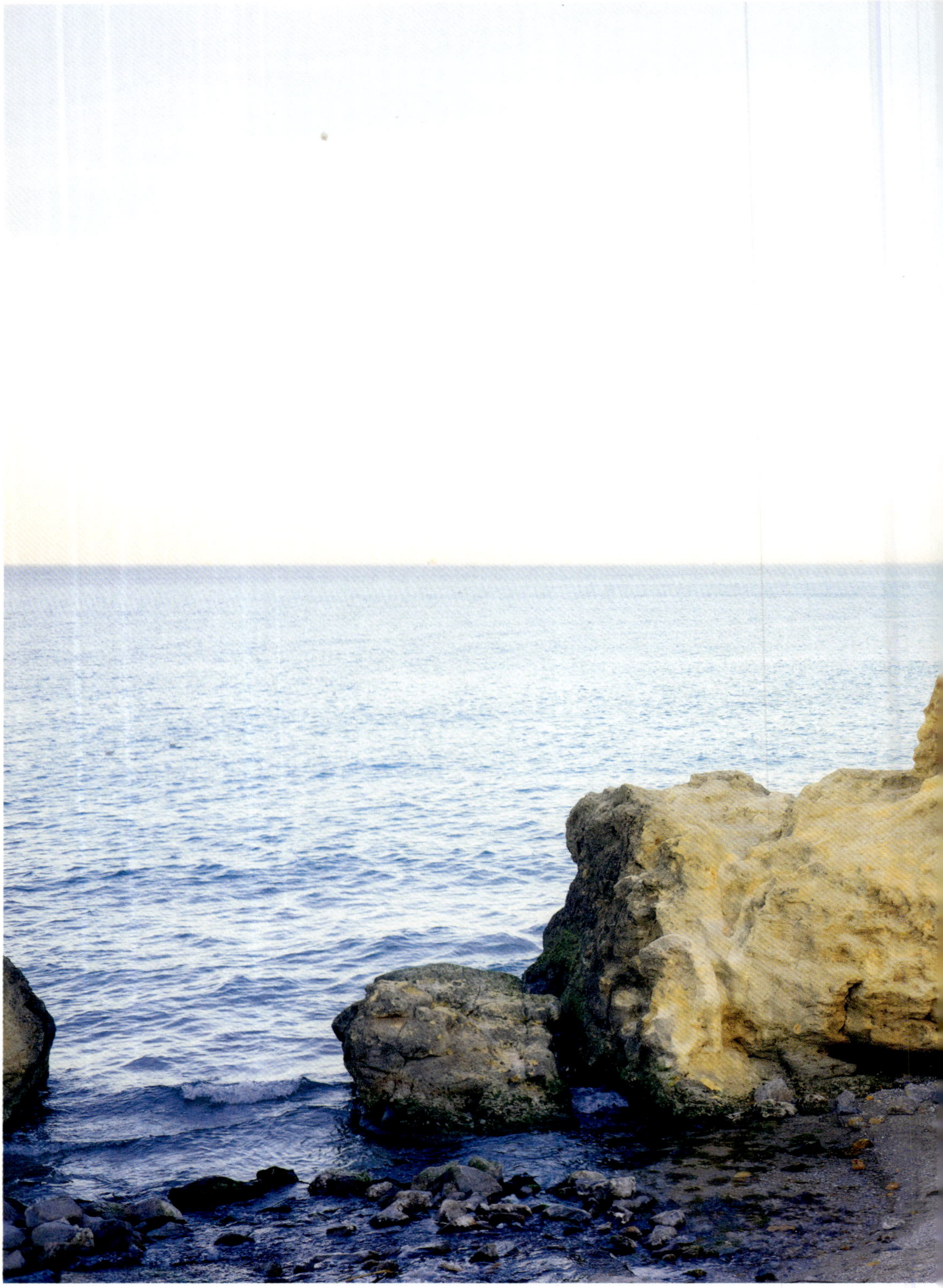

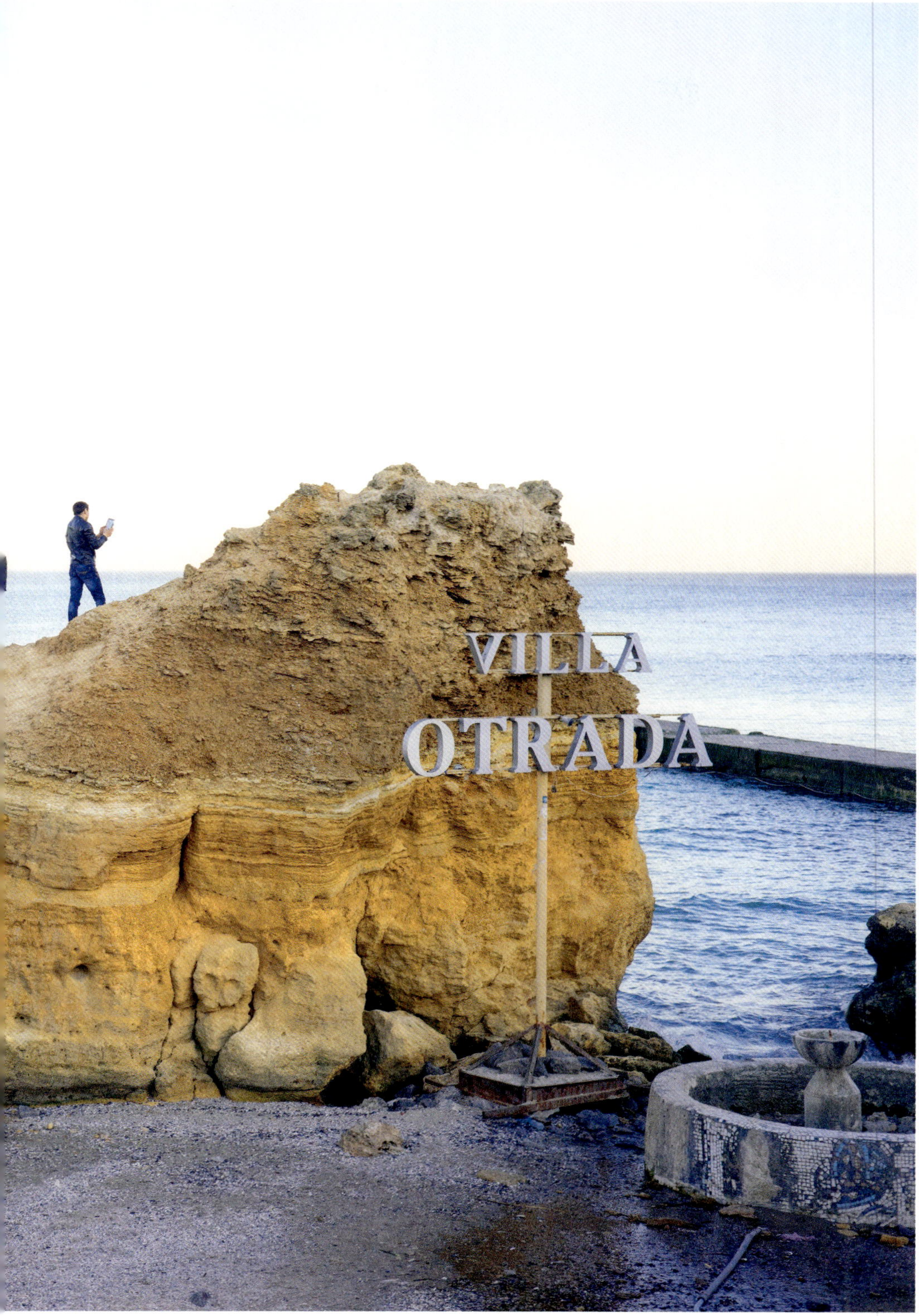
VILLA
OTRADA

Real, Soviet balustrades were cast in crudely sensuous cement and painted white. At close range, they offer contradiction—a form meant to extend light and space, expressed in the plastic language of obstructive mass. But from a postcard's-eye-view, taking in the inevitable curved sweep of any given Black Sea promenade, the balustrades collectivize to do their job: elegant, endless verandas for the masses. And it is inevitably against one such whitewashed barrier, looking out beyond the narrow rocky beach to the calm waters beyond, that two strangers meet in the chaste light of day. Typically, there is a statue nearby, to distinguish one stretch of vista from another, though a particular café might serve the reference just as well.

When I met Andrei, near the obelisk and eternal flame of the Unknown Soldier, we spent the better part of an hour looking out to sea, side by side, so that my initial impression was mostly of his surprisingly solid hands astride the broad cement railing. And even now, in memory, his face is as an accessory to those hands. He talked about his work in the hotel industry, drifting from Russian resorts to Turkey, Italy and the Mediterranean cruise-ship circuits. That career provided anonymity, opportunities for travel and occasional erotic adventure. But this particular resort town, crystalline and empty in the spring sun, was Andrei's hometown. He had grown up in this village, and now, nearly forty, lived here with his parents. It was an impossible place—too small to meet men, to provide job opportunities, to offer any kind of a life he might want. It was also the only place on earth where he was not a stranger, itinerant. Here, he was not himself, but he was home.

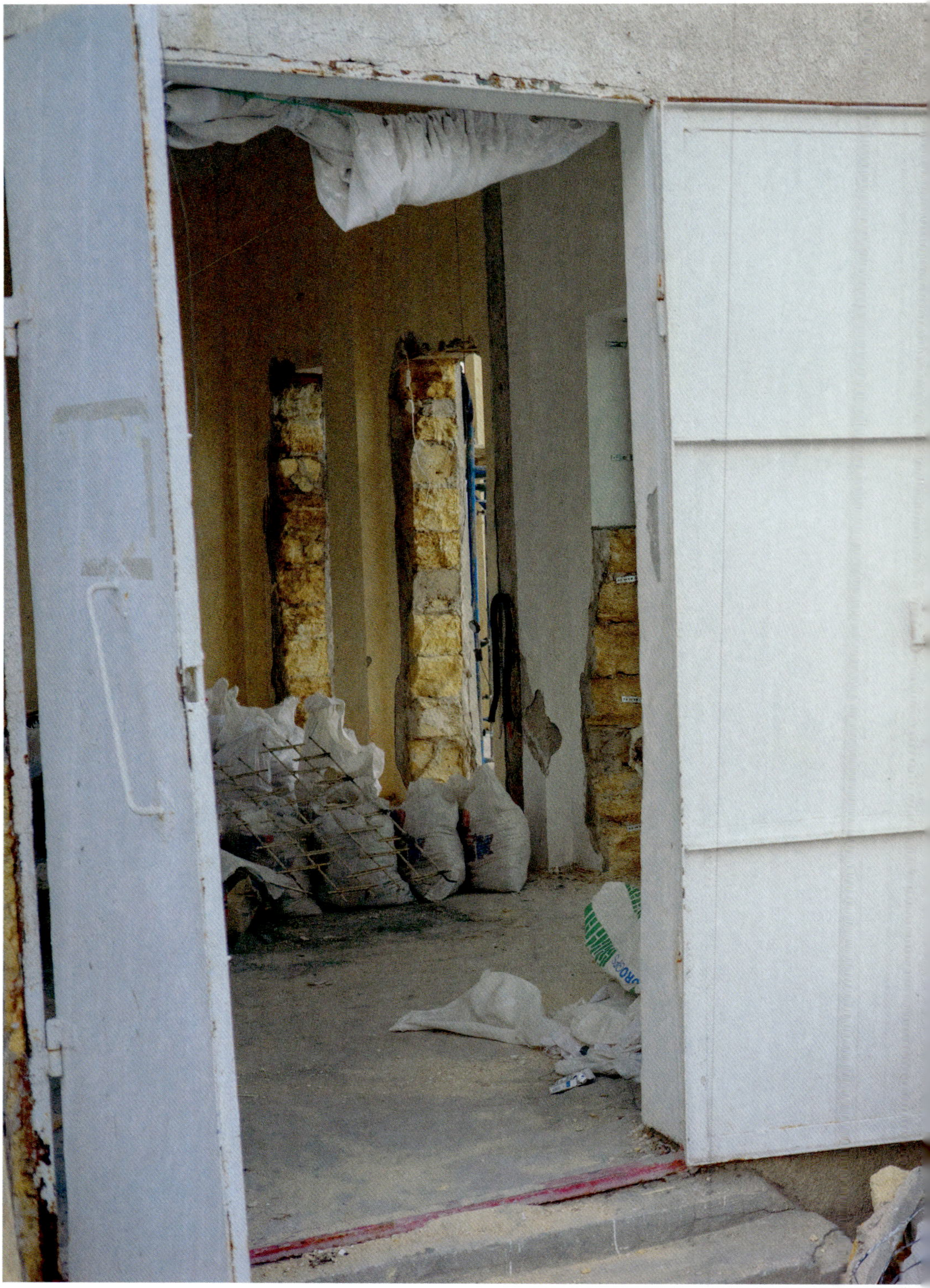

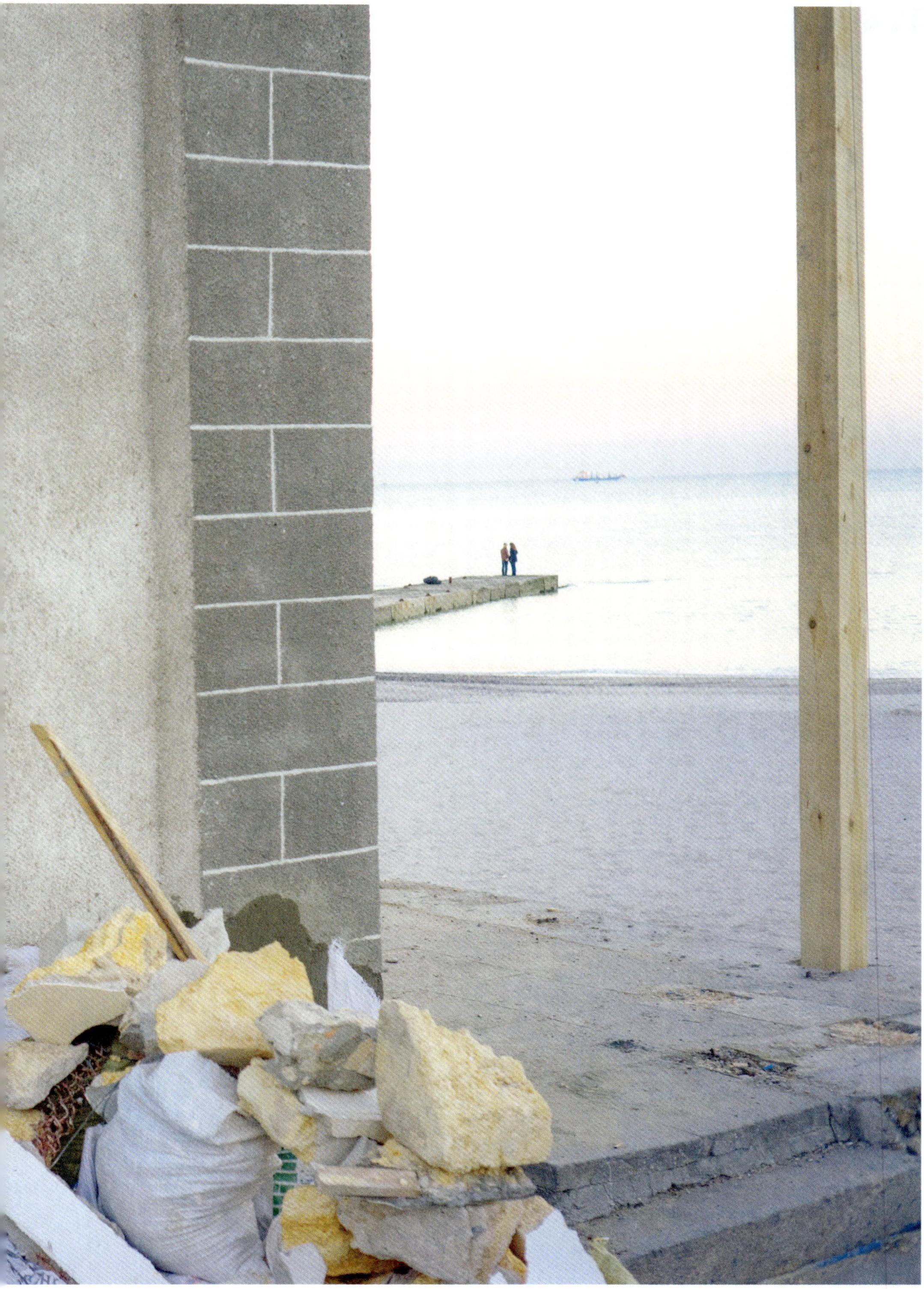

Staring out at the emptiness of our shared vanishing point, I wondered aloud where he hoped to settle himself. Andrei responded methodically, dismissing not only this, his hometown, but Sochi, Moscow, Tashkent, Paris, Athens, Istanbul and Dubai. His answer was an endlessly contracting field in a steadily expanding universe. I recognized the effect—zooming in while pulling back—my own familiar vertigo. *Some desires lack in objects.*

Eventually, we strolled south along the promenade until it abruptly gave way to a climbing road culminating in a mirrored-glass hotel. Inside, passing an idle bartender of Andrei's acquaintance, we took the glass elevator to the top. We looked out, again, at the green-grey expanse of the sea, and up at the modest blue-tinged mountains rising behind the town. The path that had brought us here unfurled below, inviting us back to the obelisk of Anonymous Man, where our friendship had started and was soon to end.

Vasya, on the other hand, chose a café just across the pine-shaded pedestrian boulevard dividing the town from the beachfront. Unlike my other meetings, anticipated by weeks or months of correspondence, we had only made contact the night before, as I scanned local listings from my hotel room.

He was curious and open, wryly aware that he did not want to be here; a worldly thinker trapped in his provincial hometown. Vasya worked for the local branch of a national media company that he knew to be at the beck and call of the Kremlin. Now in his early thirties, he'd had one long-term boyfriend, but they'd broken up after eight years of living in their parents' respective homes, pretending to be just friends. Vasya had wanted them to get their own place, move in together, but the other guy couldn't muster the courage, so it ended.

He wanted to leave. But he spoke only Russian and felt that his professional skills were not sufficient to support him in a western country, or even in competitive Moscow. So there he was, hanging out with his two girlfriends—one an equally subterranean lesbian—the only two people who knew he was gay. His family, he was confident, assumed he was sleeping with both of them.

It happens rarely that a new acquaintance seems to begin *in medias res*, as if it had always been there. Despite the circumstance of the meeting, our rapport was not charged with erotic possibility or exotic difference, but with an immediate sympathy of taste and temperament.

Vasya loved 'difficult' literature and melancholy endings. He had enormous dimples, a winsome smile, and delicate fingers. I asked him to model his hands, explaining that they made an expressive stand-in for the face I would not show the world. This metaphorical feint pleased him, and my hand-model obliged.

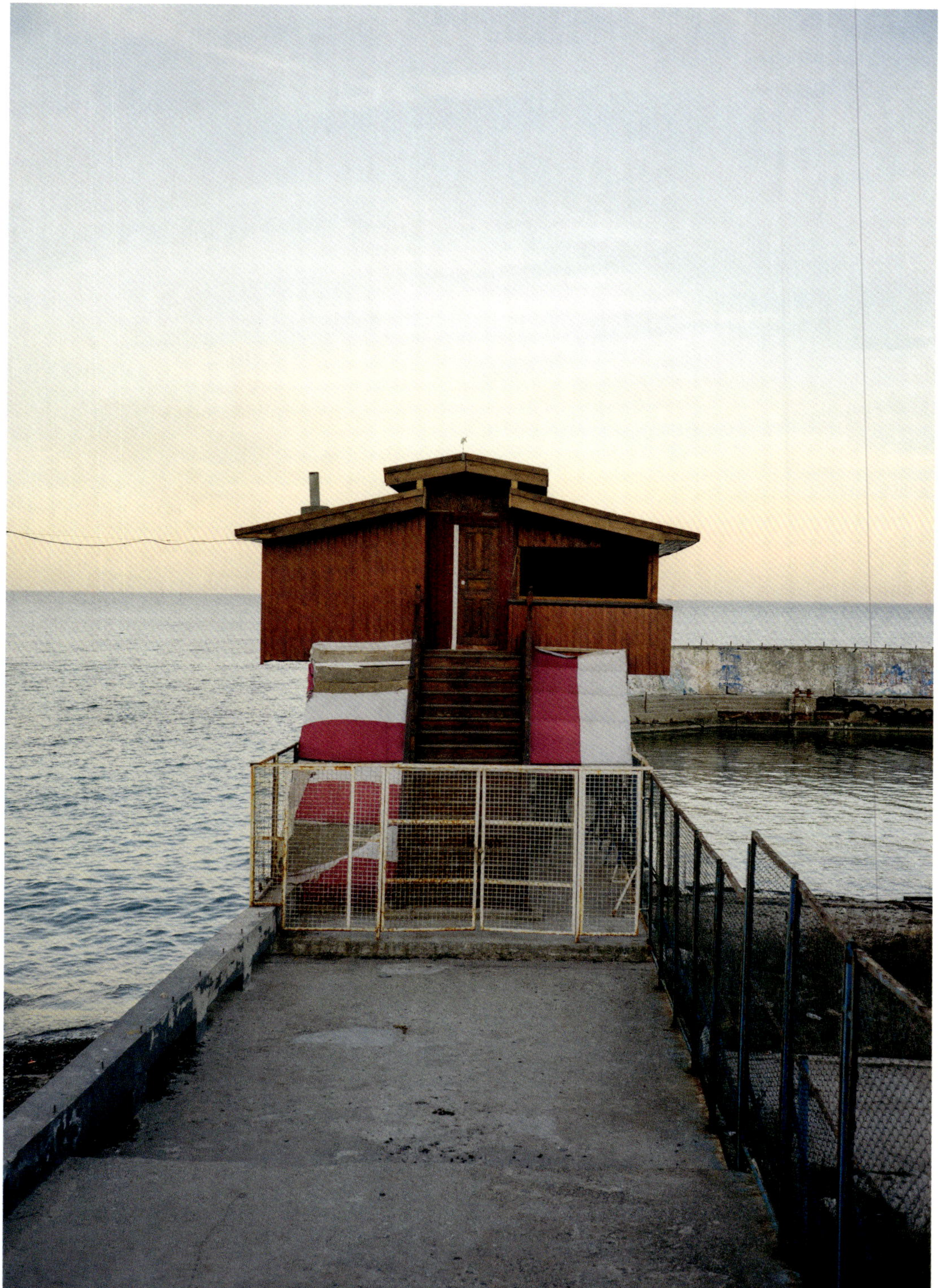

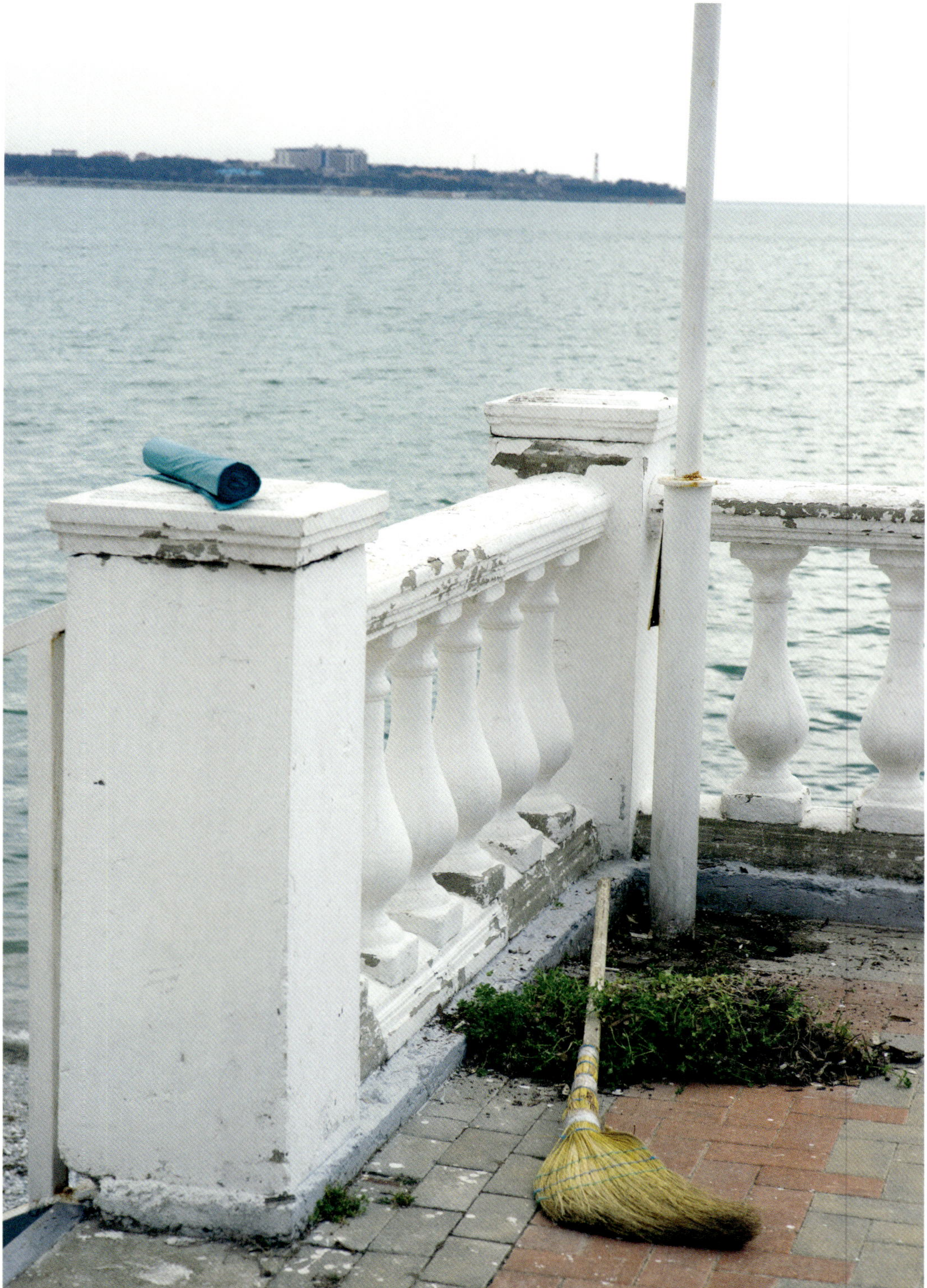

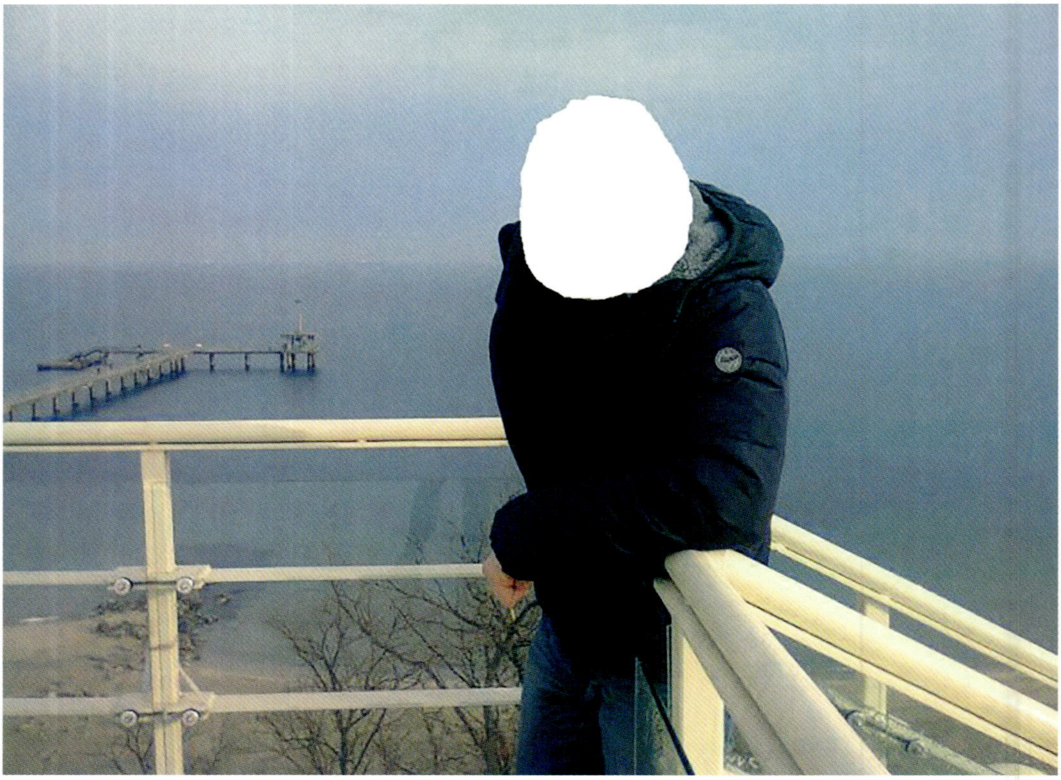

We didn't meet at the promenade, but we ended up there soon enough. I found Timur, which is not his name, outside my hotel, pacing in small circles at a safe distance from the building's colonnaded entrance. I wouldn't have recognized him but for the wild look in his eyes that was retrospectively evident in the picture I'd seen. What might have been "alertness" or "playfulness" in the image was actually a kind of antic restlessness. But other than the eyes, he was much softer and paler than I'd expected, as if he'd recently been kept well fed in a dark place. He certainly seemed pent up.

Even as we shook hands, he turned in retreat towards the water. The night was soft, moistened by two days of rain and a low fog rolling in across the vast, walled harbor. He whispered, as we walked: so where's your boyfriend? In the narrow park above the main promenade, we lingered against the railing, side by side, looking down on the nearly empty walkway below. This time, I explained, I was traveling alone. Timur had a lot of questions, and so did I.

His boyfriend is a married man with two children, and a wife who allegedly knows nothing of his double life, even though (unlike Timur, who considers himself bisexual) his boyfriend is "one hundred percent gay." They both work in "business" or "trade," terms deployed by Timur with a characteristic Russian vagueness that cast a shadow of illicit intent over the unspecified enterprises. I declined to press him further on these lines of work. But I did want to know more about this relationship, and Timur obliged, happily. They met frequently in public, at cafés or restaurants, because "here," as he put it, "there is never anything suspicious about two men dining together." They were an invisible gay couple in full public view. But it was less easy to arrange intimate rendezvous, though once they travelled together on one of the boyfriend's 'business' trips.

Our conversation was punctuated by abrupt silence, whenever strangers walked by, as if their passing bodies interrupted the radio signal between us. And when a couple paused at our railing to take in the view, Timur decided to move on; this company was too close. We descended onto the main harbor promenade, strolling slowly past shuttered kiosks.

The walkway was short, so after a few hundred yards we simply turned around and strolled back. Direction was in our words, not our walking: was it really possible, Timur wanted to know, that people at my work knew I was gay, and my boyfriend and I lived together openly? Did my family actually accept me that way? Was it true that, in America, there are gay bars everywhere? In theory (the internet), he already knew the answers to these questions. But these assertions needed to be reality-tested on a physical subject. Tested in soft voices, elbows on balustrades with our backs to the sea, looking at shops and strangers and rarely at each other. Spies sharing secrets in occupied territory.

Timur spoke in quick rushes, almost stutteringly, in heavily slanged, soft-edged, intensely profane Russian. After moving on several times from inadequately secure locales, Timur stopped into a coffee kiosk for a midnight espresso. Thus even further energized, he led me past the main promenade around a bend and along the side of the harbor, down increasingly vacant streets to an unilluminated dead-end. We shared stories of our relationships, and he spoke passionately both of his boyfriend and of his own promiscuity. His married lover is madly jealous, overly possessive, and Timur loves him dearly. To avoid hurting him, he doesn't tell him about any of his assignations, usually internet-arranged: in closed offices, strangers' apartments, occasional hotel rooms and, most frequently, in parked cars. He wouldn't even tell the boyfriend about me, he confided, because it would only enflame his envy.

By now, we had arrived back at the promenade behind my hotel. I told him it had been great to meet; it was unfortunate I was leaving in the morning; I was tired and had to go to sleep. His antic eyes lit up again. Timur informed, with unmistakable intent: "I have slept in your hotel before." We both smiled, and shook hands, and clasped shoulders, as men do, and went on our street-strolling ways.

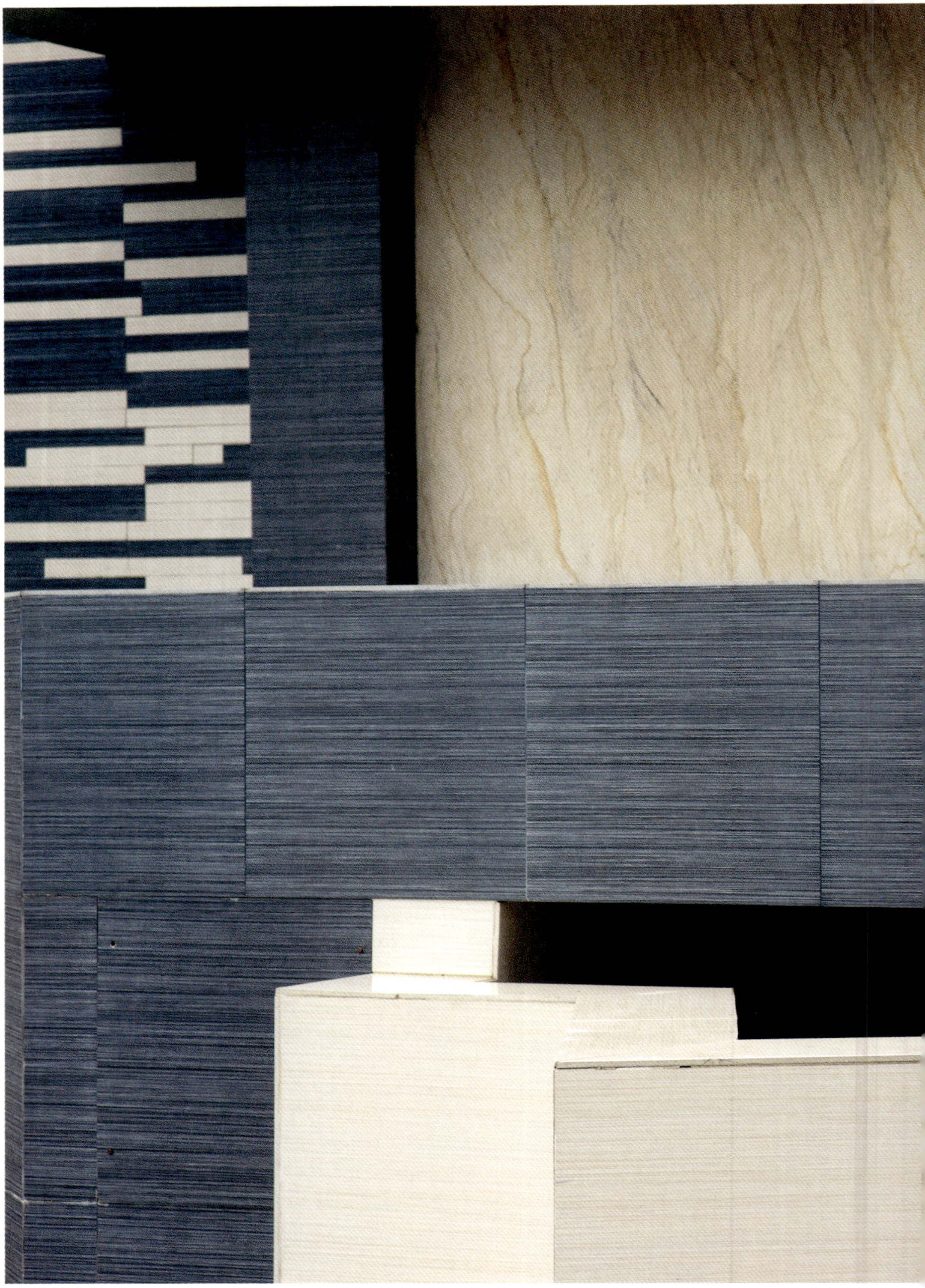

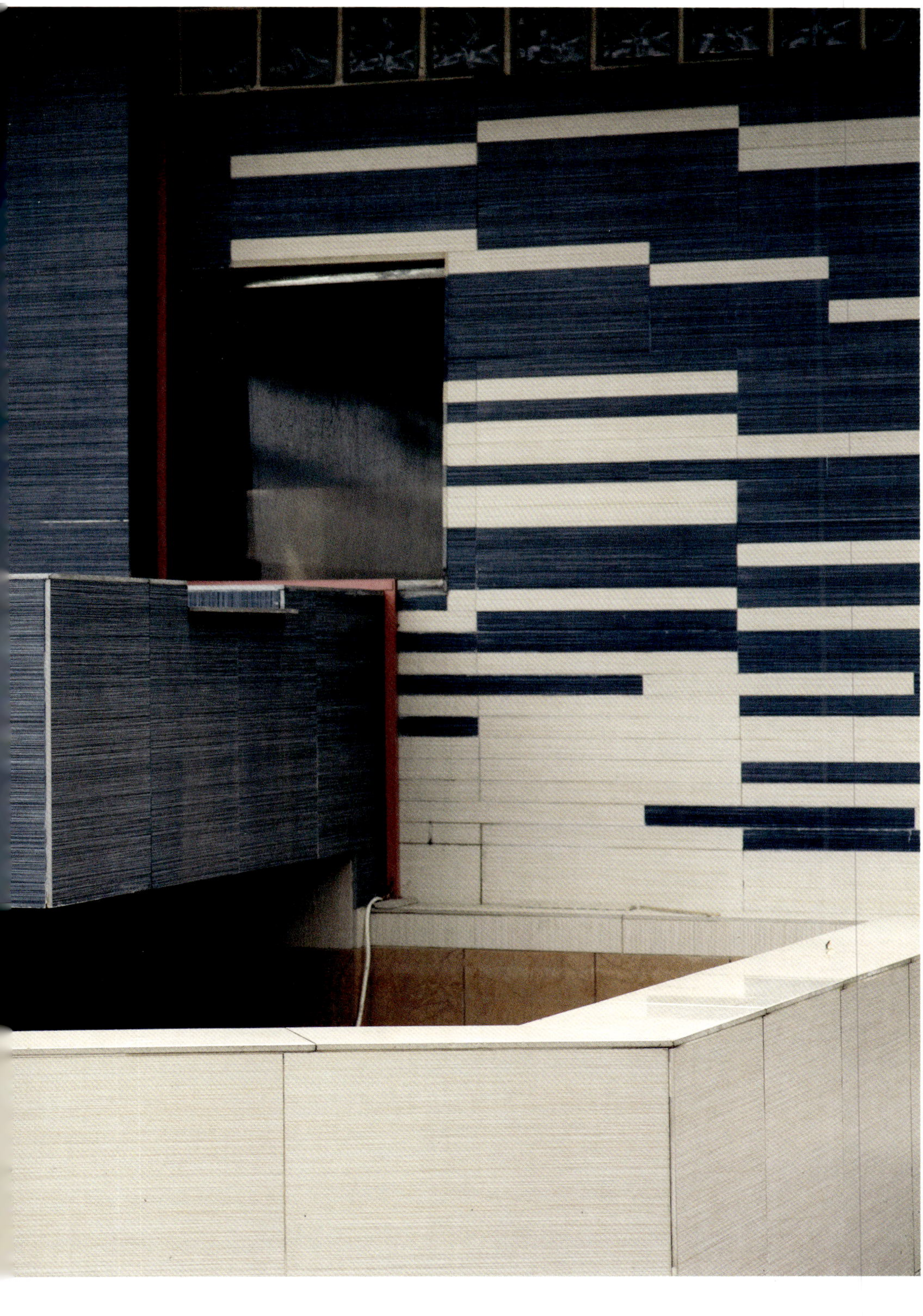

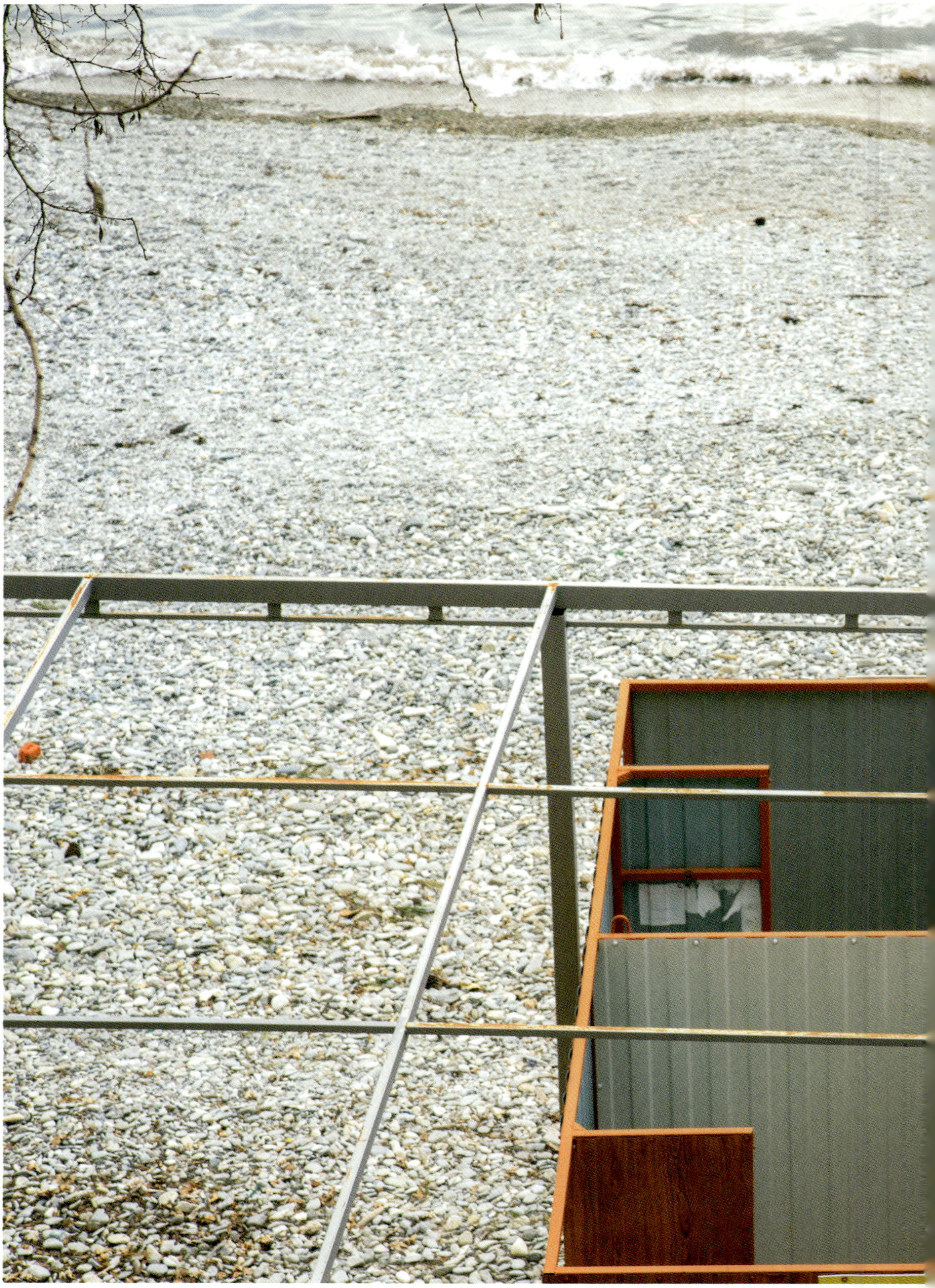

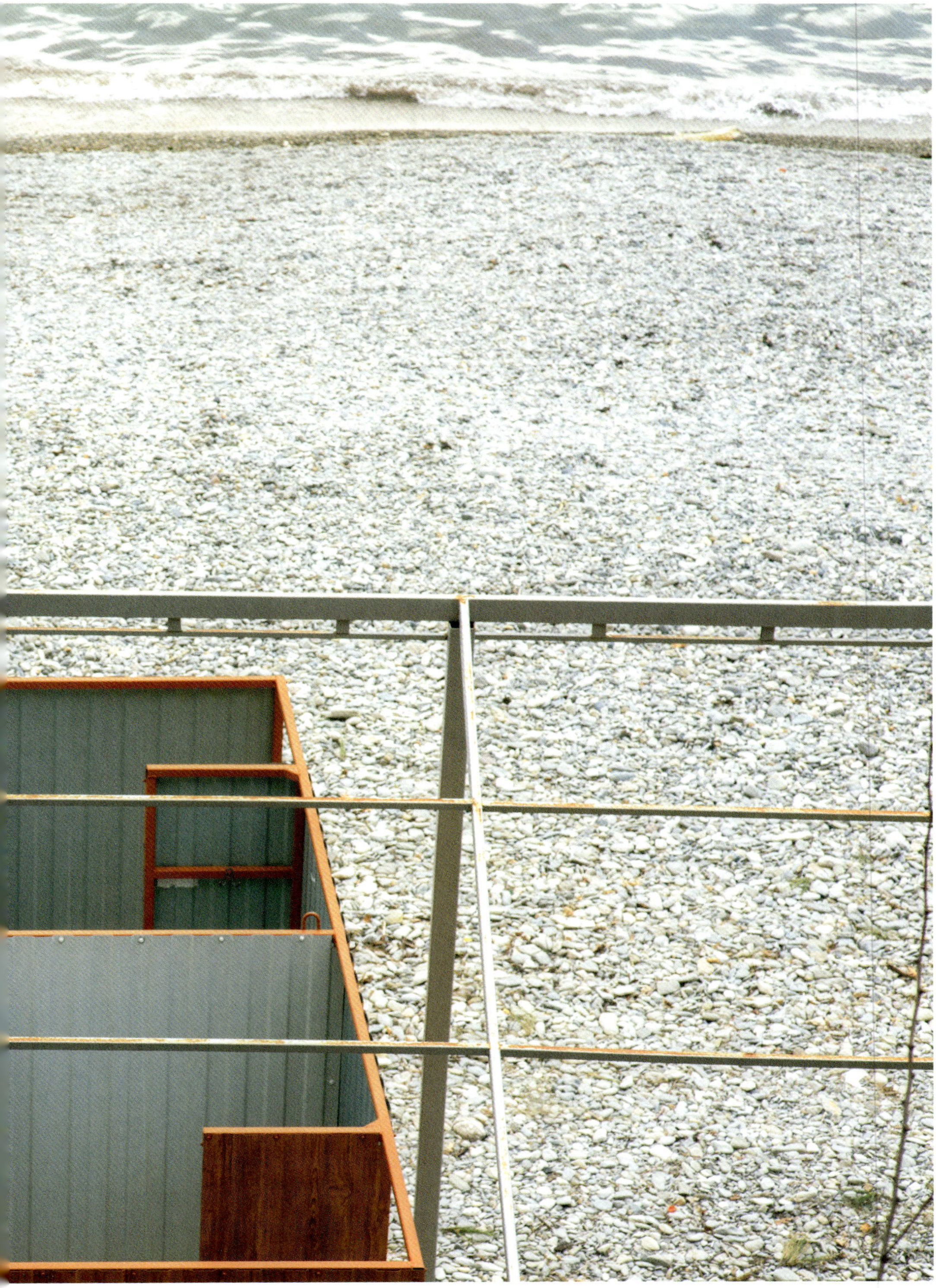

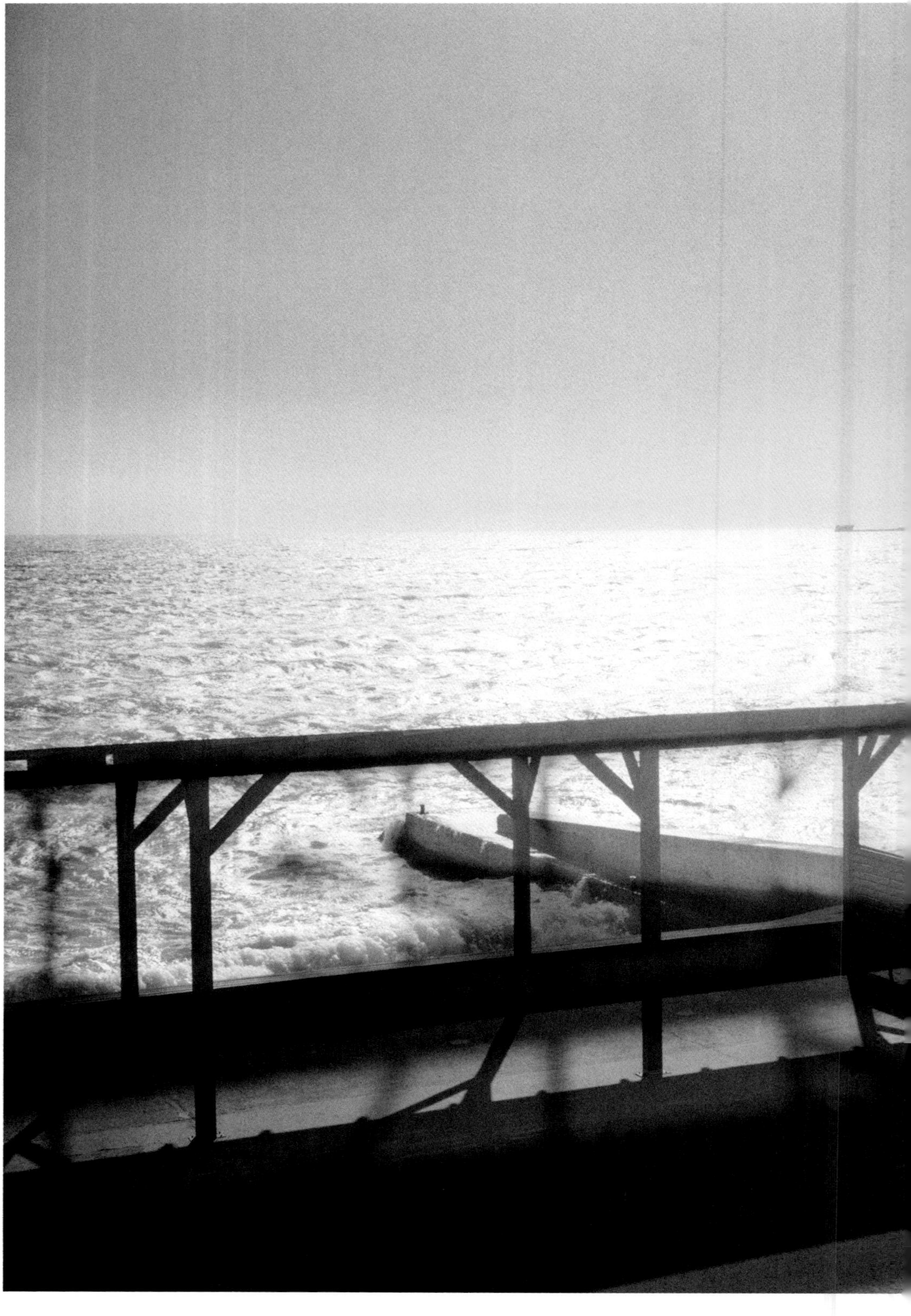

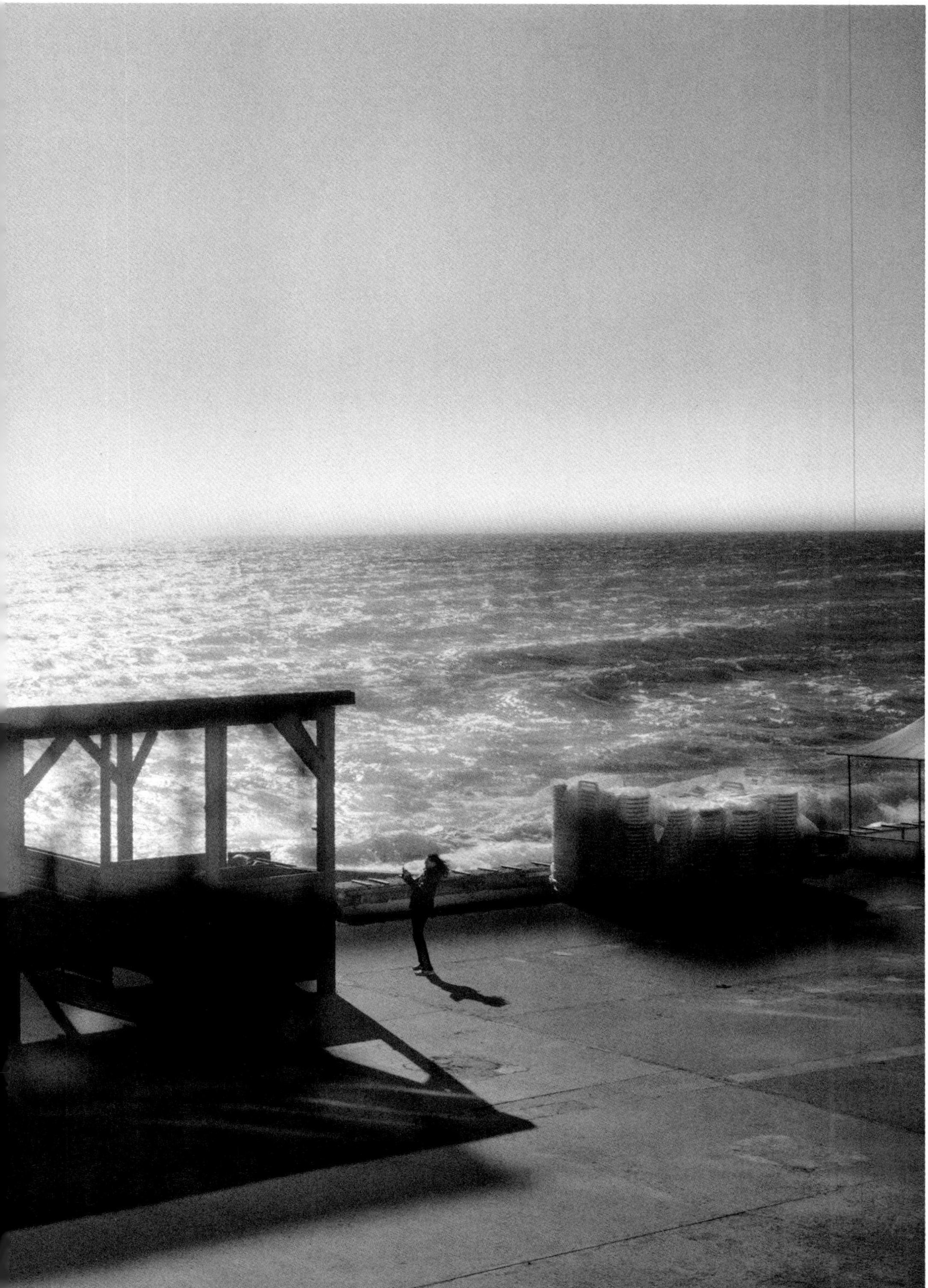

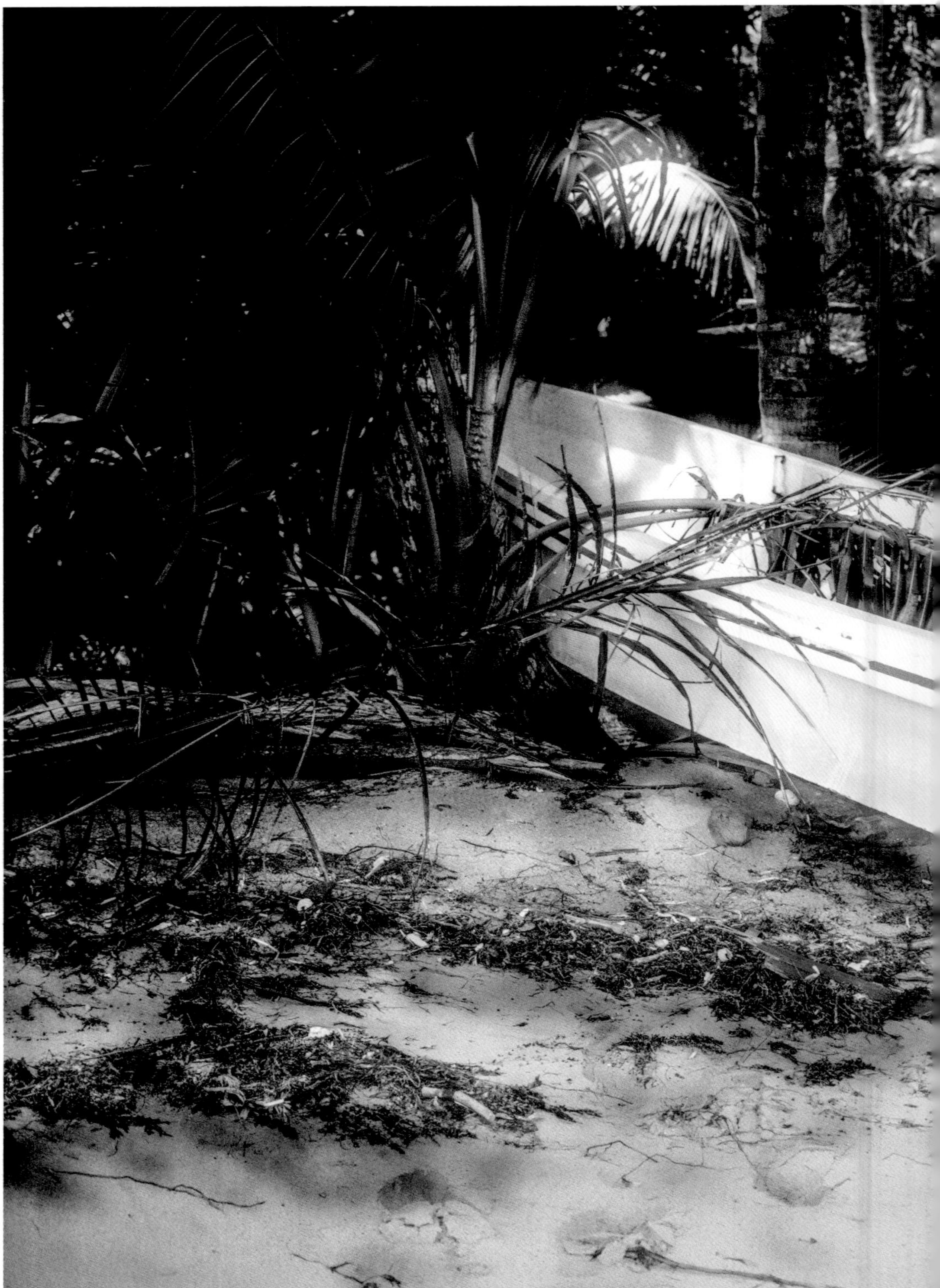

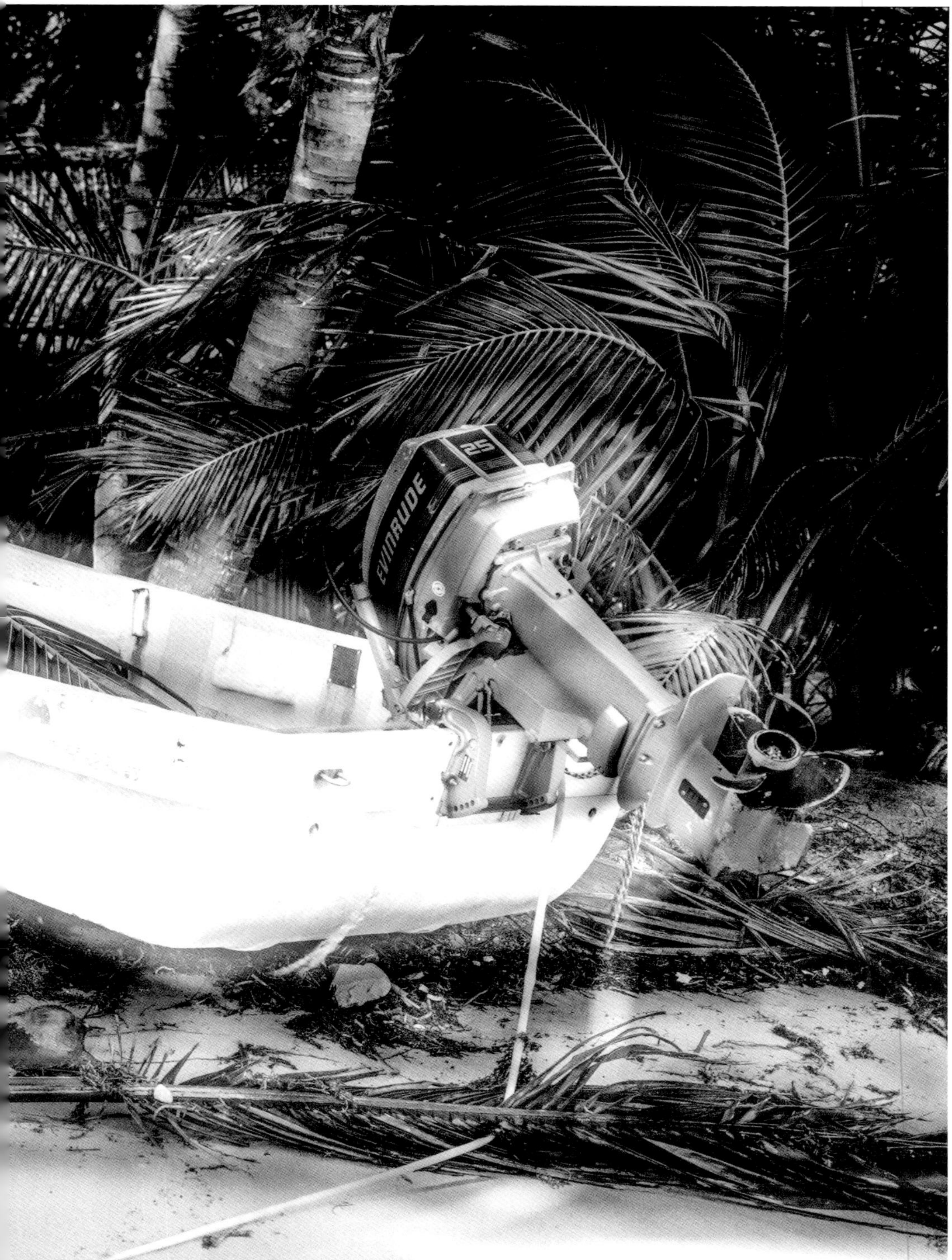
25
EVINRUDE

Chapter II

To call it a wind would have been inadequate. To call it a storm would have missed the point. But it was coming, and Isabel, suddenly distracted by the rustling of leathery leaves, let the small coconut drop from her idle right hand without a break in stride. It bounced once, rolled across a few feet of gently sloping Bermuda grass, and came to a stop against a broken conch shell, chipped open to reveal the workings of its many-chambered heart.

Four hundred years before, in nearly the same spot, a daydreaming Carib Indian, dazzled by the sea, probably dropped another coconut in much the same way. Because even then, three centuries before the island had been systematically reshaped into a palm oil plantation, and just as thoroughly abandoned, there were always more than enough coconuts.

The agitated current of air that had captured Isabel's consciousness was particular in its moisture, its energy, the noise of its incipient threat. It was also, in all of these ways, familiar, suggesting with the irascible certainty of the tropical that something fiercer was not far away. But Isabel continued across the grass, small coconut abandoned to the anonymity of dropped coconuts, without the slightest acceleration of pace. As she approached the house—or rather, the agglomeration of huts and shacks and thatched shelters that made home—the unhindered air that rushed across the clearing whipped up urgency only in her hair. For a moment, a gust from behind threw her uneven, dense tresses up into a fan at the back of her head—the temporary crown of a tropical bird, passing above the anxious scatter of actual chickens seeking shelter all around.

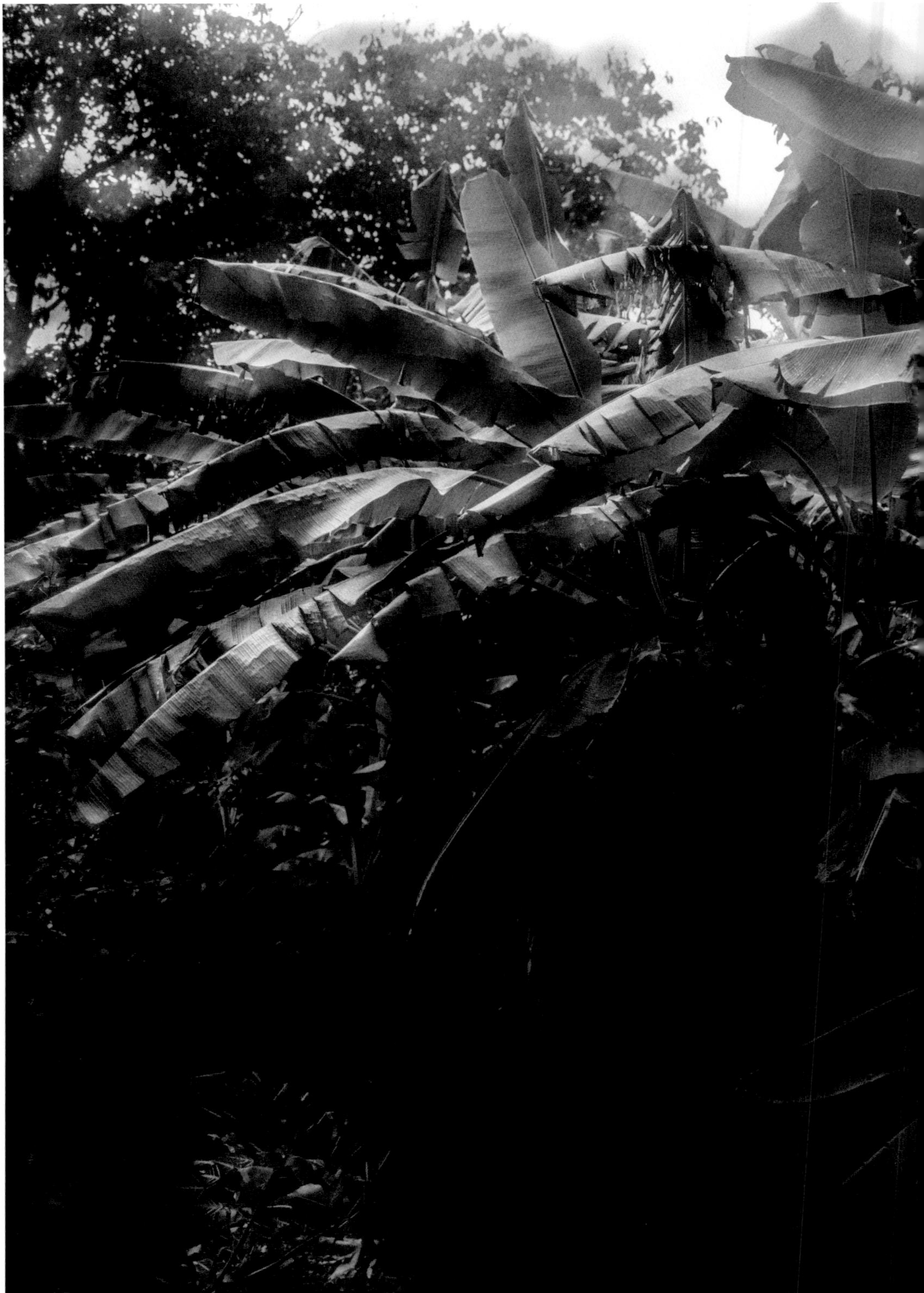

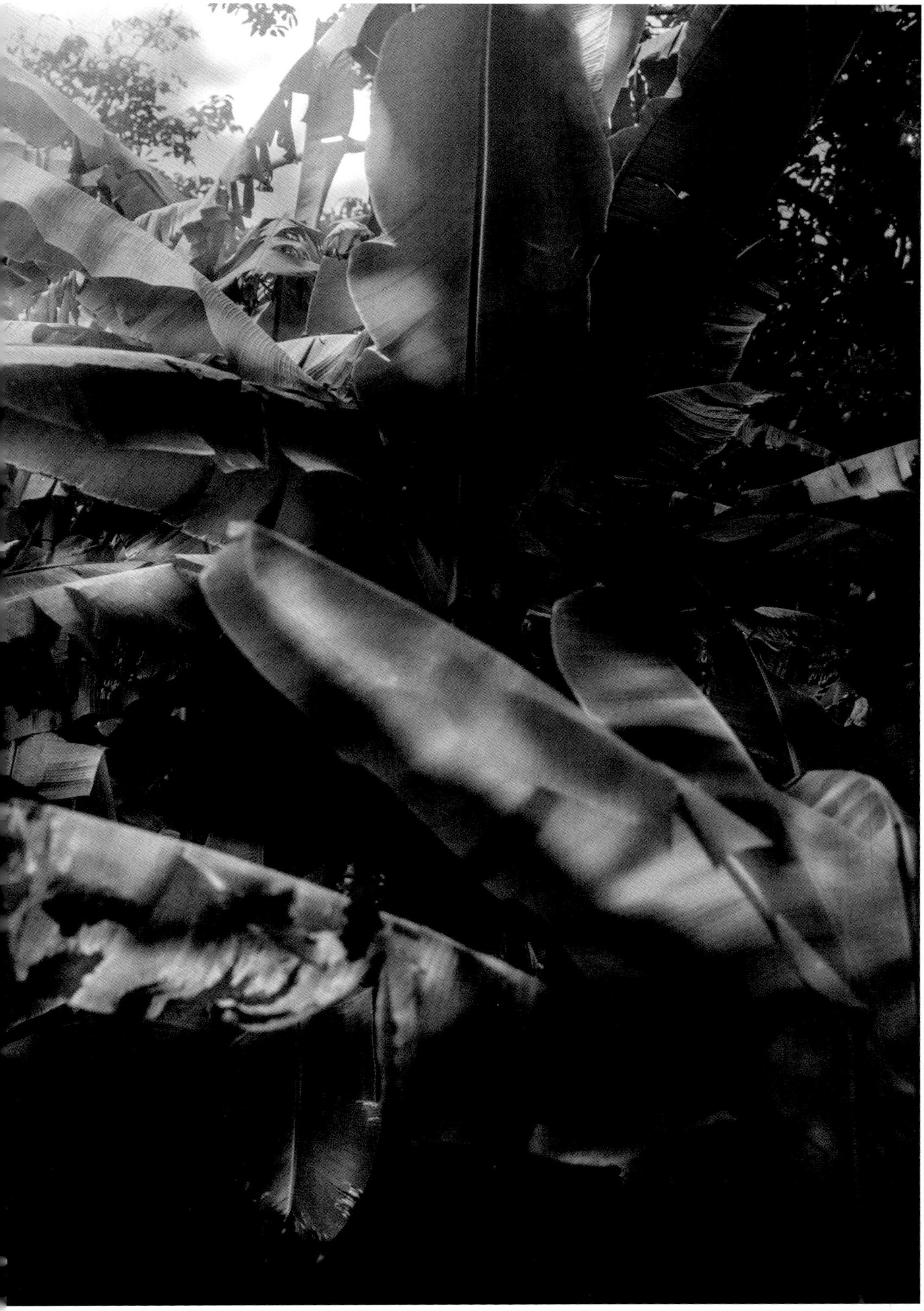

I could not say with certainty that she waved to me at that very moment, though it seemed possible from the sudden jerk and arc of her leeward arm, backlit against the ocean. Was that a hailing gesture or a wind-tossed limb? She certainly did not slow or detour to clarify. Set back under the thatched overhang as I was, it seemed doubtful that she had even seen me. But I had by now become accustomed to this uncertainty of her attentions. Such were the sensations of Isabel: interest and neglect, intrigue and oblivion.

Spurred by the sudden winds, Isabel's route, previously unclear, took on focus. Her trajectory sharpened. Perhaps, in fact, the gesture that I had hoped to read as a wave was simply an offshoot of this sudden turn—arm whipped out centripetally as she tacked east across the lawn, inland.

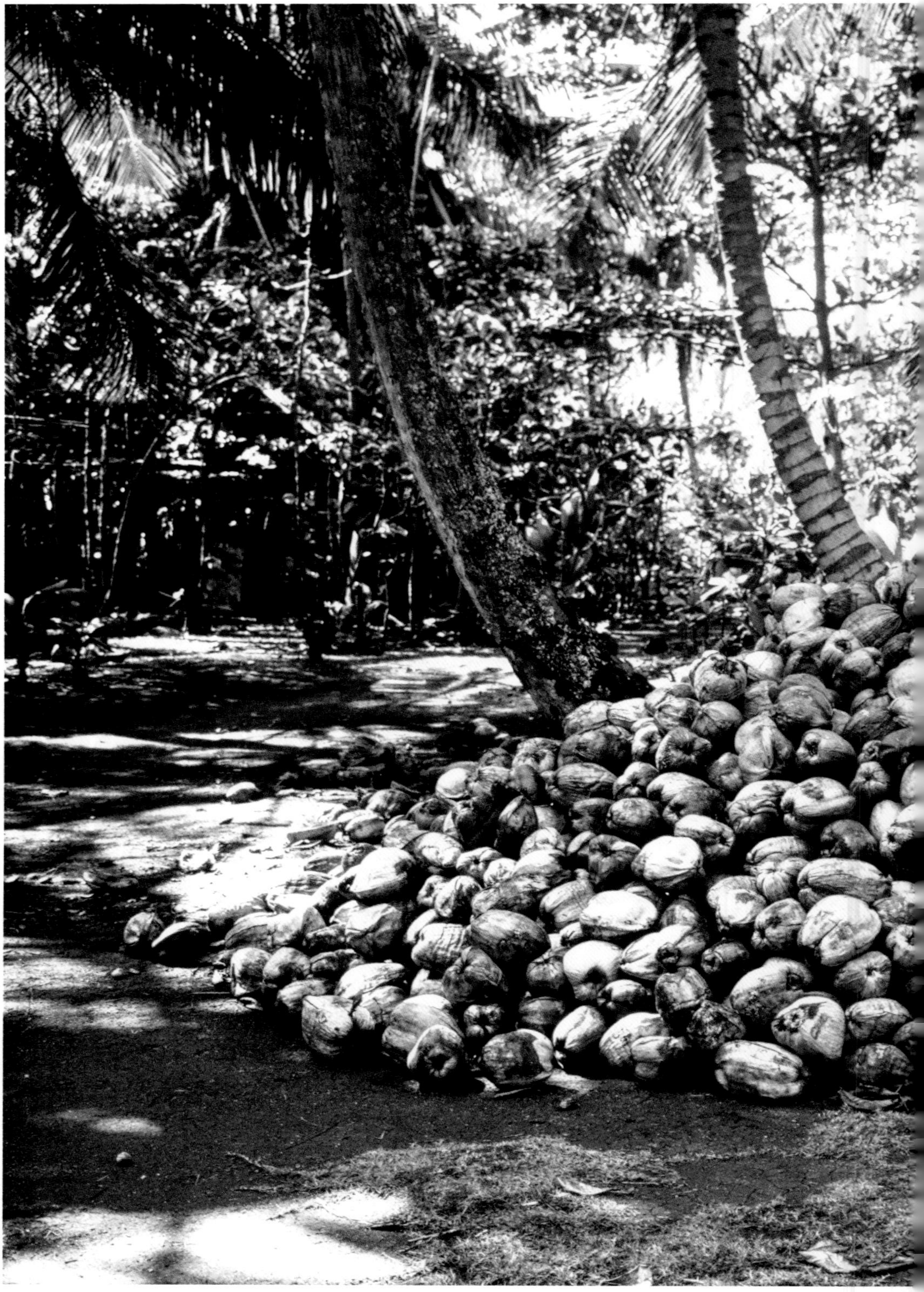

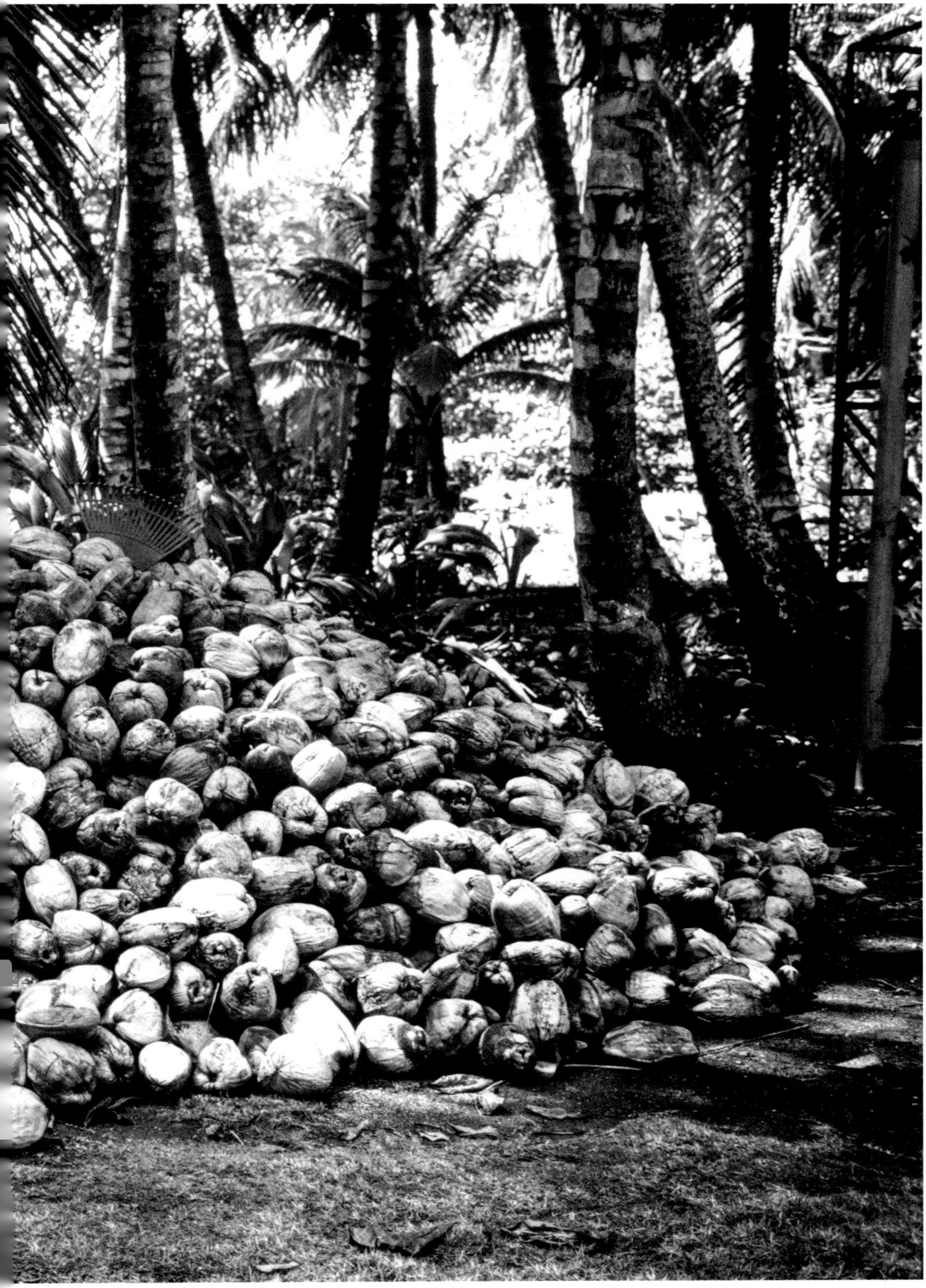

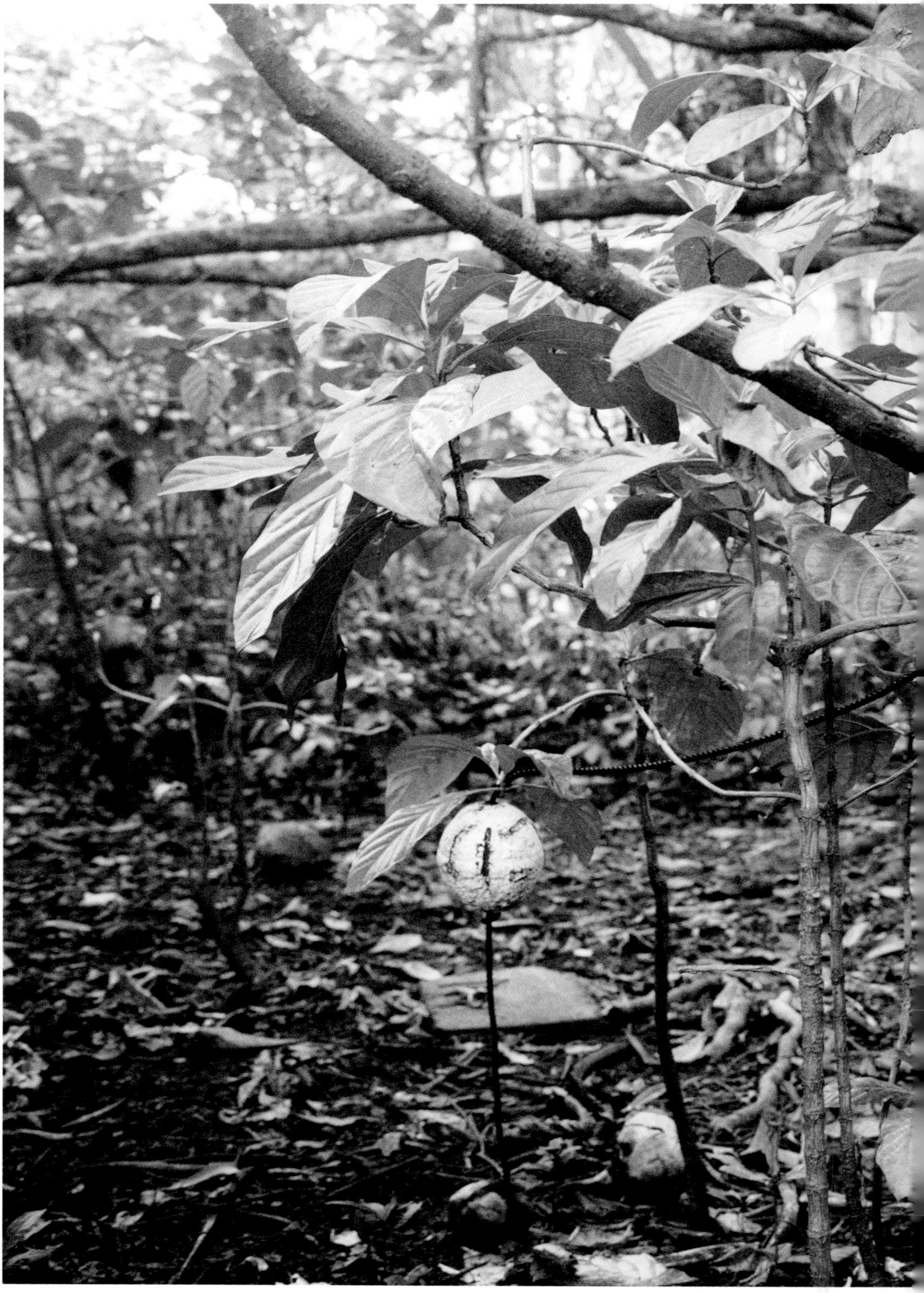

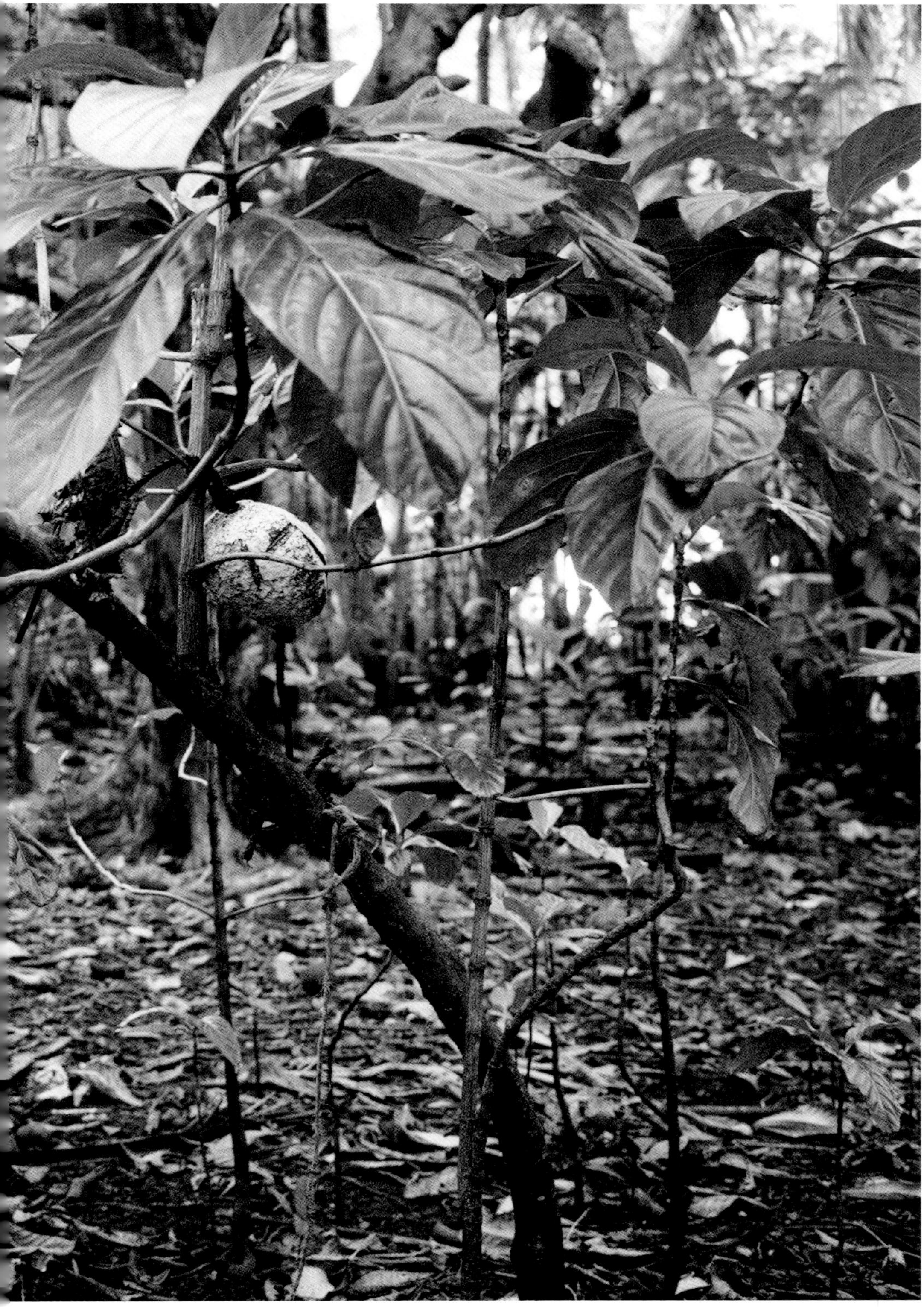

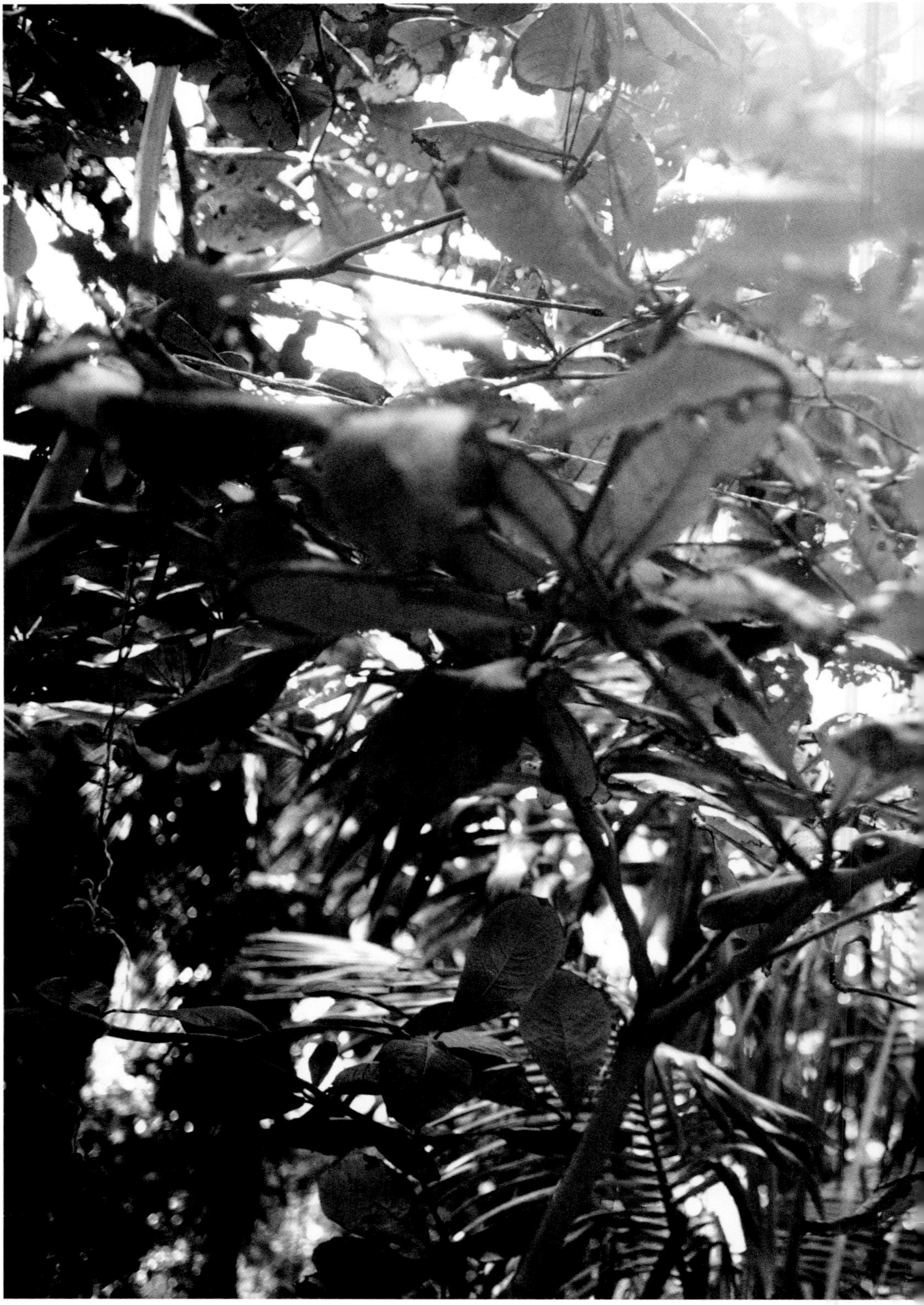

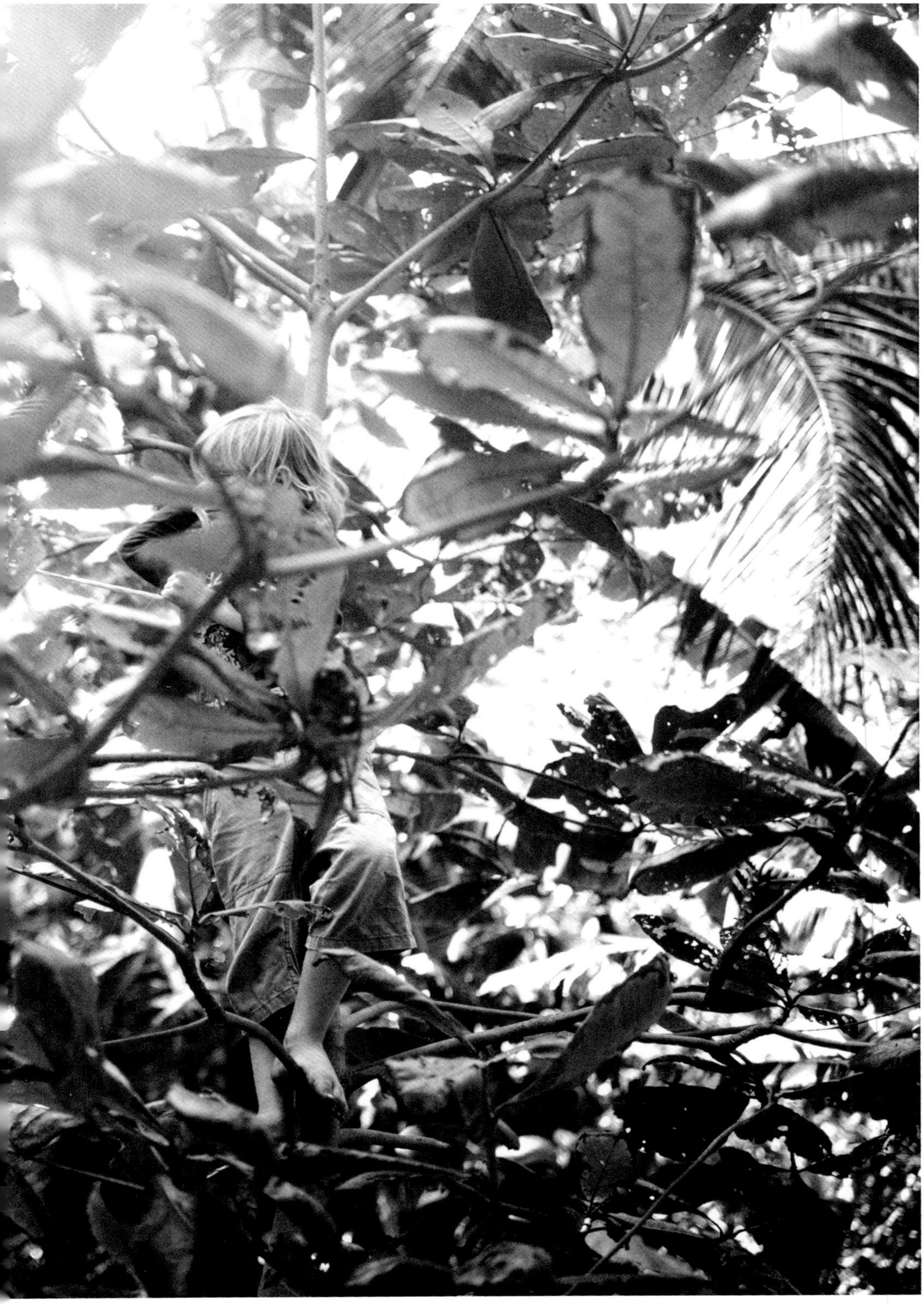

Another afternoon: Isabel on the lawn, alone on a towel in the dappled sun. Crossing the grass to refill my water bottle, I widen the arc of my path to steer closer. She sits upright, an open notebook and pencil at her side. Lola, her parrot, hops along her left hand, pecking, with lazy aggression, at her forearm. Slackening my pace to observe the scene, I realize that she is trying to teach the bird new words. *"Mano,"* Isabel repeats, several times, holding open her right palm in a gesture of surrender. Lola cocks her head and shrieks *"Hola!"* It is a mean parrot, either stupid or stubbornly perverse. To my knowledge the bird knows two words, and this is one. Meanwhile, from within the jungle, ambiguous cries drift across the lawn: the sounds of another Lola and her brother. Isabel's daughter and the parrot share a name. The children's noises, by now familiar, encompass both desperate wailing and shrieking merriment, unified by an undertone of bored desperation. Isabel, attending to her pupil, seems not to notice.

The parrot's other word is "Lola," and she deploys it freely. It is impossible to know if the bird is announcing itself or calling for the child. Likewise, when Isabel calls for Lola, I can rarely say if she is summoning the daughter or the bird. Most often, neither arrives. Only one thing is certain: the girl, when she screams "Lola," is always speaking of herself.

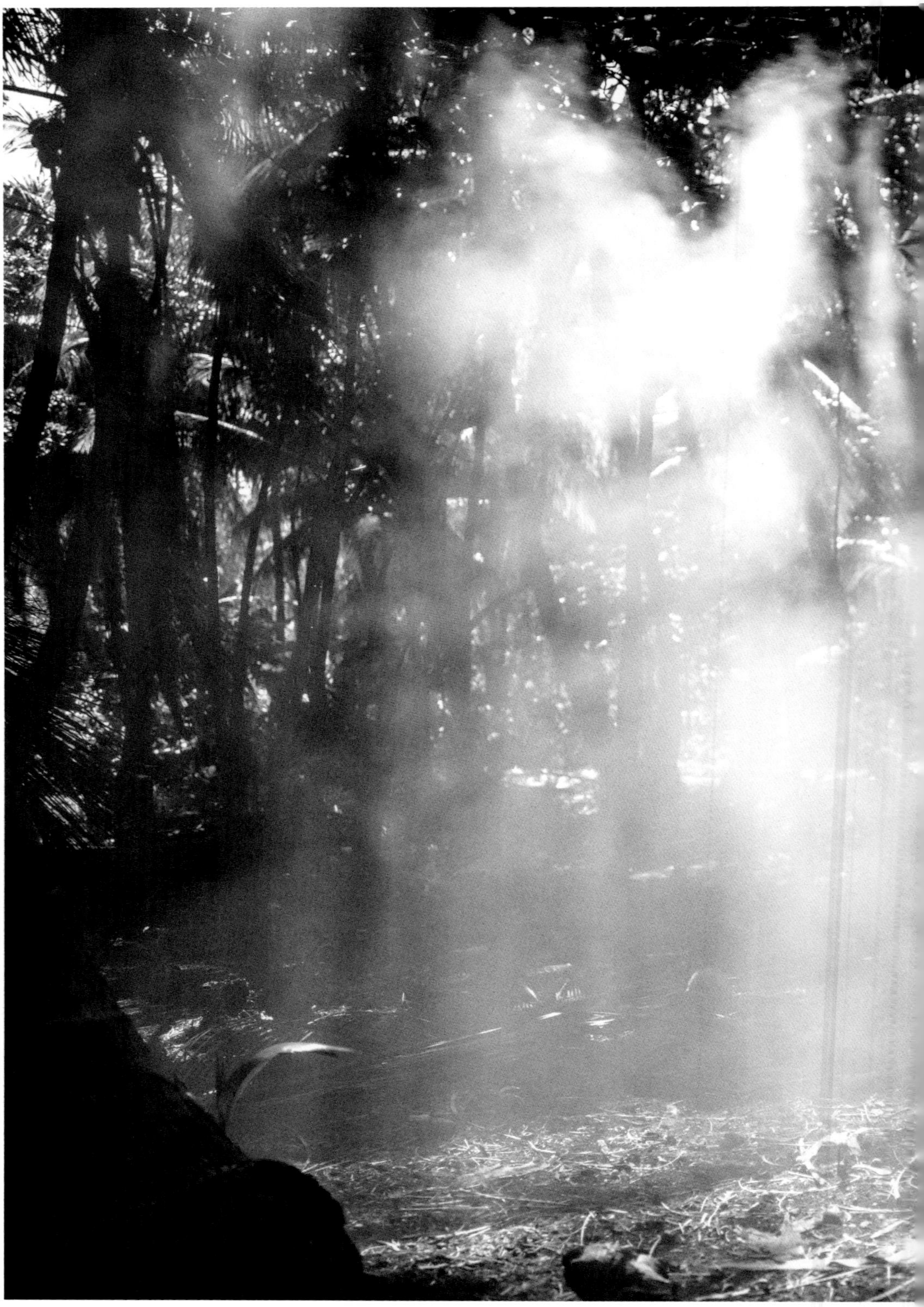

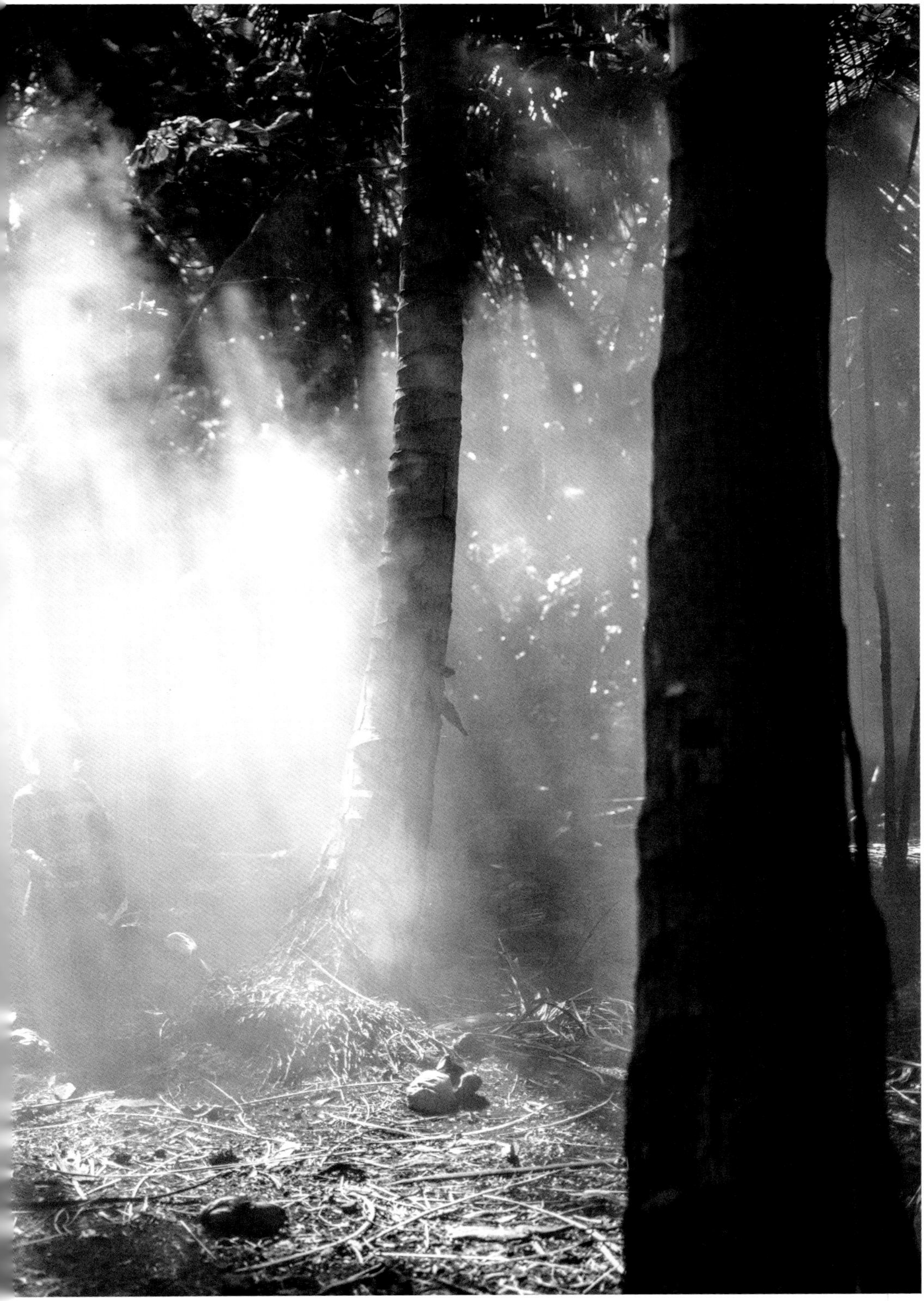

Isabel pointed urgently towards the ocean horizon. But I saw nothing.

She insisted: out there, one after the other, small waves of men, signaling from distant boats. Too far to hear over the engines and the surf, their semaphore unfamiliar. Is it a hailing, a warning, or a call for help?

Her husband had gone to the village docks, to pick up the day's new guests. It was early, and she was alone, outside her hut.

Where, I asked, do you see these men?

She looked at me as you would a mad fool, and walked away.

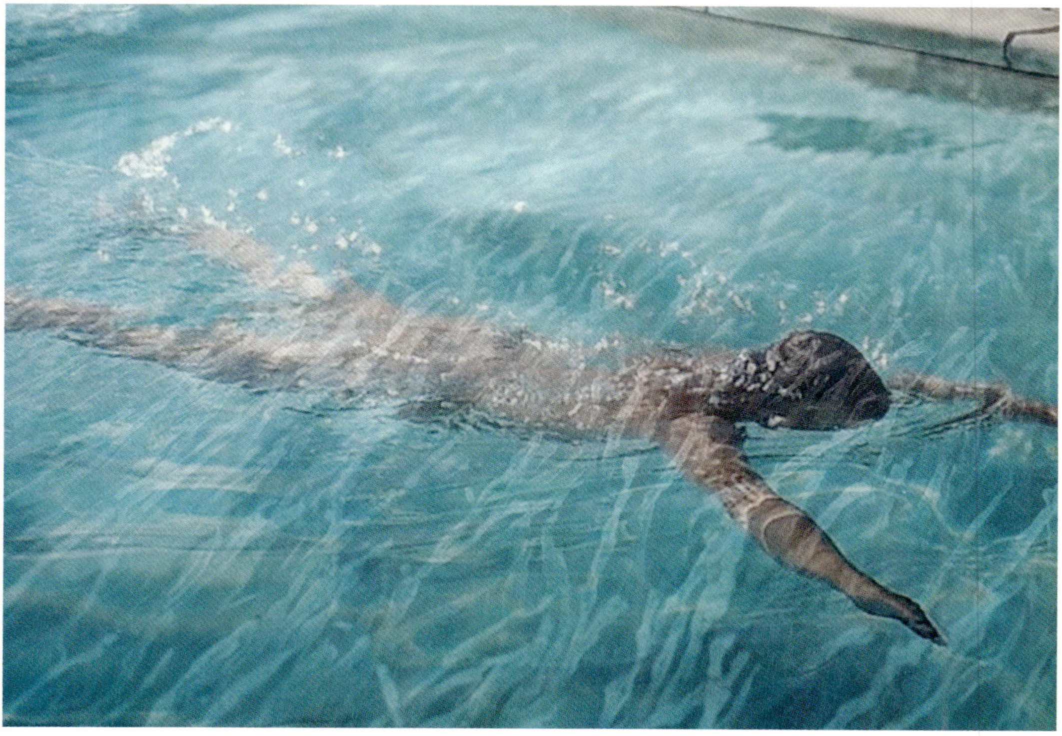

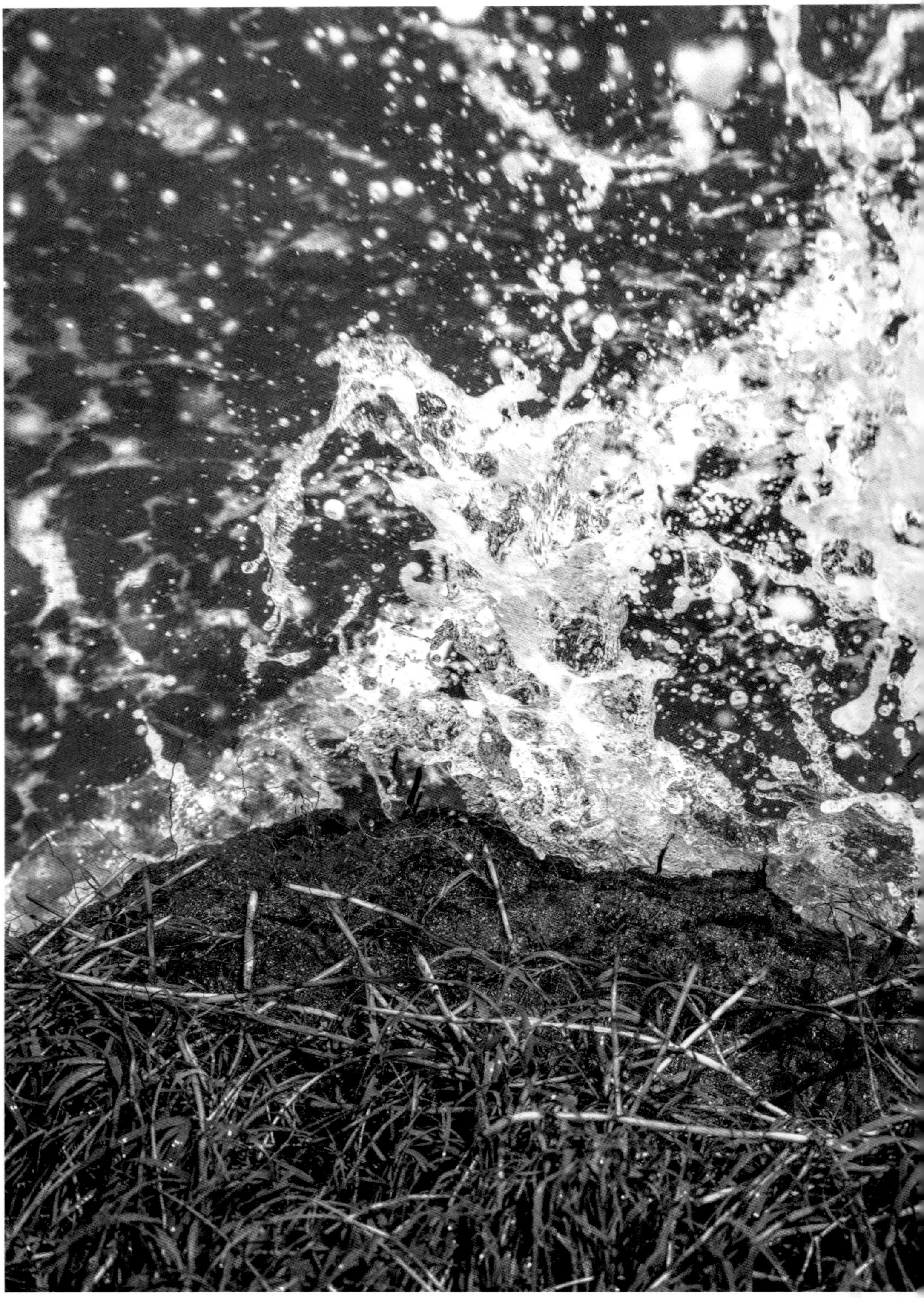

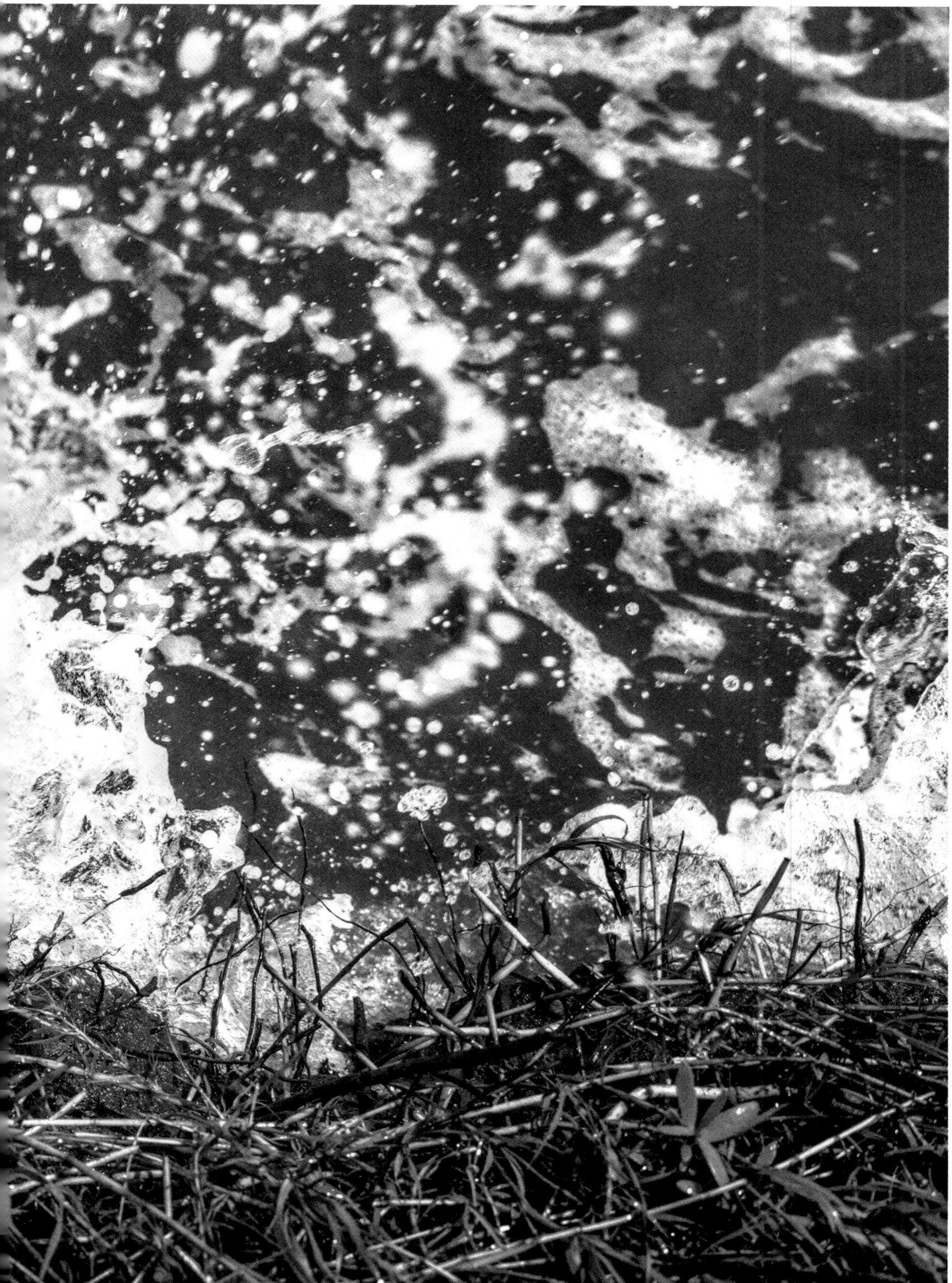

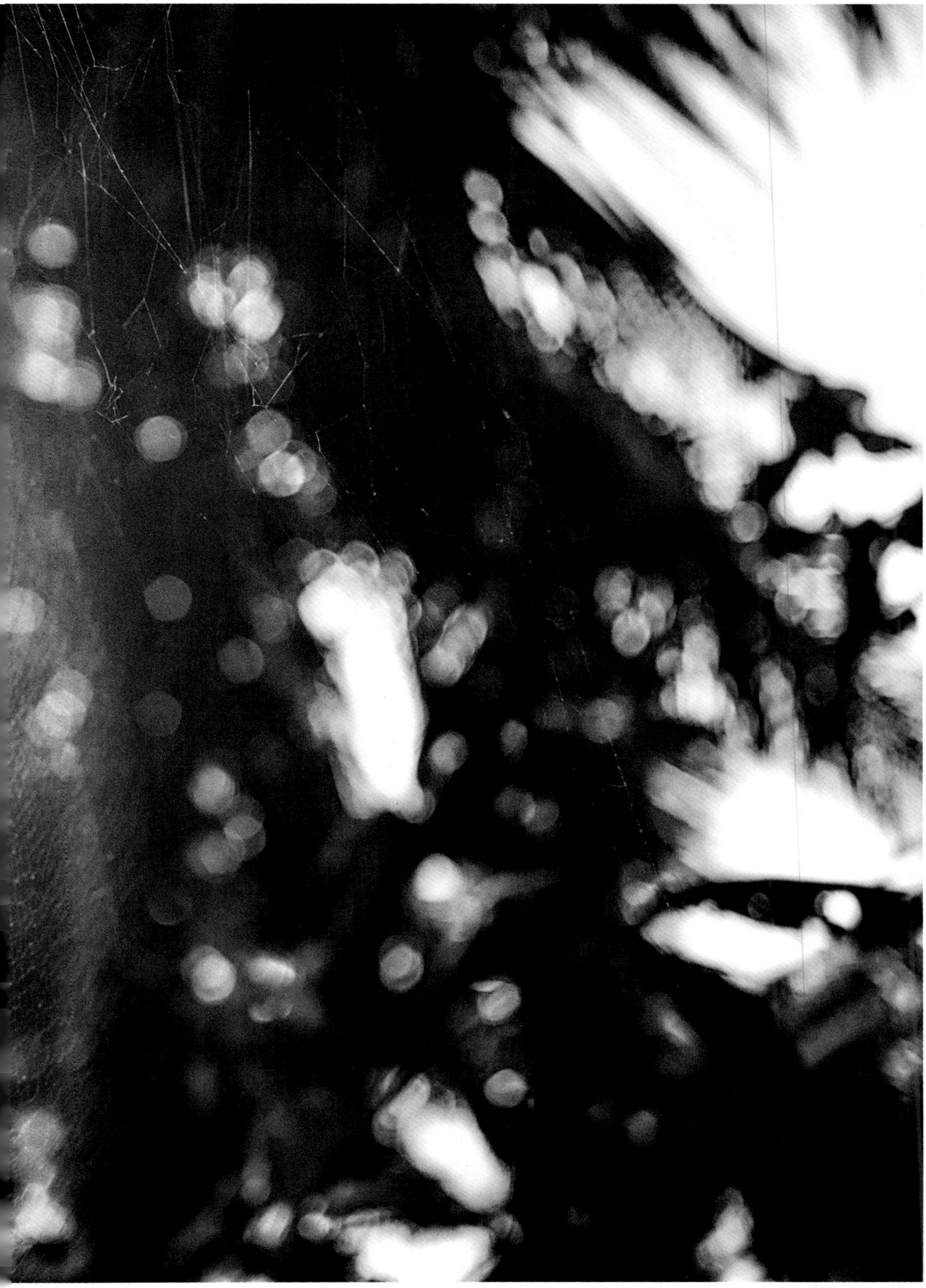

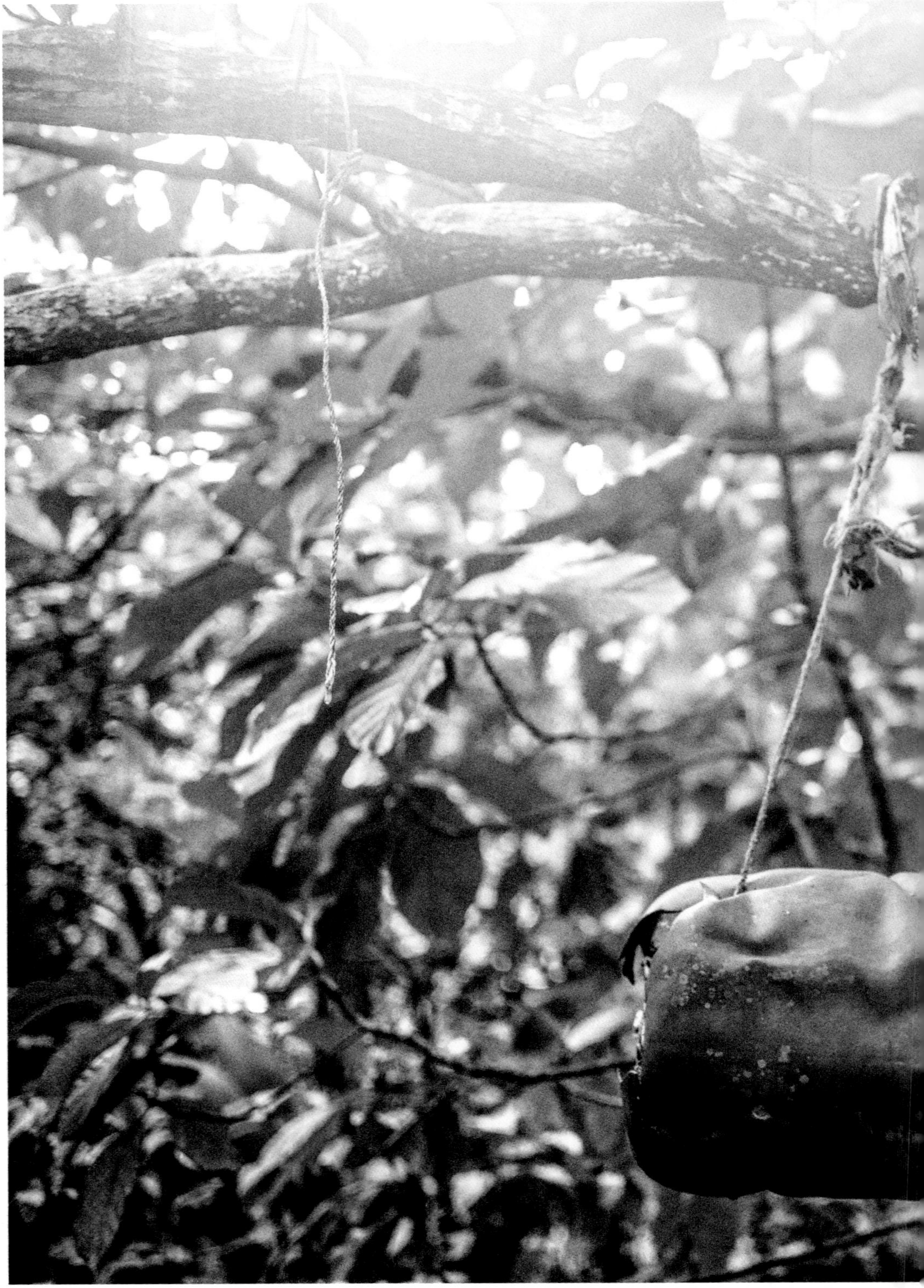

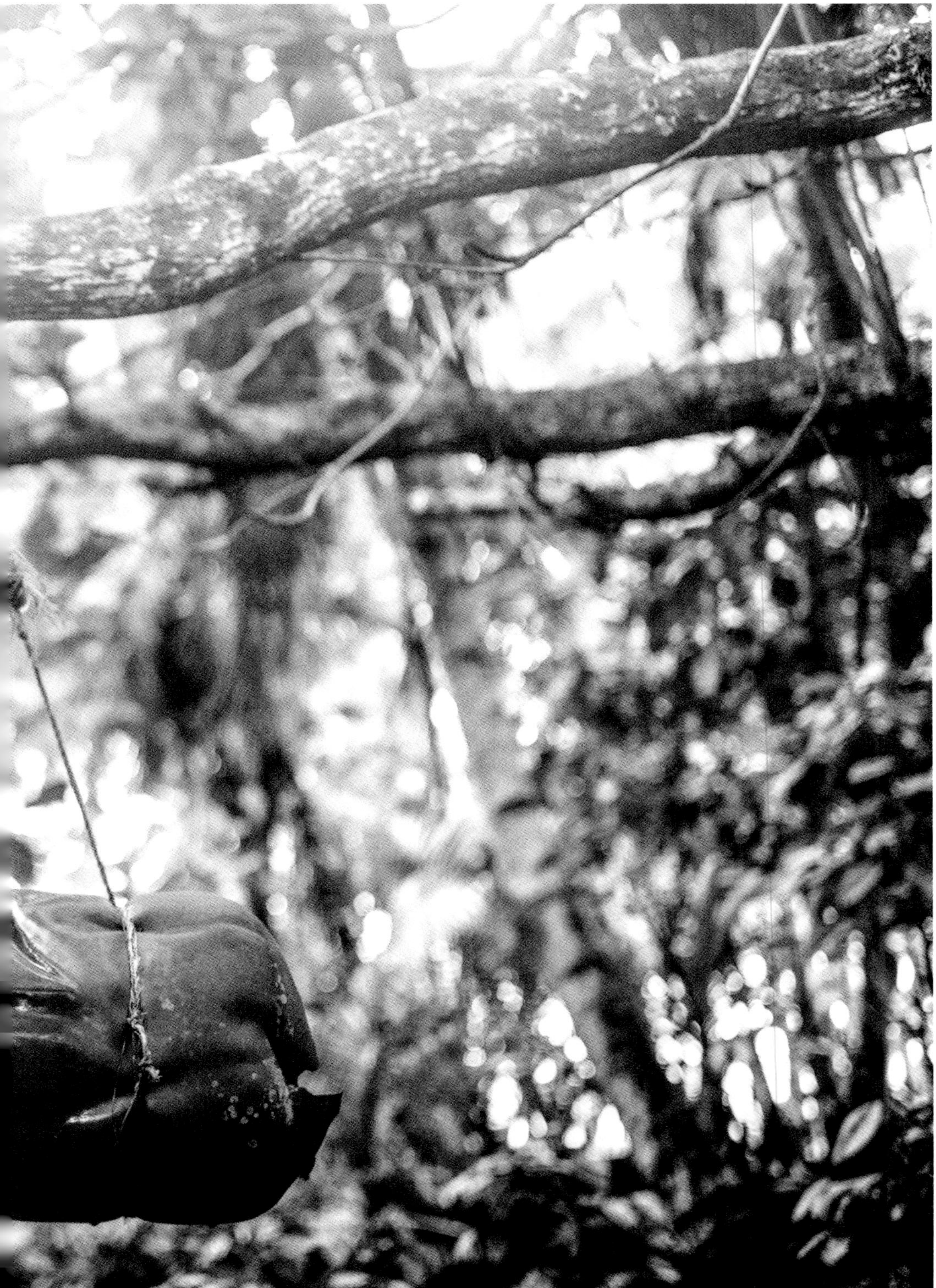

Too late for breakfast, too early for lunch, I woke up in a sweat. The sun had shifted while I dozed, pitilessly revealing me. In search of a cold drink, I wandered into the large open-sided thatched structure that everyone called "the pavilion." Here I found Isabel, seated alone in the middle of the long table, arms stretched forward and resting with open palms, as if the master had painted the martyr, but the pupils had yet to fill in the supporting cast. She was oriented westward, towards the ocean breaking over the reef beyond the lawn, Europe emphatically to her back, but the distance of her gaze was unknowable—somewhere between five feet and infinity. She smiled when I entered, but didn't seem to adjust the focus of her eyes. I might have been her child, or one of the stray dogs that wander the island, in search of shade and crumbs. She wouldn't know.

I sat down nearly opposite her—unavoidable—and smiled back. Her jaw relaxed slightly, perhaps in reply. I couldn't say. There was a dull pounding in the distance: two men working on the family residence. I asked if they were making repairs to her home.

"That is not my home," she answered. "We are making a new home, in the hills." And here she lifted her left hand off the table, rotating the whole limb from the shoulder to make a languid flourish, topped with a suggestive dropping of the wrist, to indicate the indefinite jungle behind her back. "The first real home," she added. "The other ones..." (here, her descending hand thrust itself forward and out to the left, dismissively indicating the cobbled network of huts where the hammering came from) "the other ones," she repeated, and then broke into a grin, having found the English word she sought, "were mistakes."

Isabel then described her new home, never looking my way, and never truly focusing on the sea before her. She was seeing the house, prismatically reflected down to her eyes from the mind above. The core, she began, is a large, round cylinder of stone, filled with clay and plastered white on the inside. The exterior is thatched over with a rough layer of coconut husks, so that it appears a crude shack from the outside. Isabel wanted this room—the circle of hidden stone—to be empty of everything, though it is the center of everything: the place to which the family will retreat, as to a fortress, if anything happens.

Other rooms will radiate out from the round house, connected by covered walkways—"as in Africa," she said, grinning mischievously. These rustic satellites will be built of whatever materials come to hand, natural or salvaged: driftwood, palm logs, strips of fiberglass and sheet metal, parts of old boats. The stone house will have a second floor above the citadel walls, with two bedrooms and round windows. Above that, a brightly painted mirador from which to survey the whole island. She pronounced mirador with a special relish, enjoying, it seemed, both the drama of her native Catalan and the spectacle of such visual command. Threading through the center of these three floors: a spiral staircase.

This was to be the first of several accounts of the house I heard—some directed at me, again, as for the first time; others intended for more recent arrivals at the compound. Details would shift—the number of rooms and outbuildings, the materials used for the thatch and auxiliary structures. But two things were always the same: the soaring mirador, and the perfectly round room below. The circular stone core at the heart of the home must remain, Isabel would insist, "completely empty, so you know it could always be anything."

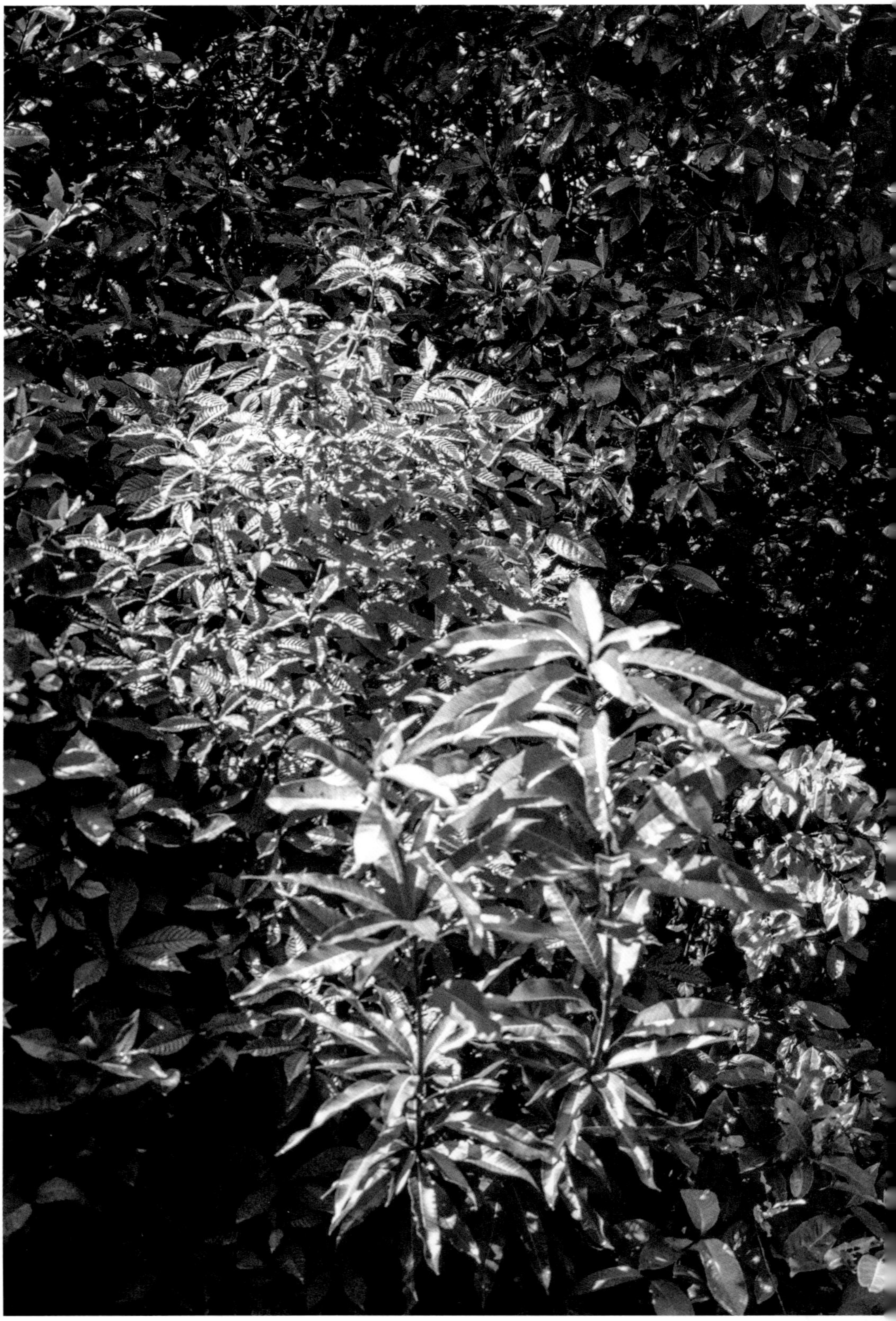

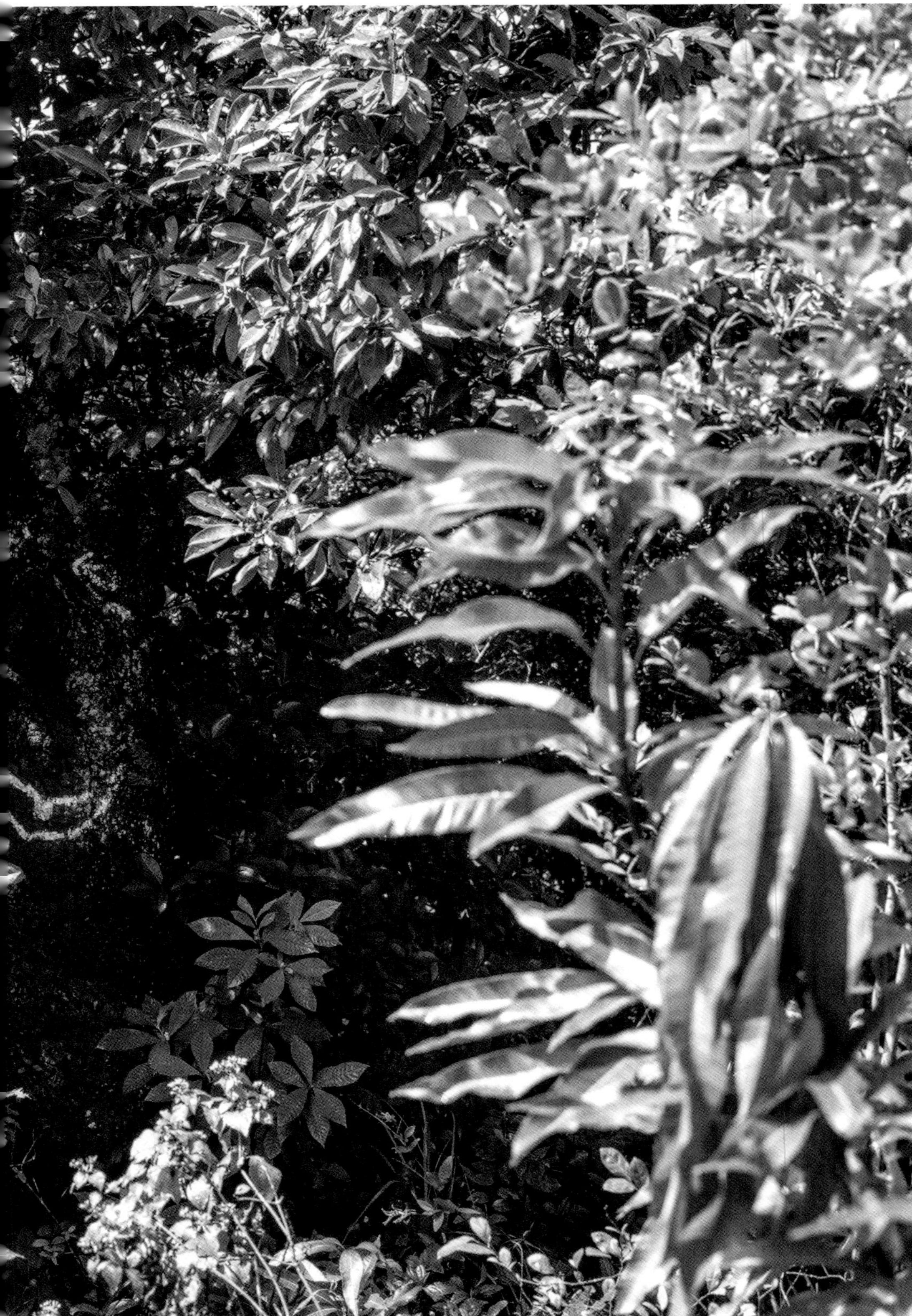

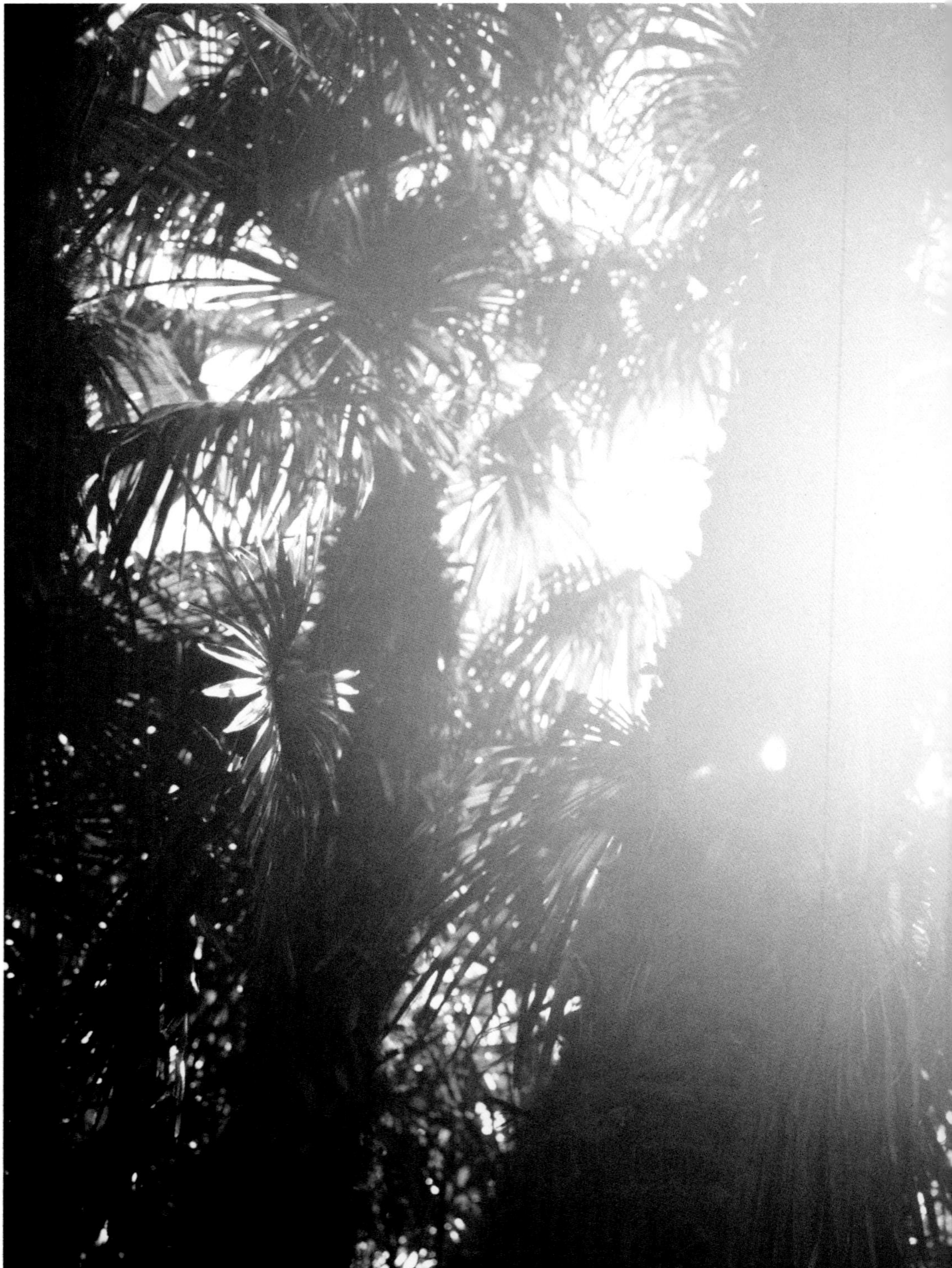

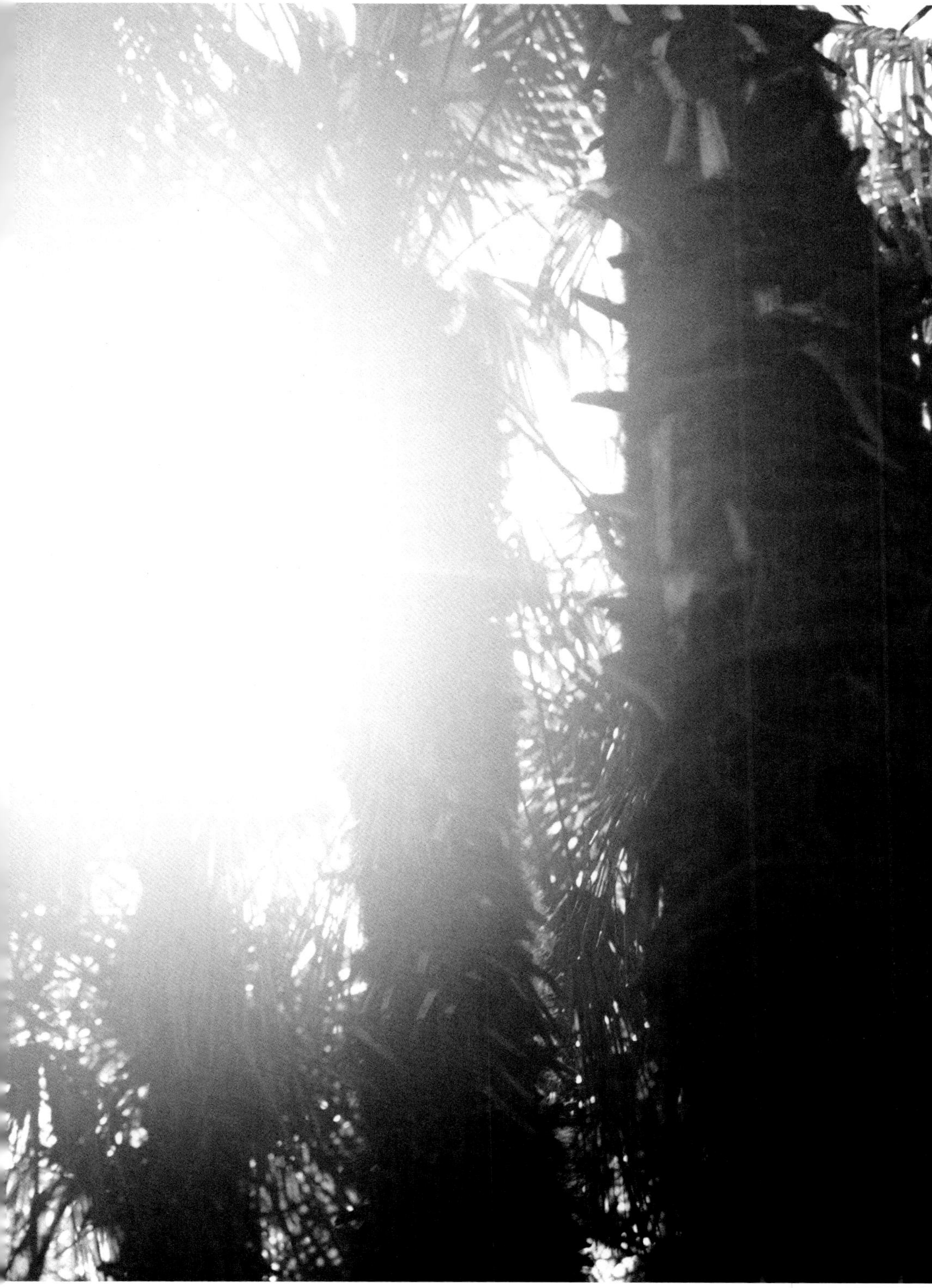

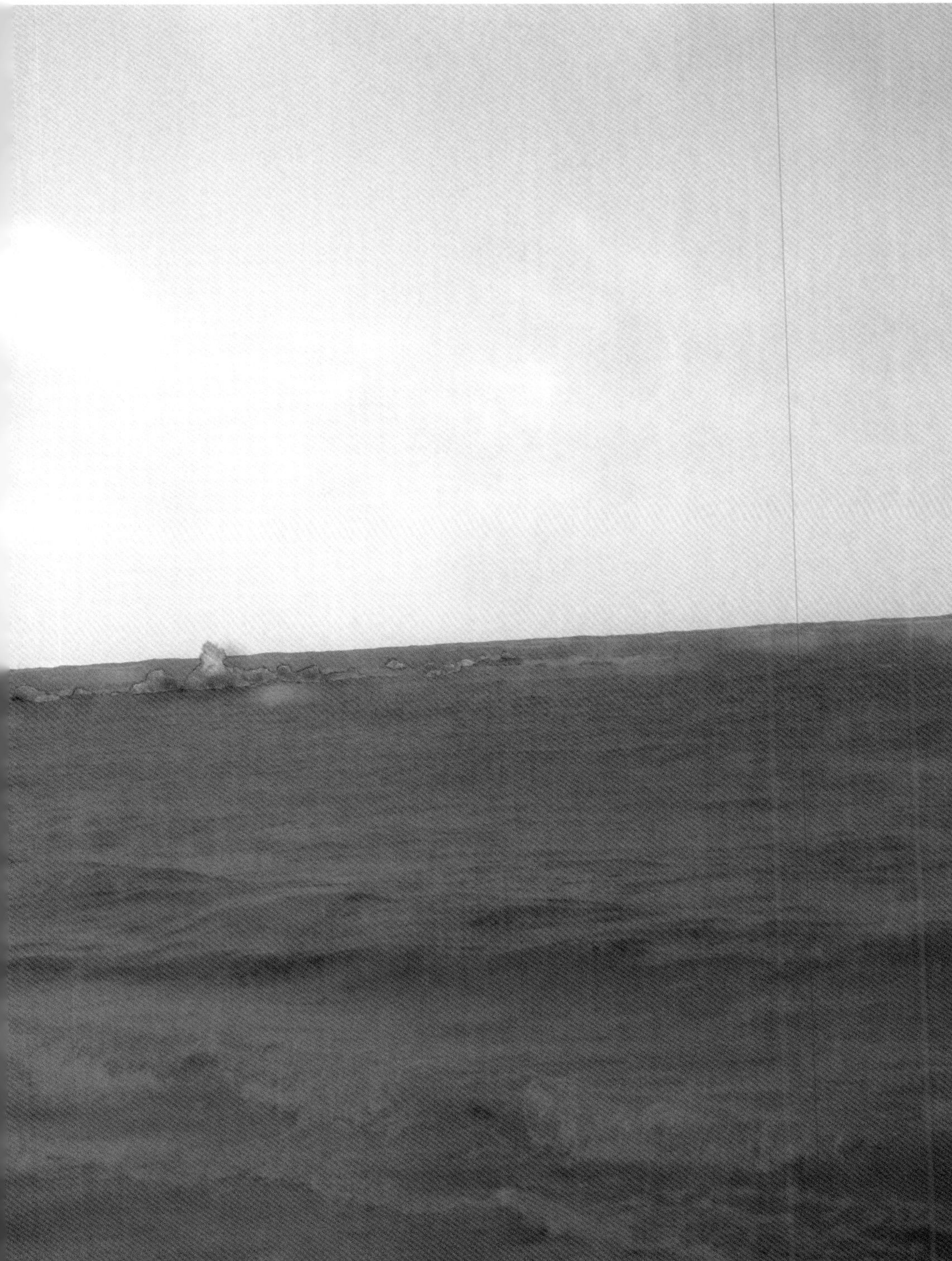

Yet again, in the quiet hours after dawn and before breakfast, I see Isabel on the beach. She is frozen, outstretched arms held out slightly from her sides, in an attitude of stylized astonishment. Suddenly, her left arm swings up to a perpendicular and her right arm rises to arc left above her head—as if a yoga instructor had taught a cheerleader how to spell a word. She holds this ambiguous shape for several seconds, returns her hands to her sides, and sinks to her knees on the damp sand. Only the ocean can see her face.

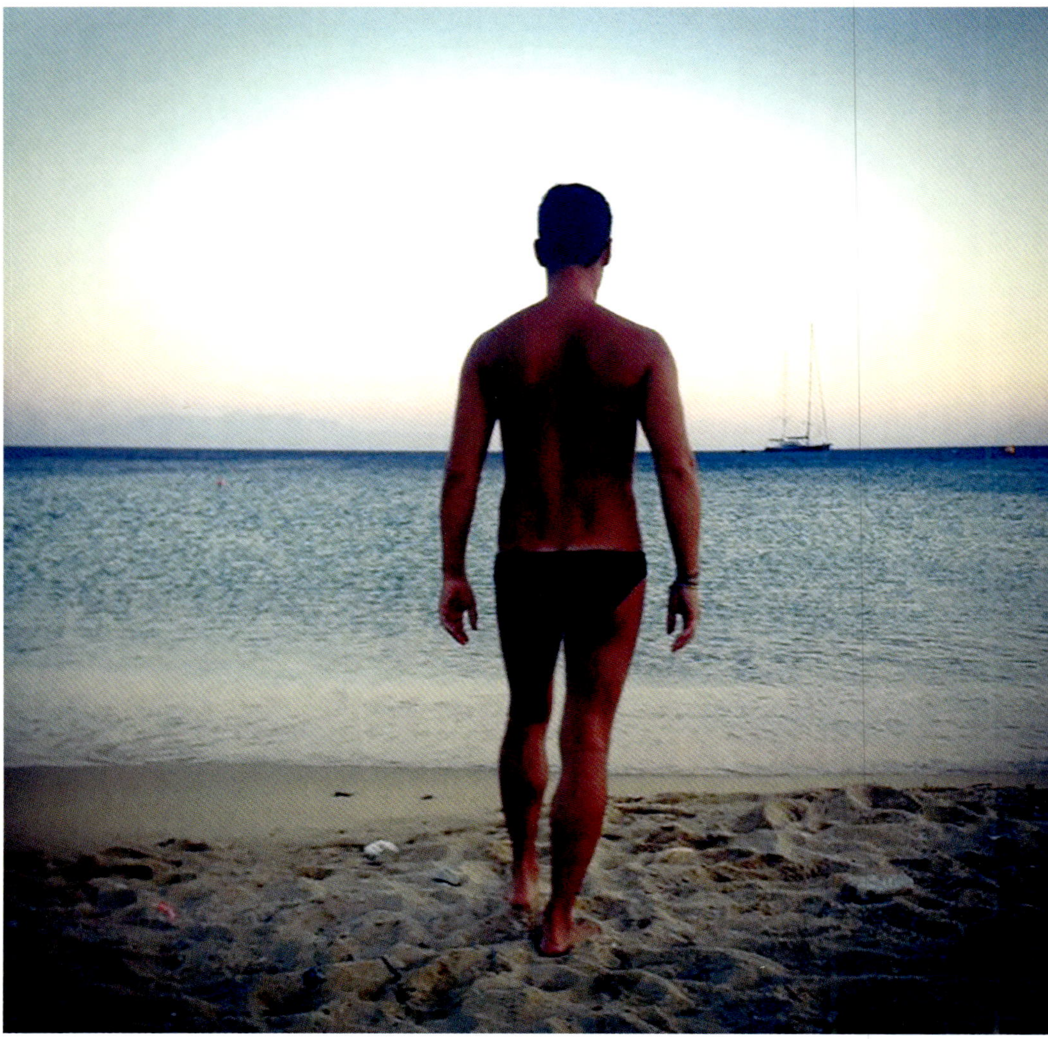

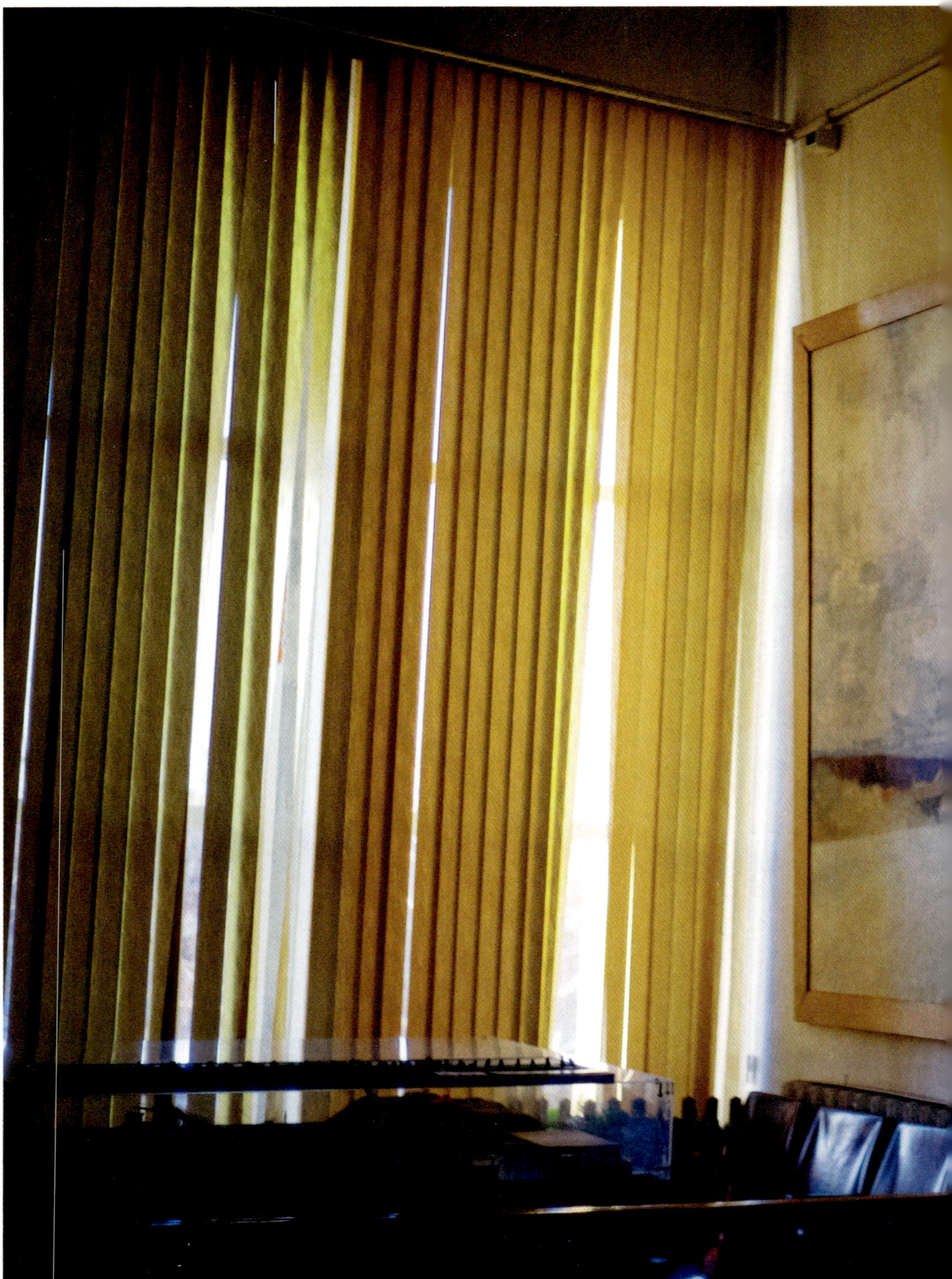

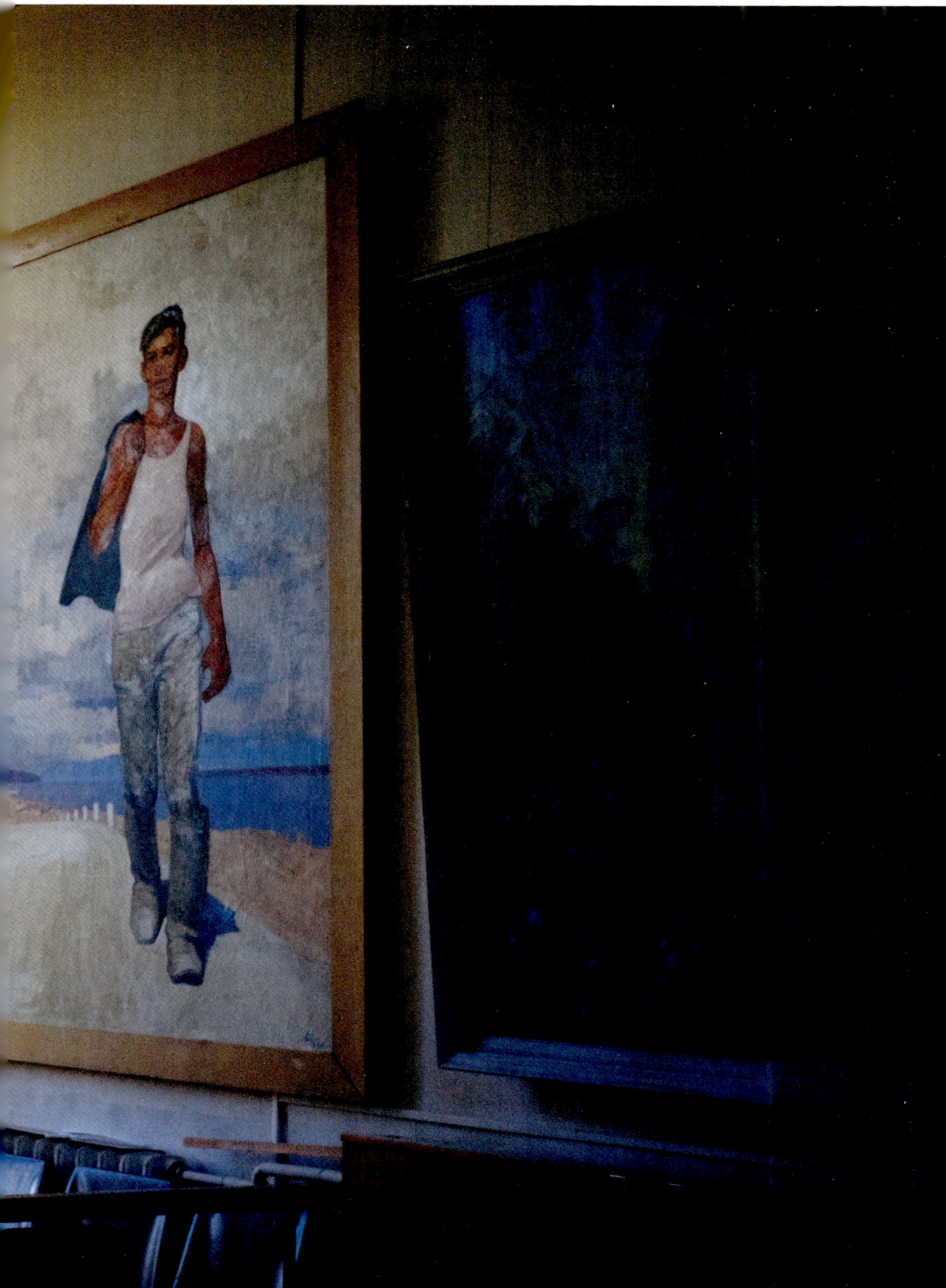

Chapter

III

We were sitting at his laptop, chairs side by side at a tiny corner table, in the kitchen that doubles as his bedroom. It was evening, and in the apartment's other room, his mother was watching a variety show on TV while his son sat enthralled in a multi-player game on a big desktop computer in the corner. Artem was showing me YouTube clips from a famous Soviet-era cartoon, the best one, his favorite one, one that everyone should know. We chuckled at the physical humor. I admired the wit and beauty of the animation, with its elegant high-folk simplicity. We giggled. And under cover of this multi-screen domestic distraction regime, we spoke quietly of other things.

He began: when you and your friend are together, what do you do? Who is *aktiv* and who is *passiv*? I tried to explain—my Russian equally ill-equipped to discuss explicit sex as contemporary politics—that there are a broad array of additional activities beyond those specific roles. He took in this intelligence thoughtfully, a cartoon wolf in overalls reflected in his large grey eyes. He asked: with your hands? I elaborated: hands, mouths, and then I momentarily forgot the word and stuck out my tongue, pointing at the awkward organ. "*Yazik*," he supplemented, in the voice of a former schoolteacher. I said yes.

By now, Artem was playing scenes from another Soviet children's classic—a live-action fantasy film populated by flamboyant bearded wizards, over-rouged minstrels, young men with blue faces and long ears, or in short dresses and pastel tights. It was a mincing woodland-camp wonderland.

I asked: is there someone he has been together, in love, with? No, he answered, nobody at all. I asked if he had been with men just for sex, and he said no, never in his life. On the laptop, singing wizards pranced around a steaming cauldron. How long, I asked, have you known that you were gay? He answered, firmly: a very long time.

That morning, Artem and Sasha had met me at the bus station in L., as if that was the usual thing: being picked up by my two pals. Except that I knew Artem only as a correspondent in a messenger window on a gay website. I had only one picture, in which he appeared distant in the frame, his face almost unreadable, his right leg lifted with the foot resting jauntily on a post, right hand on right knee in a faintly readable gesture of self-confidence. Or so I conjectured.

As for Sasha, I didn't know him at all, or even to expect him. I had imagined that Artem would be alone. Our brief phone conversation, while I was on the bus, might have told me otherwise. Beyond confirming my arrival time, he asked also, furtively, that I not tell anyone where we had met. But he hung up before I could ask: how, then, are we supposed to know each other? Seeing two men approach me, this question took on greater urgency. But somehow, miraculously and then obviously, nobody ever asked.

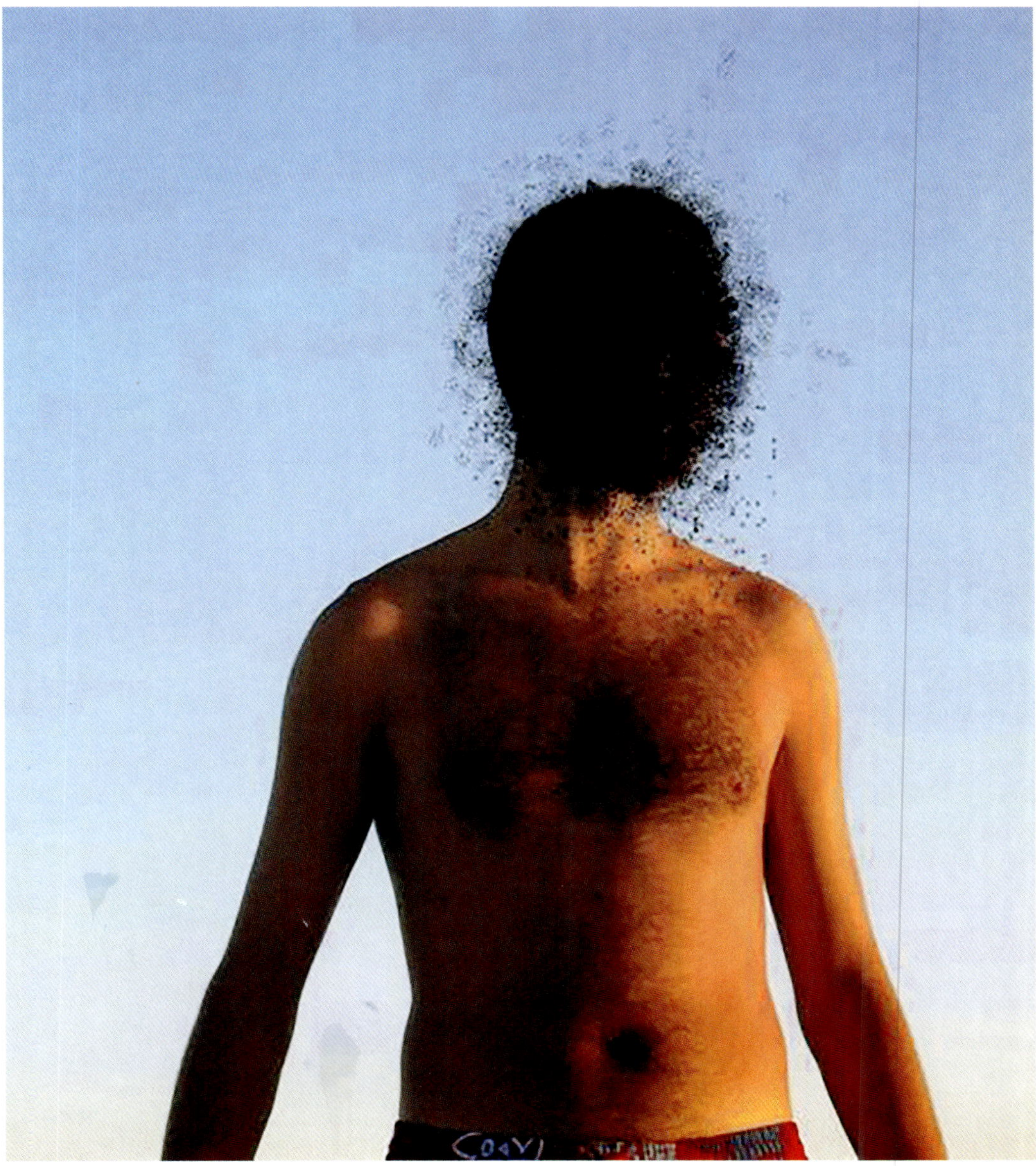

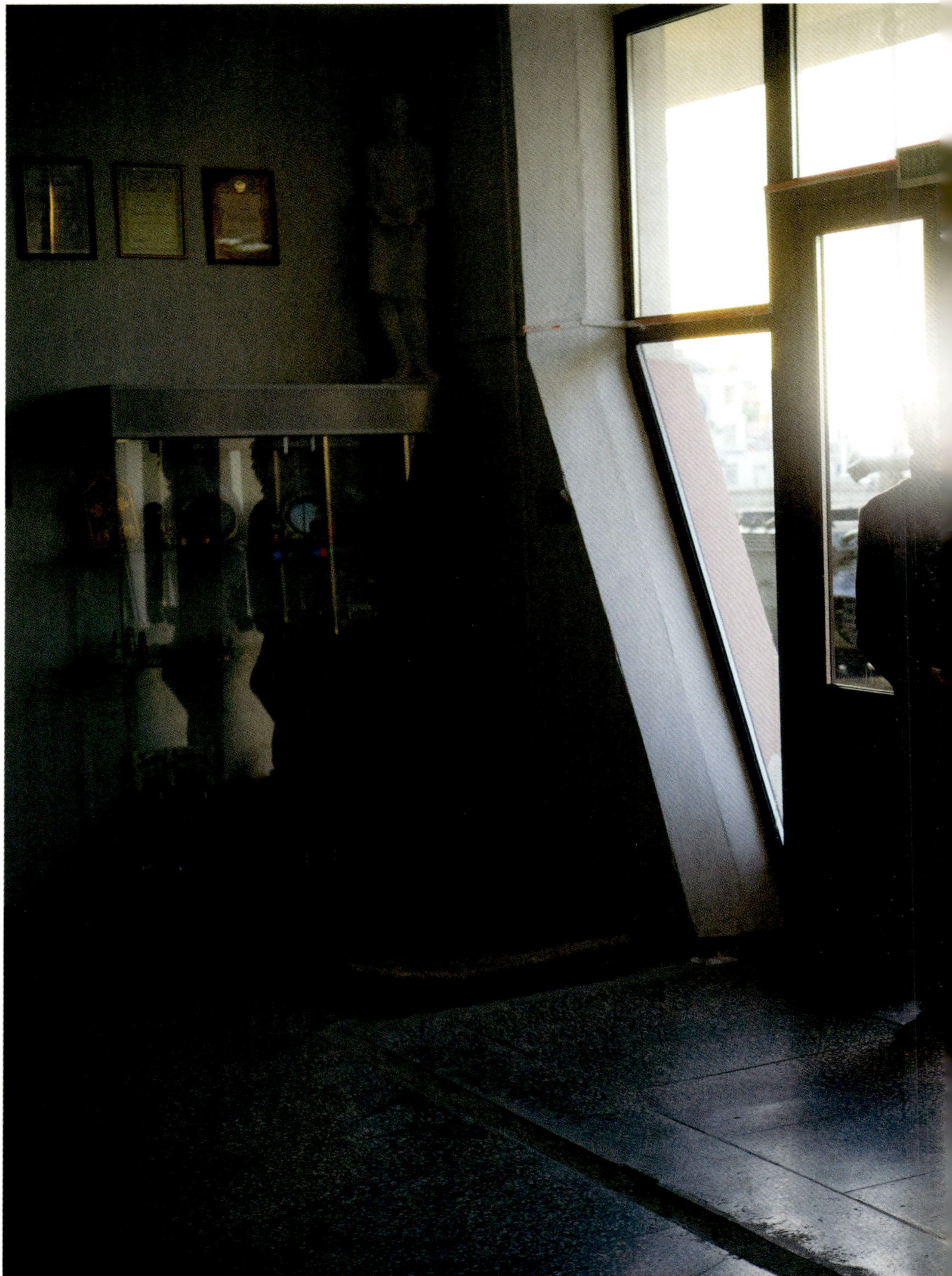

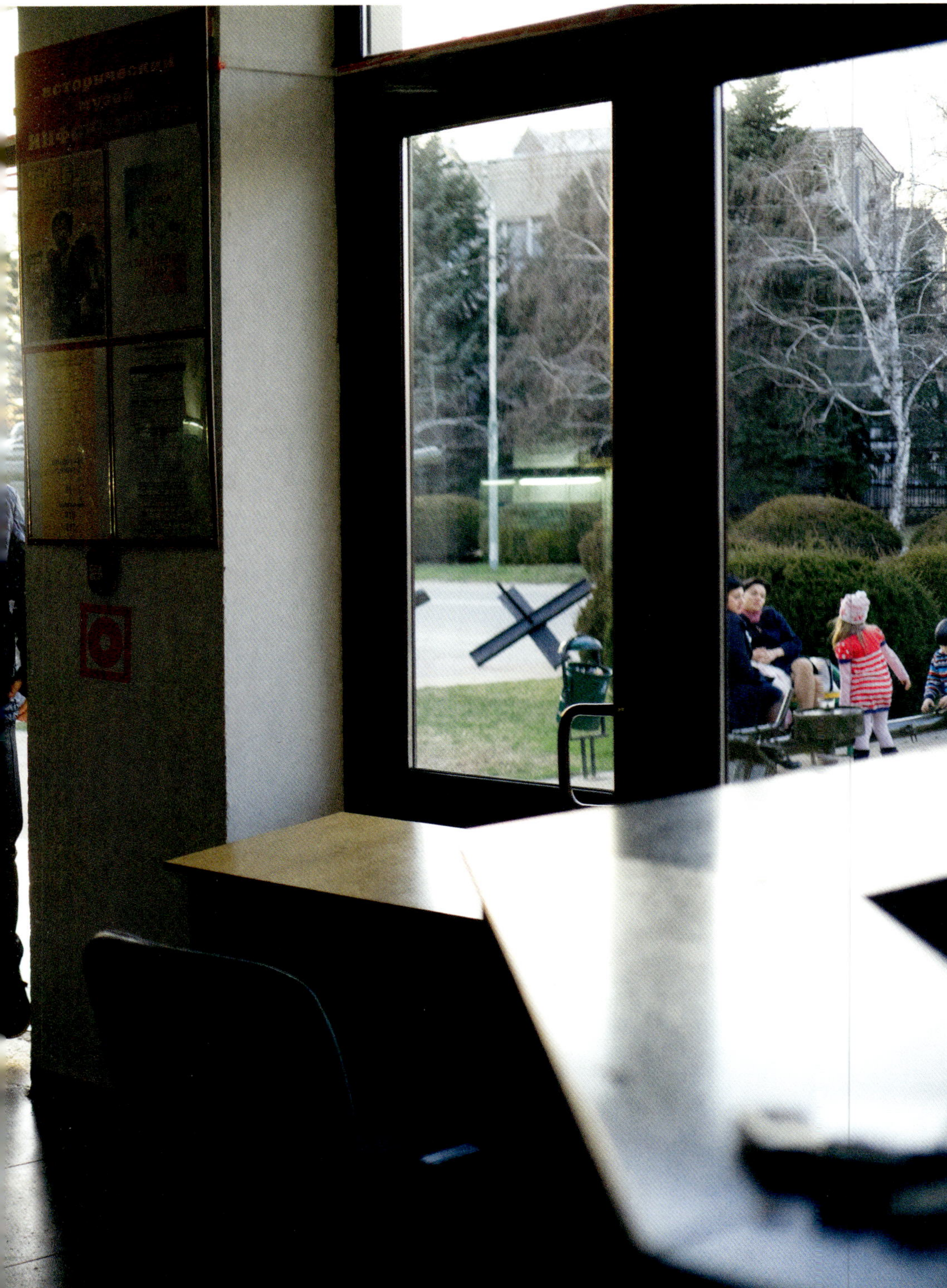

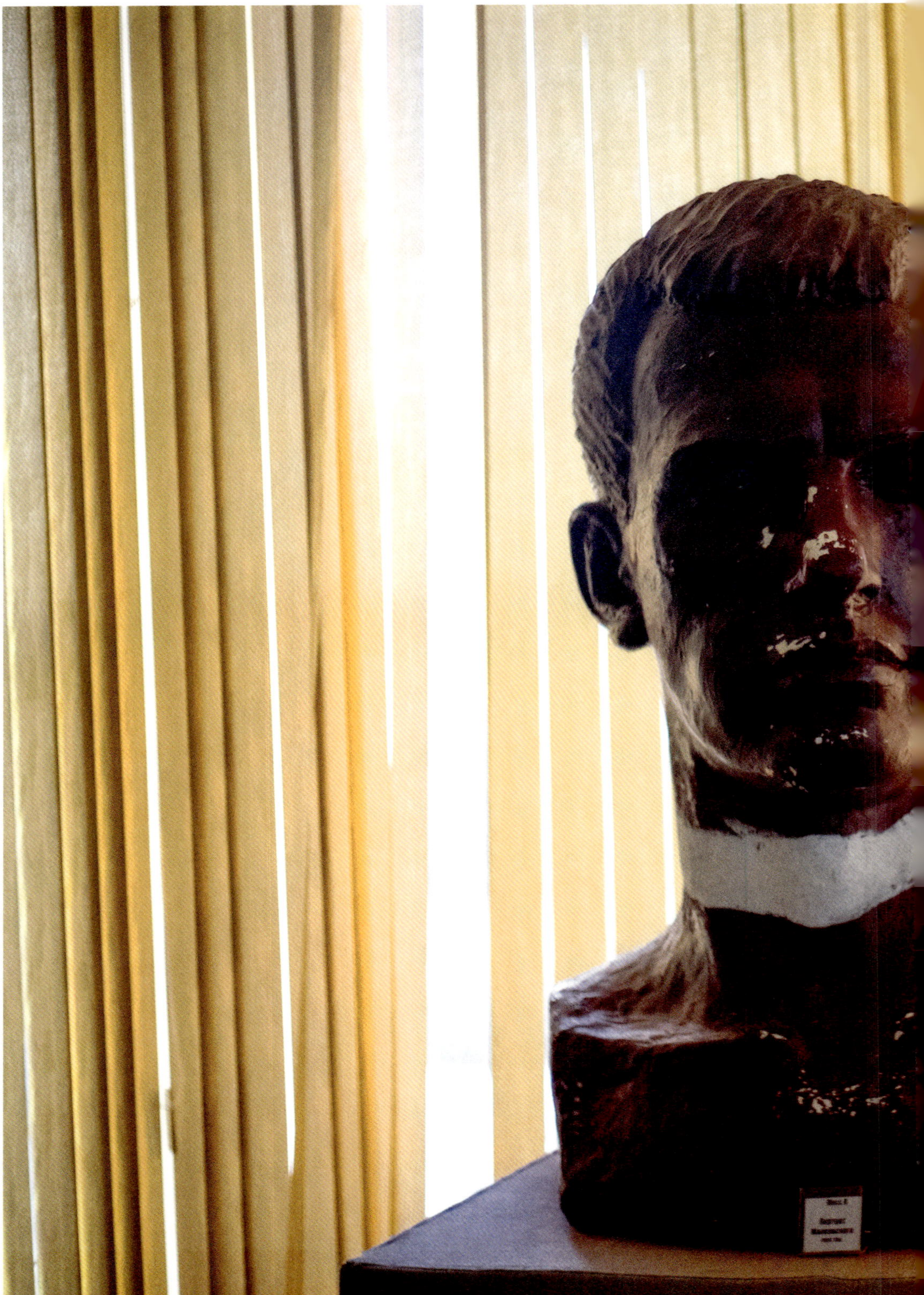

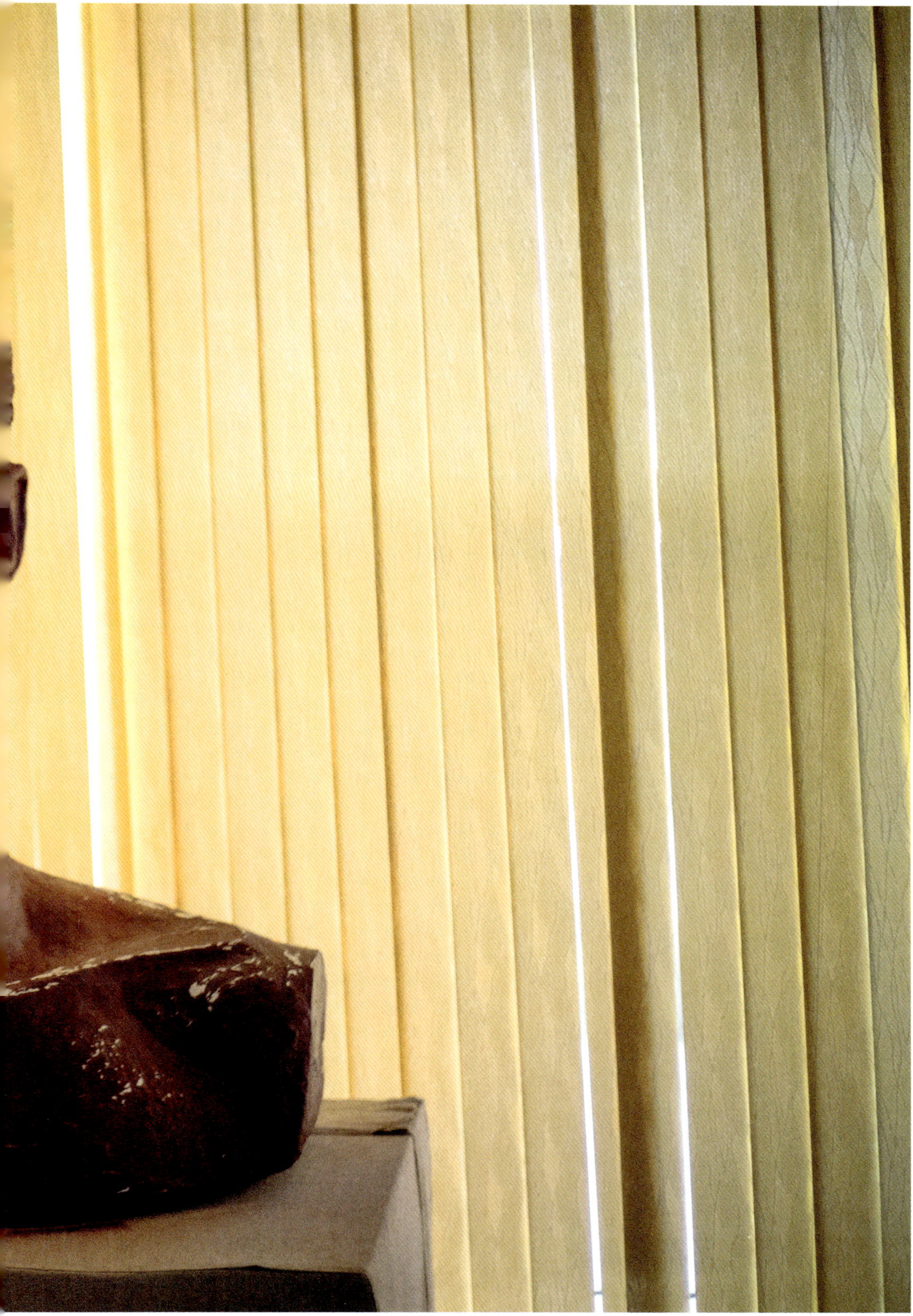

Our first stop was the town's museum, a lovely cement-and-glass Soviet structure, with a sleekly modern façade. We were met by Sasha's father-in-law, a retired Soviet army officer who appeared to be the institution's primary keeper. The exhibition swept from the stone-age to the collective farm in one room, before giving way to extended displays of regional oil painting in several more galleries. Our guide informed me, proud hand on round belly, posing beside an imposing Cossack uniform that would most likely fit him: "many of my own works are in the collection."

After the tour, my hosts magically summoned a small picnic from the shadows of the darkened museum lobby. As always, I had been roused by the sculpted musculature and seductive utopianism of Socialist Realism. Now, between crisp pickles and vodka toasts, secretly gay Artem spoke longingly of the lost communal intimacy of Soviet times. Perhaps, it struck me, I was not the only one so ravished by this art.

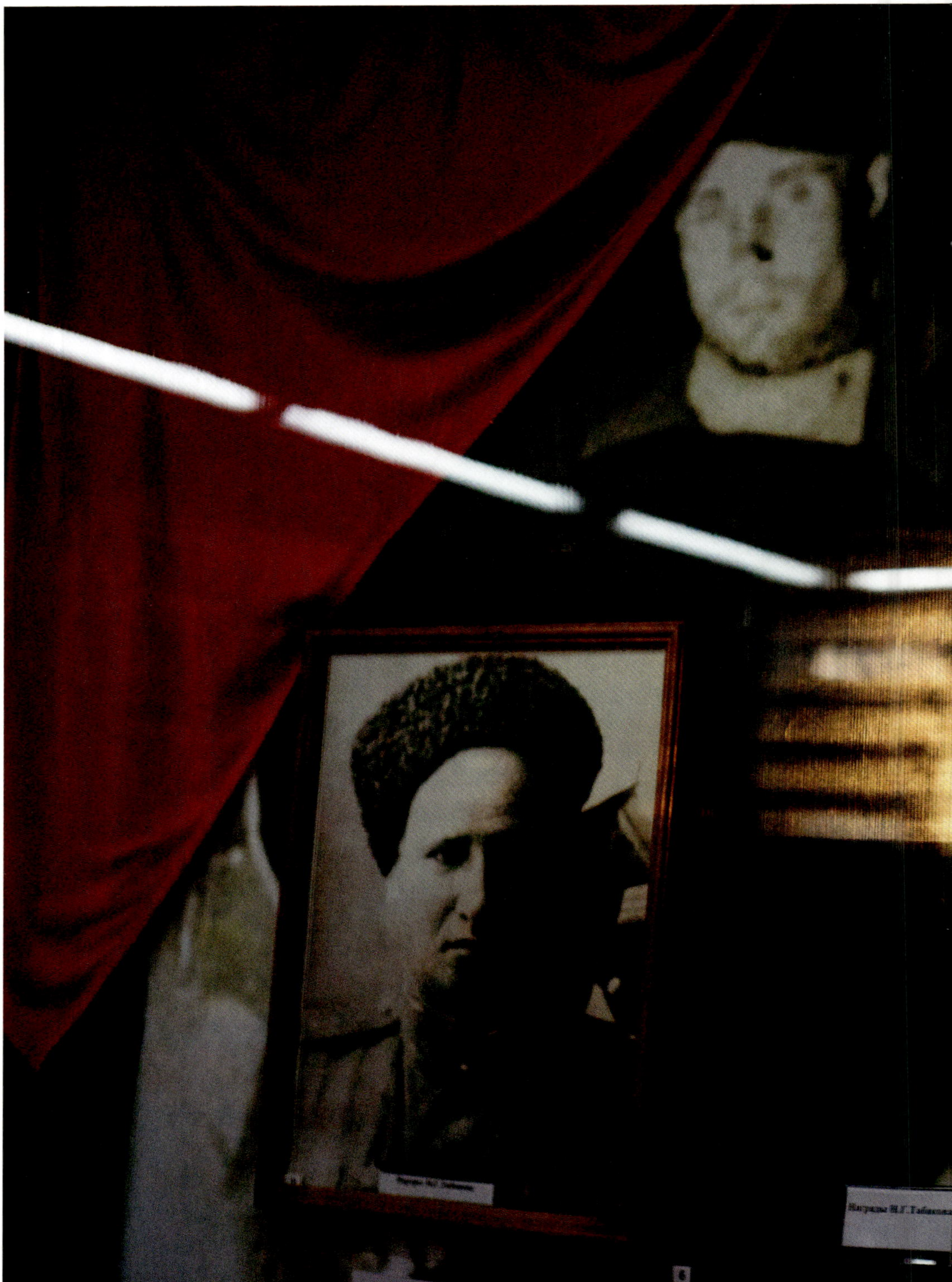

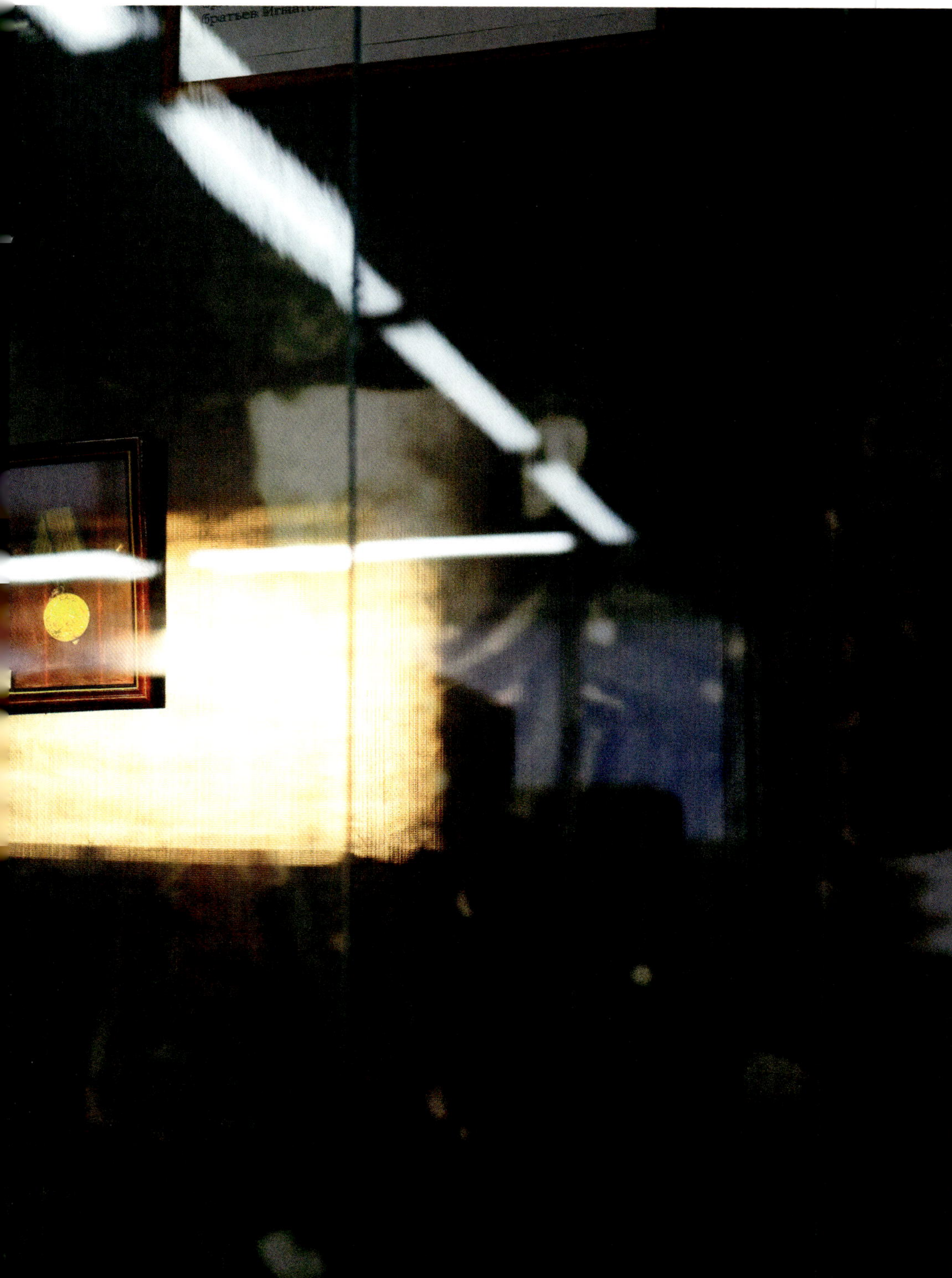

That afternoon, at the former officer's dacha, I learned to shoot from the man who raises rabbits for food but does not have the heart to kill them. The animal lover, discussing the cultural attitudes that make this region distinctive, explained it thus: "for example, if homosexuals arrived in our town, we would kill them."

Later, alone together for two minutes when his friend stopped to pick up milk for his wife, I leaned forward from the back seat, and asked Artem: when your friends say things like "here, homosexuals would be murdered," does it hurt you? He nodded, brows arched gently above bright eyes, and whispered, simply: "yes."

Moments later, loose-limbed, open-faced, ever-grinning Sasha jumped back in the car. As we drove back to Artem's place, a surging romantic Russian rock ballad came on the radio. They both loved this song. One of them cranked up the volume, and together, full-throatedly, the two men sang about love, all the way home.

We stood between his apartment building and the small dusty playground, waiting for my taxi in the dark. I picked up the thread of my questions: do you want to find a man to be with?

Artem answered: it's impossible. The town is too small. It is too dangerous even to try.

"What about going to R.," I asked. "It would be safer there."

He just shrugged, almost imperceptibly, the gleam of a yellow street light in his eyes standing in for sadness. "I have my family, my job, my friends … and at night, I have the internet."

This statement was a summary and a conclusion. Already, the taxi had pulled up, headlights raking admonishingly over our bodies, side by side. There was never enough time to ask the next question. He walked me to the car, instructed the driver and insisted on paying for my ride. I hugged him, briskly, lightly, as much as he would allow. He settled me in the back seat, asked if I was okay. I watched him wave in silhouette as I rode away.

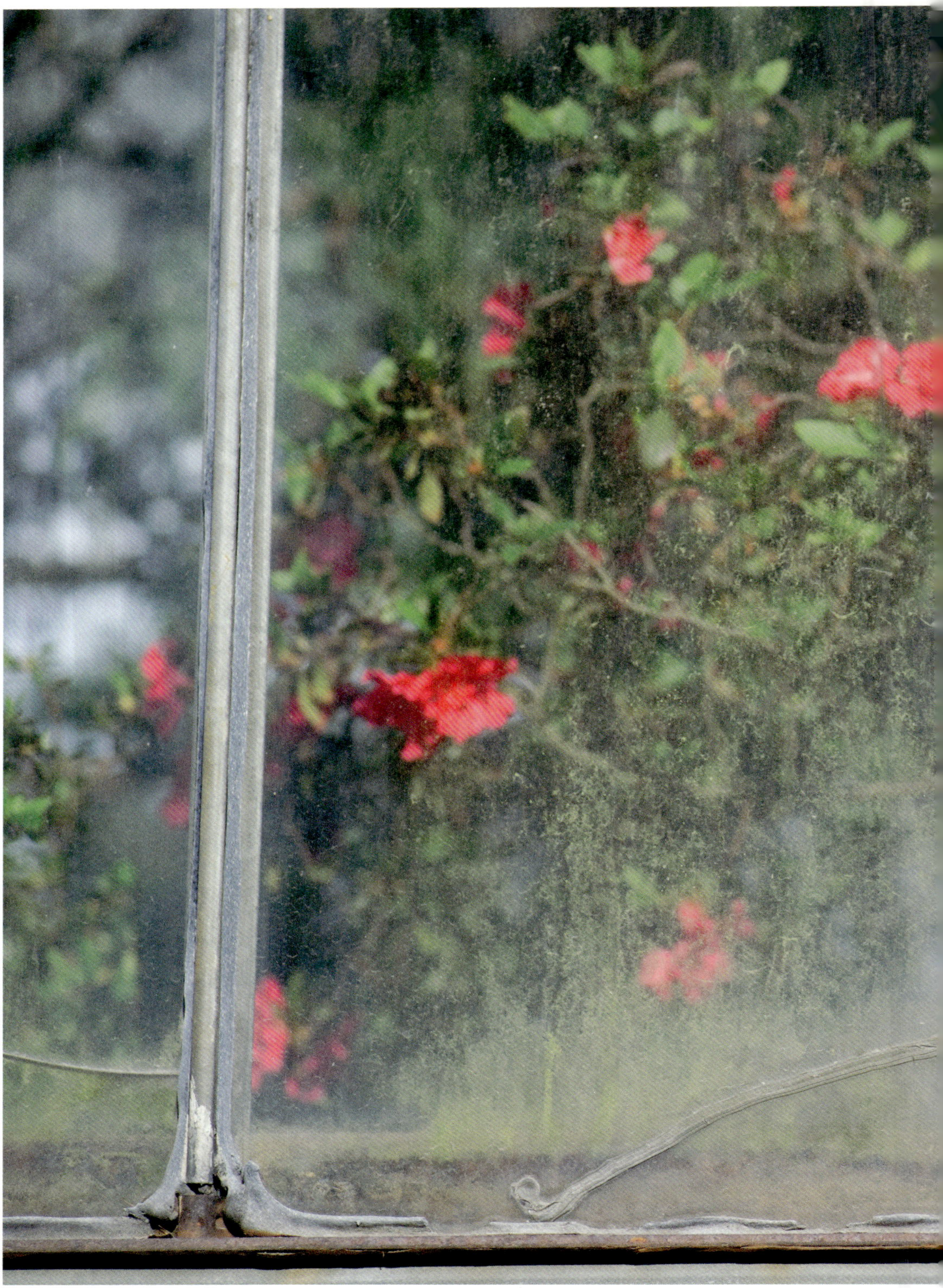

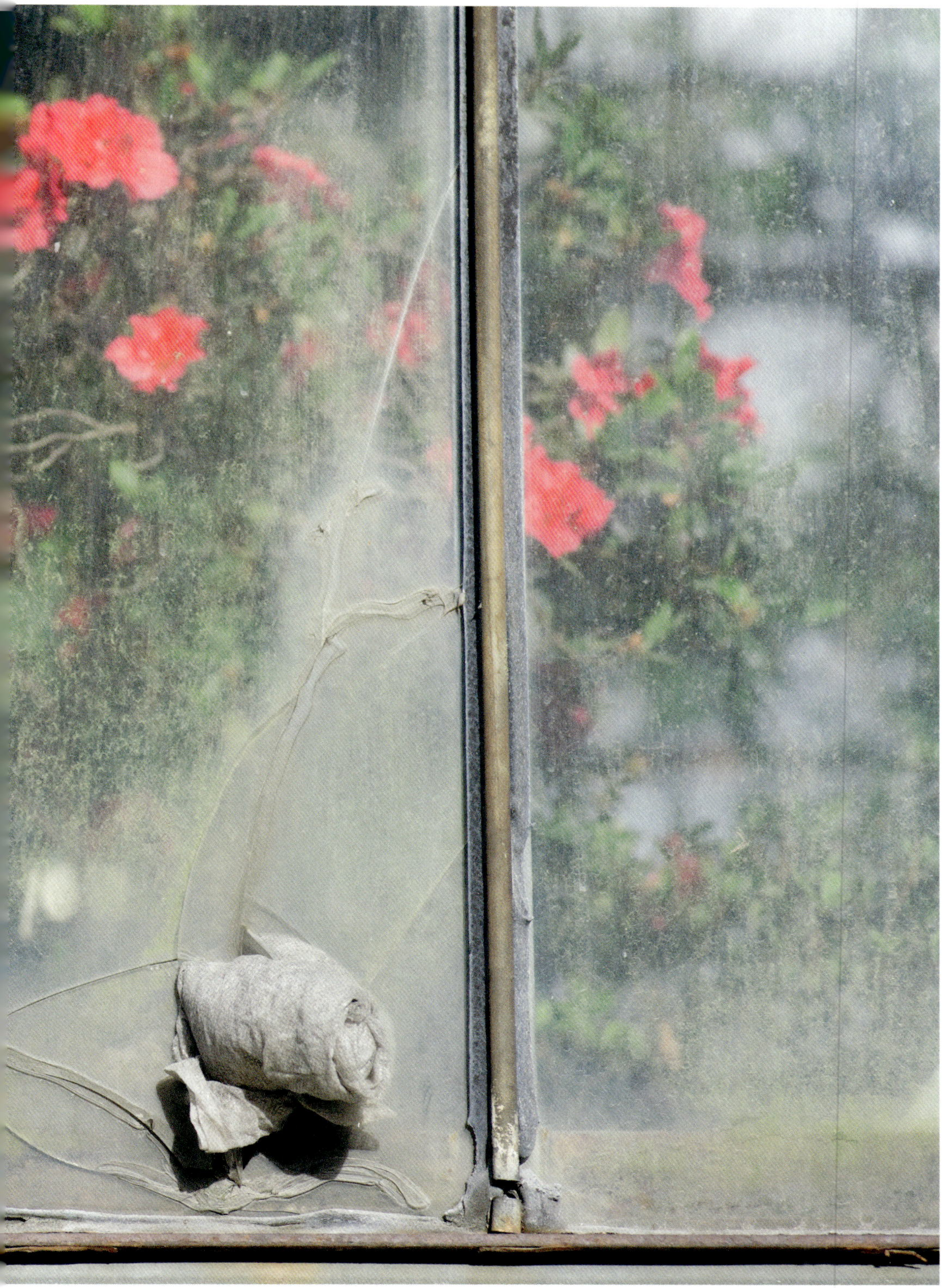

Artem imagines a home in the future: a dacha, just outside the town, quiet, with a car to drive there, and a room for each of them. There would be a field for fruit trees, starting with the apricots that bloom first in spring. He would build his own greenhouse, to protect the defenseless early shoots. No more trusting those grassy seedlings and striving sprouts to the hands of others, or to the predations of this untamed environment. At the back of the garden, trailing over the fences, blackberries, currants and gooseberries: a perimeter of well-defended fruits. In the front, facing outward, will be roses.

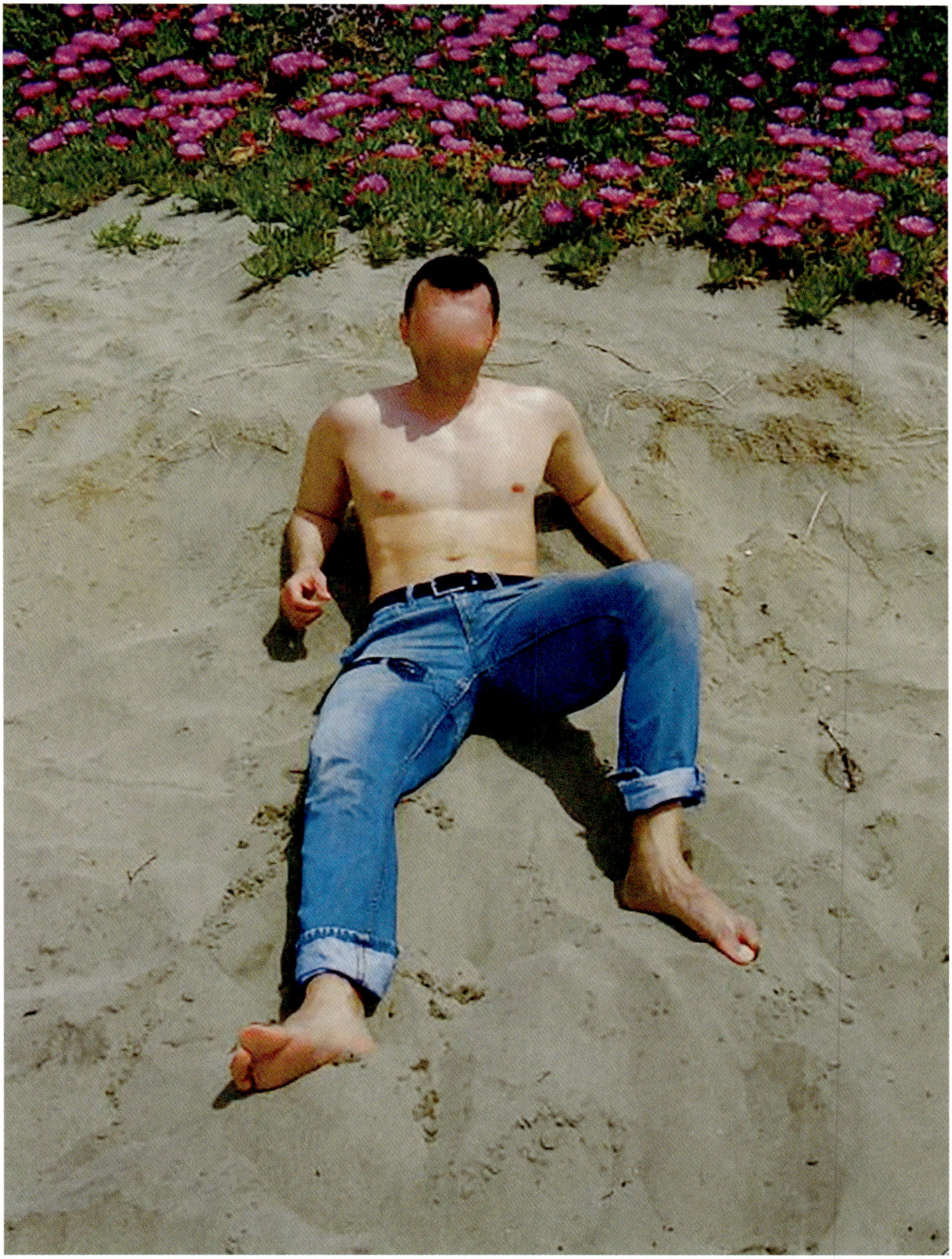

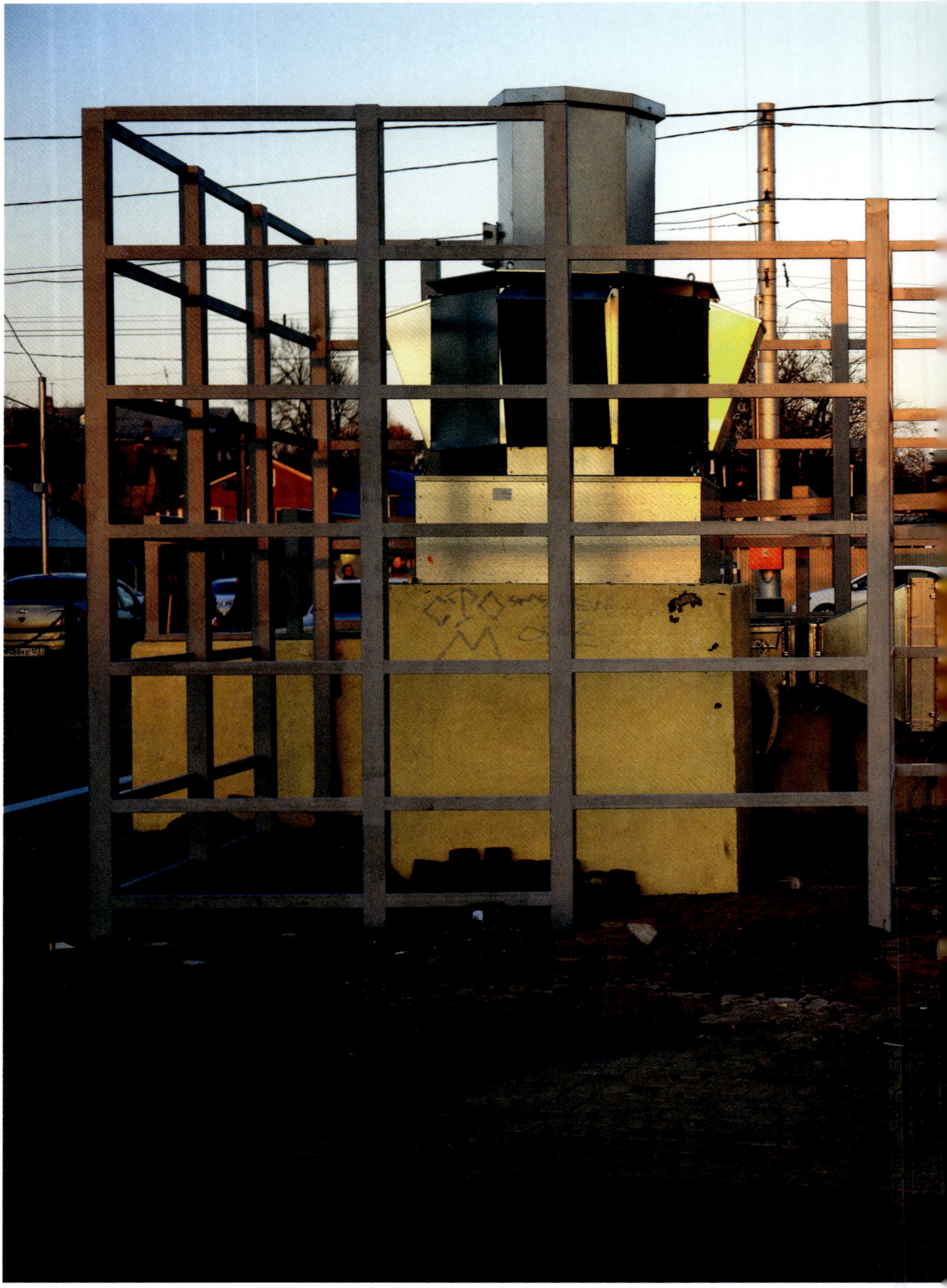

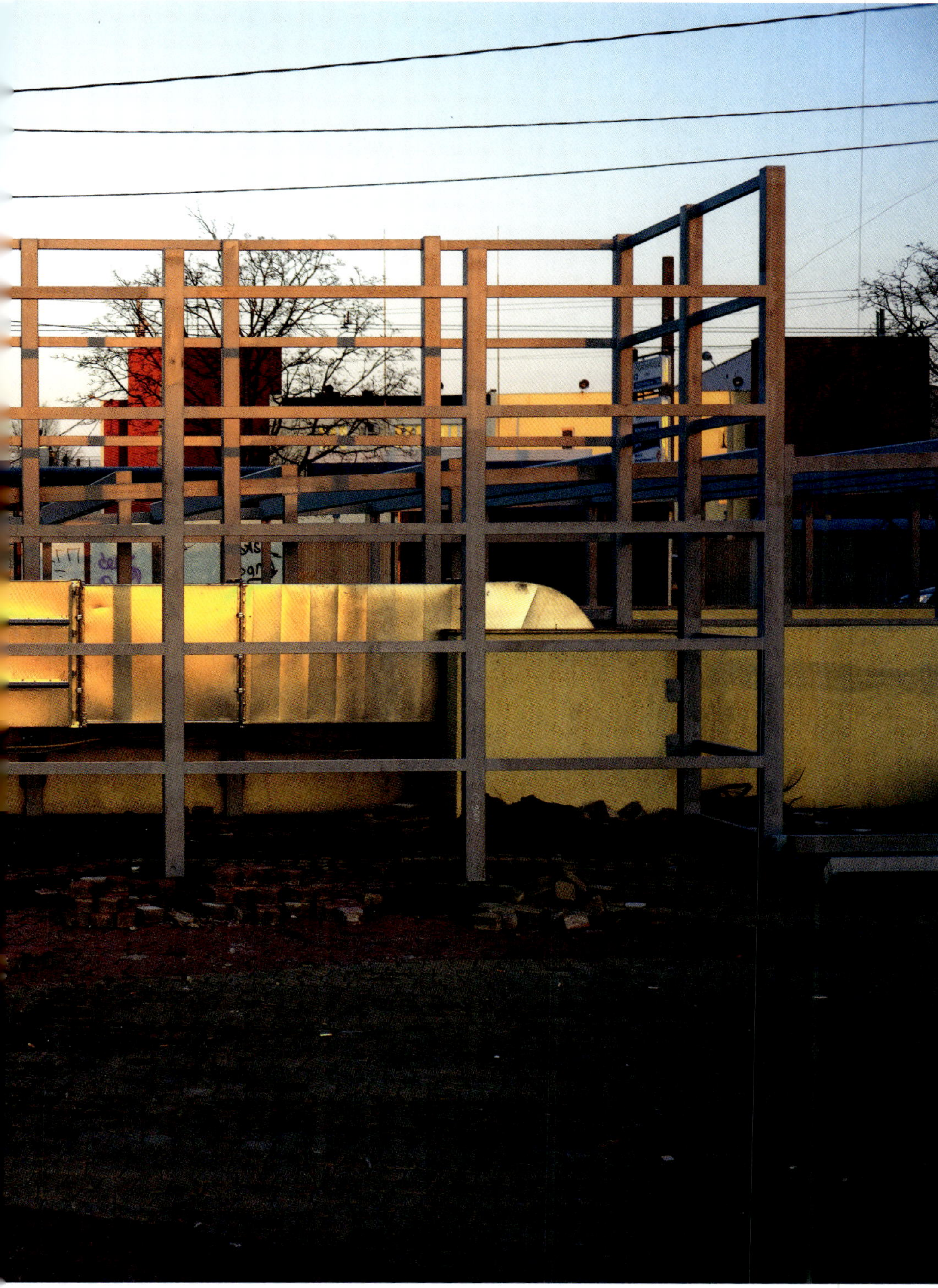

Chapter

IV

Photography is not allowed at the N° 8 public men's bath in the provincial capital of the Cossack heartland, in southwestern Russia. But I will tell you what I saw.

To begin with, I will say this: the place evoked fantasies of Zurich or Casablanca during the wars. Everyone seemed to know everyone else, but it was difficult to know who was a spy, or where anyone's sympathies lay.

The baths staged a kind of naked set piece, with a cast of stock characters, variously troubled or beaten or wounded beyond their years. Waxen-faced Ivan, my first internet contact in R., a full 15 years older than his profile promised; vain, bigoted and prone to leather raincoats and excessive cologne. His pal Volodya, a jolly, straight banker by day, and a sadistic boy-chasing drunk by night. Downstage, at a square table, naked under loosely draped towels, Volodya and a young thug ripped into rank-smelling smoked fish with fat fingers and bared teeth, washing it down with warm beer.

At the bath's social periphery sat a slow-moving, drawling hulk of man, with a boy's face and the body of a giant baby. One might cast him as the village idiot, the kind of fool who is kind, and shy. He put his hand on my arm and told me he liked me, looking dewily into my eyes. I assumed it was just his simple touch. Later, Ivan told me that he was also gay—a fact never acknowledged between them, but learned from furtive sightings at the club near the edge of town.

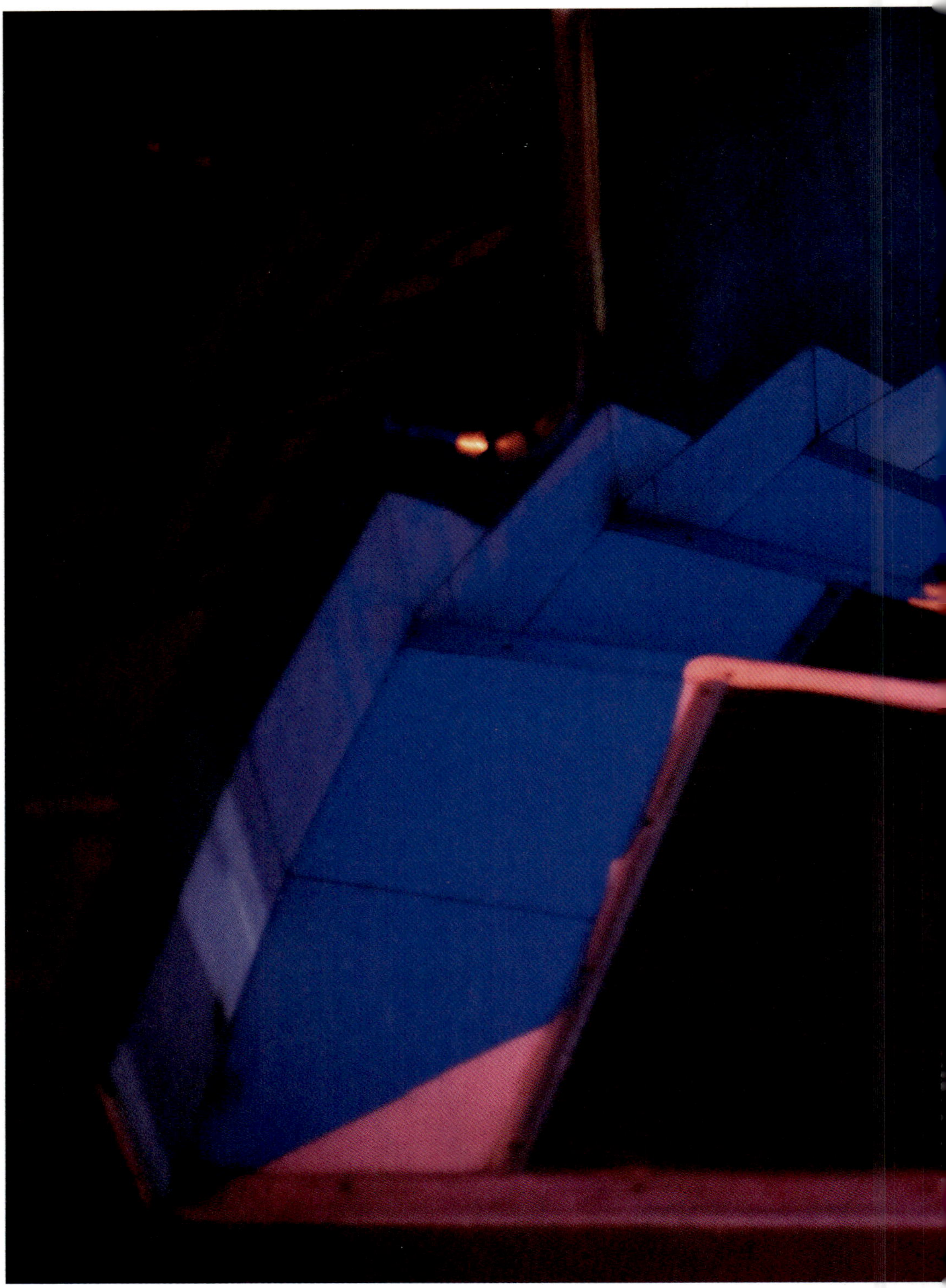

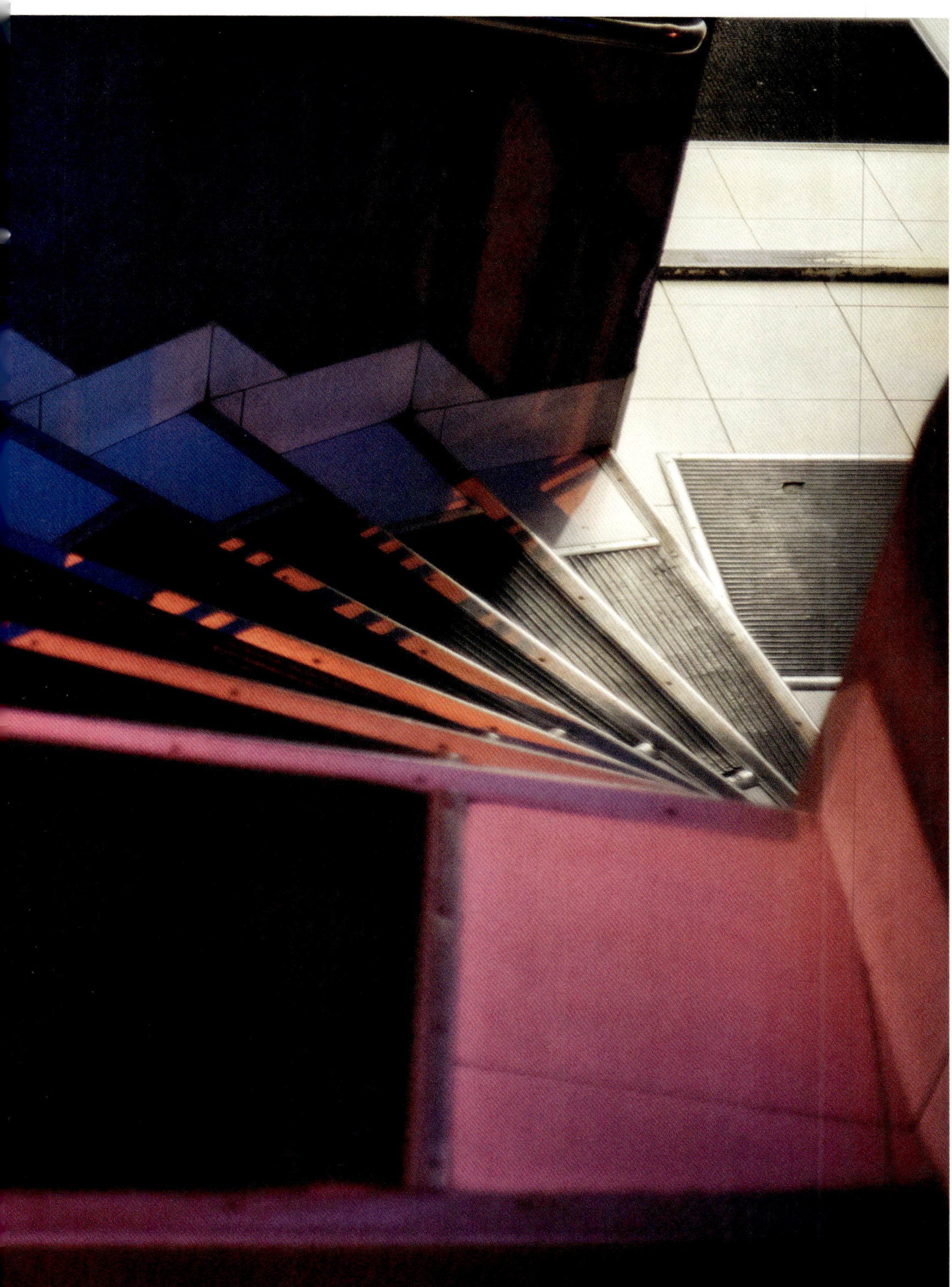

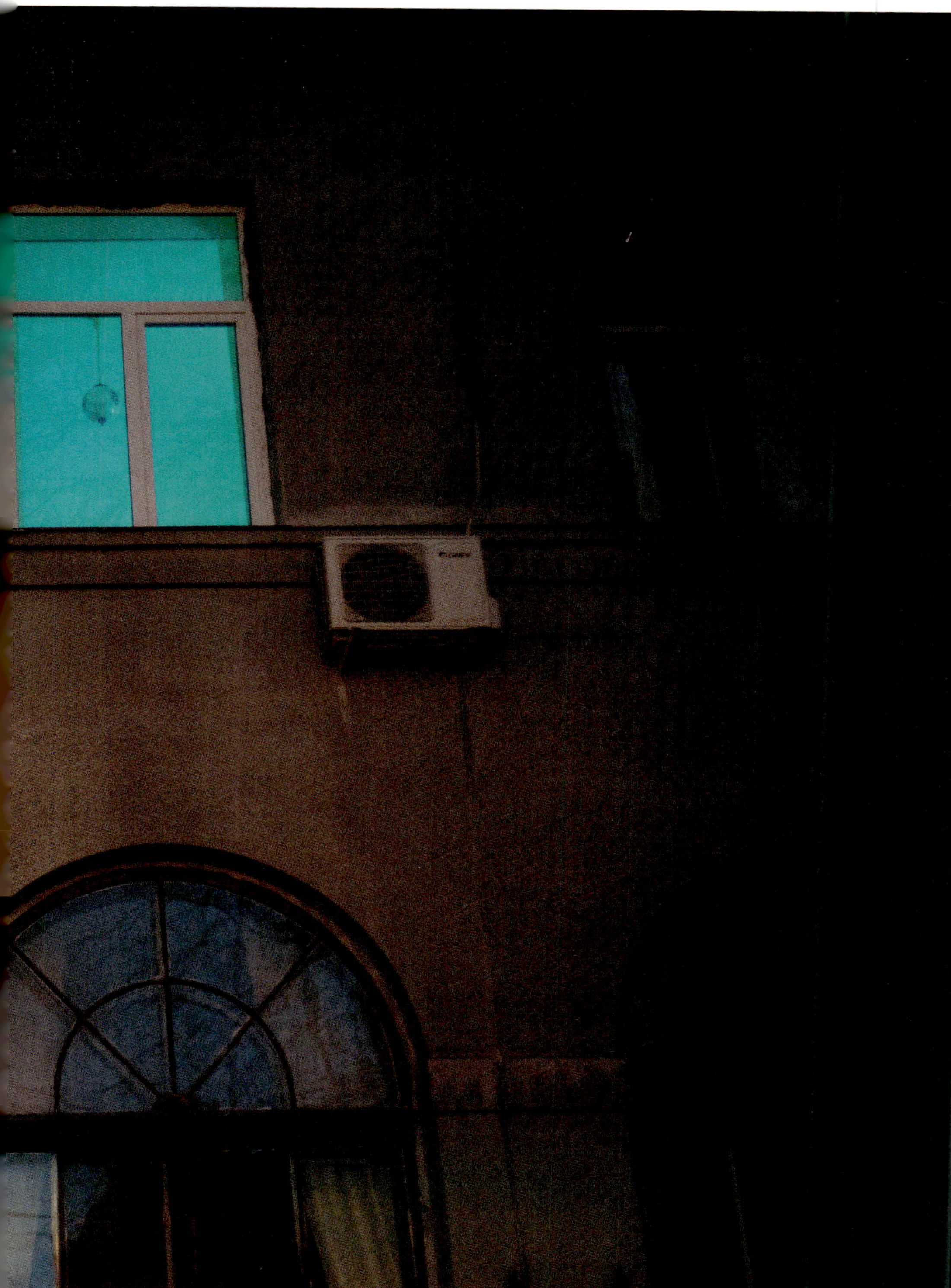

Finally, there was the boxer in tight underwear, mournfully parading his paradigmatic young body in an extended ritual of withdrawn self-display. He dreamed of fighting in America; he wanted to be a pro. The staging notes might say: spotlit by his splendor in the chiaroscuro of steam and poor lighting. The character notes, perhaps: beautiful and already heartbroken, as if the limits of his life hung across his field of vision like an idea in tatters.

Inside, everyone else passed from naked to variously wrapped and covered as they moved from the bath to the showers to the locker room and back again. But the boxer stayed always in his wet pink briefs and nothing else, through all environments, and almost always on the move: stretching, bathing, pacing, posing. I tried not to look directly at him, as the closeness of the space, the seriousness of his purpose and the beauty of his body made me nervous. One time, I entered the sauna to find one older man, seated, contemplating the boxer's turned back. The fighter stood alone, one pale, muscular arm raised above his head towards the steam-diffused light, languidly and solemnly shaving an armpit.

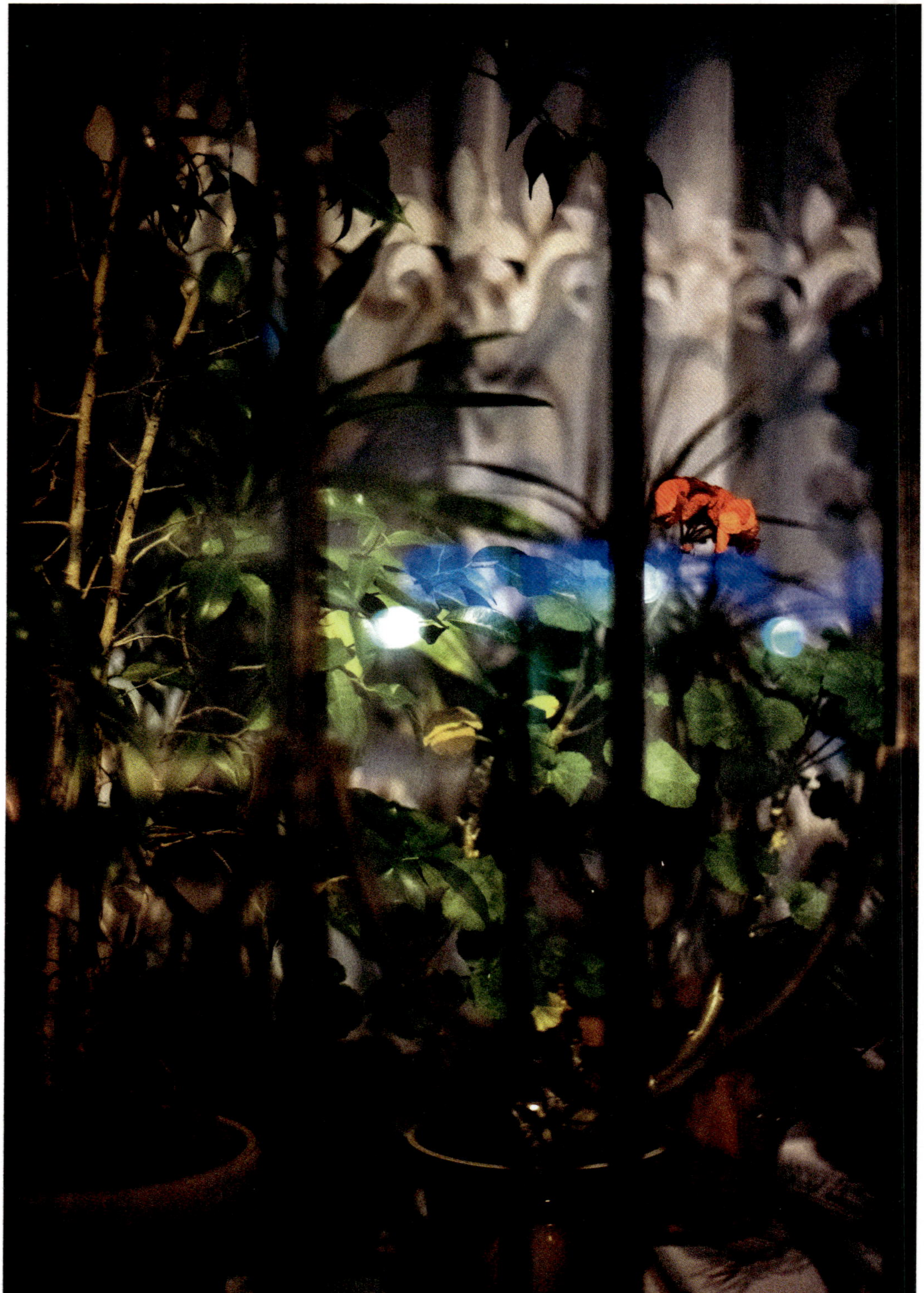

In these misty chambers, it was impossible to say who was gay, and who was simply naked and melancholy. Just as, in high school, the word 'faggot' functioned as a general marker of suspect nonconformity, so my unfolding encounters with the complex of gay male suffering, beauty and evanescent hope drew a kind of dragnet around a broader male condition. Sometimes, in search of a particular suffering, a larger field unfolds. In Russia, the inability to live up to one's dreams of self—a fantasy life hiding in each of us, but particularly cracked open by the half-promises of a post-Soviet world—seems to have released a particular form of male sadness, traced with disappointment and violence, humming under everything, blooming like radon from the dusty ground. Infected, for me, by the eroticism of the impossible, all of this hurt began to feel queer.

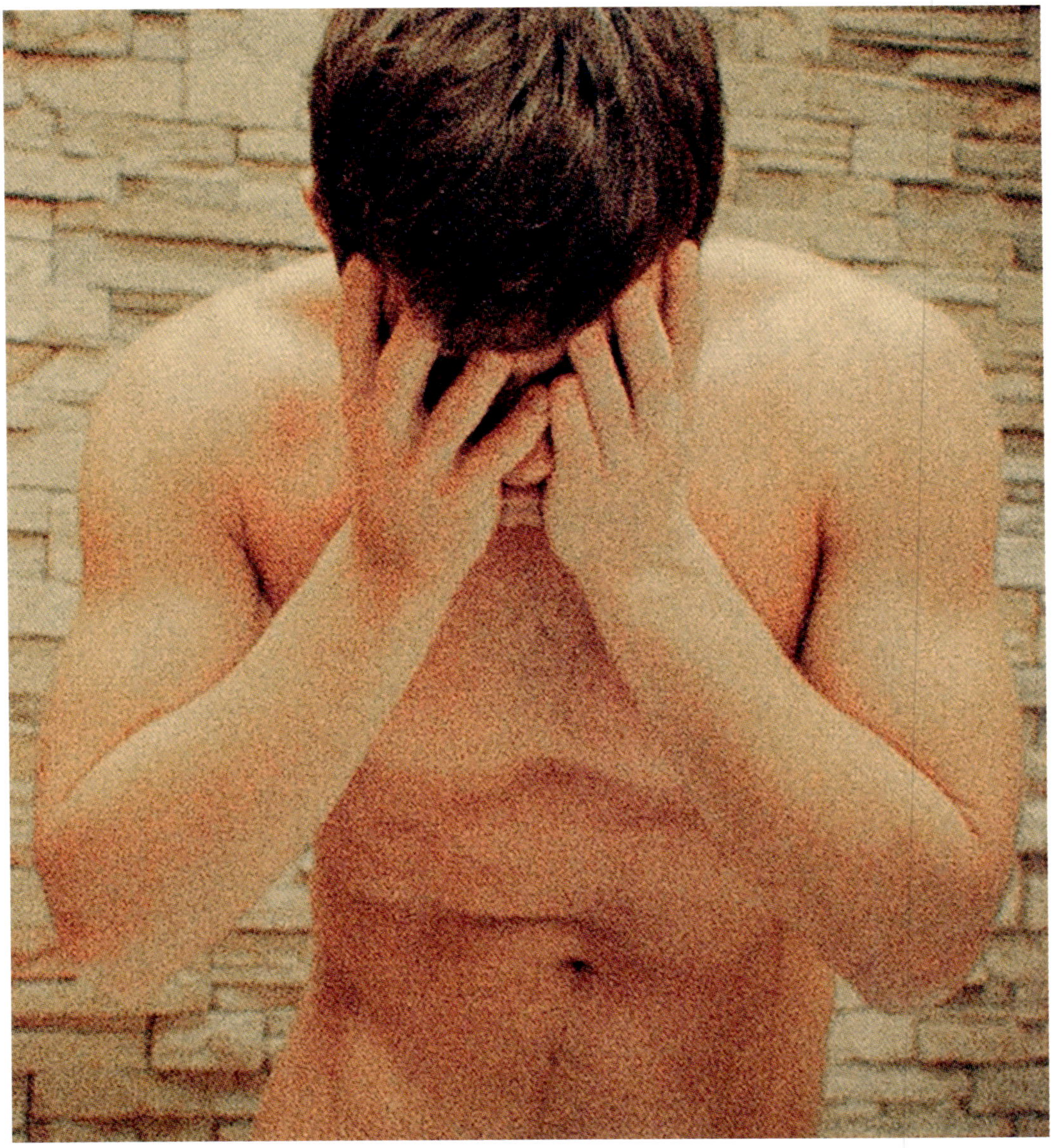

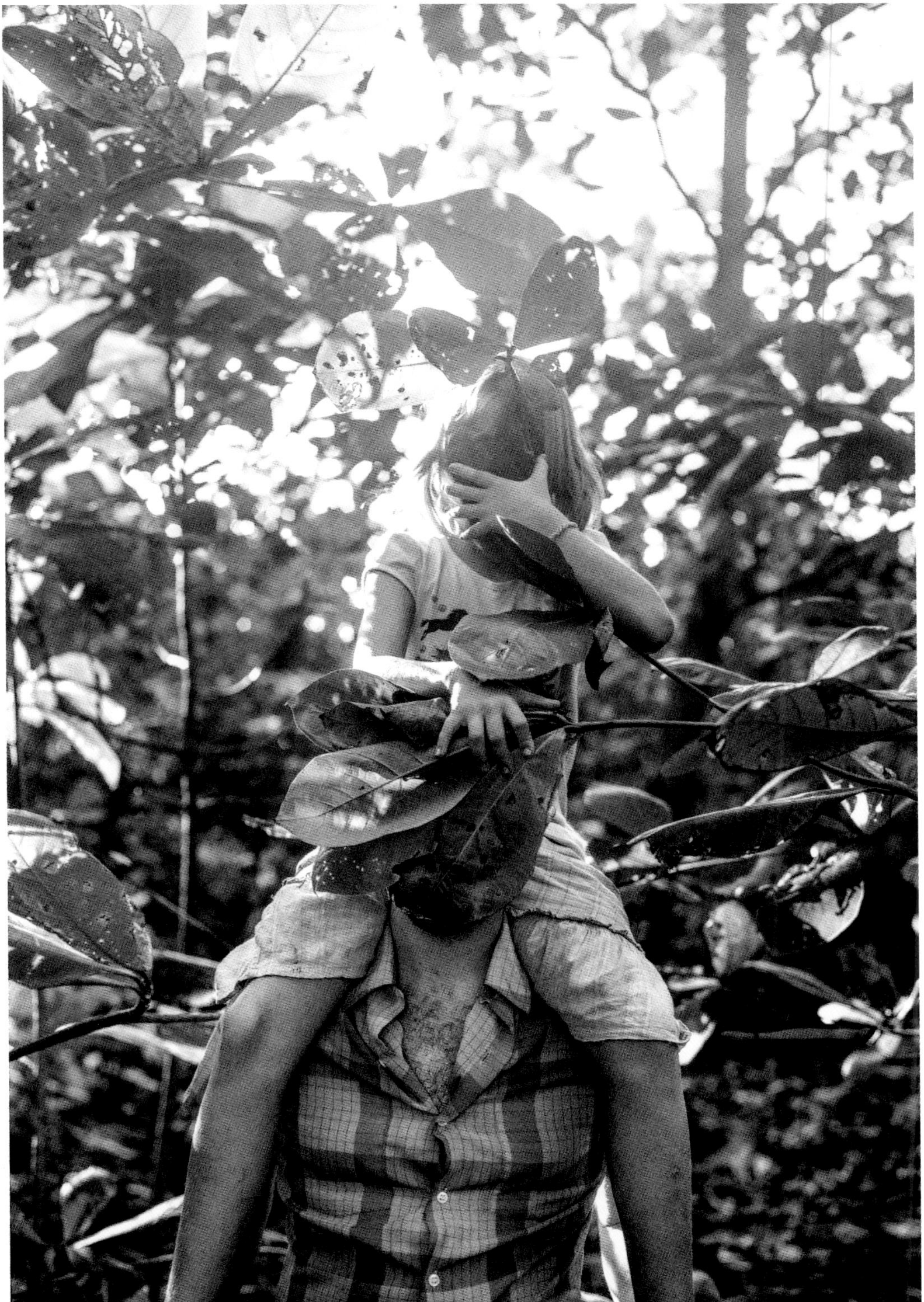

Chapter V

Lunchtime on the island, and I had yet to leave the compound, having spent the morning pretending to read in the pavilion. Mostly, and without leaving my chair, I had been discovering and cataloguing the half-finished and abandoned craft projects that littered this living and dining area: bits of weaving, stranded half way between potholders and small carpets; scraps of unraveling patchwork that may once have aspired towards quilting; numberless and mostly desultory dream-catchers, propped and hung and impaled on rusting nails, the faded yarn smudged and sparkling with salty rime; and various ill-executed driftwood stools, ostensibly for children.

The wind was picking up, violently tousling the taller palms, and I resolved to take a walk before the weather sharpened more. Behind the washrooms was a westward-seeming path I hadn't yet explored, its entry marked by what appeared to be a disintegrating punching bag, half-gutted and hanging sideways. A few yards further back, under mangrove leaves, the hull of a half-finished dugout canoe lay rotting. Pure jungle closed in on the neglected trail. It grew darker, buggier, and more fragrant with decay. Within a few minutes, the path was nearly impassable, and the slick mud, increasing with the incline, slurped desperately at my flip-flops with every step. Only a rigid contracting of the toes held them to my feet.

A moonlight green land-crab, claws raised in alarm, scuttled across the path in front of me, its hard black eyes suggesting defiance in the face of attack. But the crab marked an improvement in the scene, as if its little upturned mandibles had meant to hold a tiny sign announcing "better times ahead." Within moments, the land leveled and dried and the path improved. Sun broke through the canopy. The same wind that had propelled Isabel across the grass was now sending a field of improbably low and fast-moving cloudlets across the sky, creating streams of light and shadow. This movement, combined with the urgent motion of wind-tossed palms, created a nearly stroboscopic effect on the trail. The ur-disco of the island.

A small clearing appeared. At its near edge, half a dozen untended pineapple plants struggled to be themselves against the incursions of rude vines. The clouds were gone now and the wind had stopped dead. I scanned the overgrown boundary of sun-swept shrub and surging convolvulus, looking for a way forward. Ahead of me on the right, one large tree, set back in the brush by several yards, cast a circle of shadow under its oaky canopy. Its smooth trunk bore an image of wide-open eyes and an indulgent, toothless smile, drawn in loose brushstrokes of white paint. The tree smiled at me from its place of self-produced shelter. I grinned back.

This moment of contemplation was broken by another apparition: the head of a horse, seeming to float, collaged, above a dense hedge to my right. It hadn't moved into the scene so much as materialized. This was the German woman's horse.

I had seen her riding at the edge of the beach two days before, her skirts hiked up and her jaw set forward, looking every bit the Boer pioneer woman returning to her settlement. As always on this island, I had no sense of distance. Was it possible I was near her guest-house, clear around the northern tip? The horse did not say, or even move.

Unfamiliar as I am with the side-mounted eyes of equines, it seemed possible that it was surveying me with its left eye and the smiling tree with the other, deciding which being, if either, to trust. As we all stood there, uncertain, the light turned a sudden violet, shifting to bruised blue, and an immense storm cloud entered the palm-bracketed sky above. An audible drop of rain fell on the horse's muzzle, right between the eyes. It blinked but did not budge. I turned in retreat.

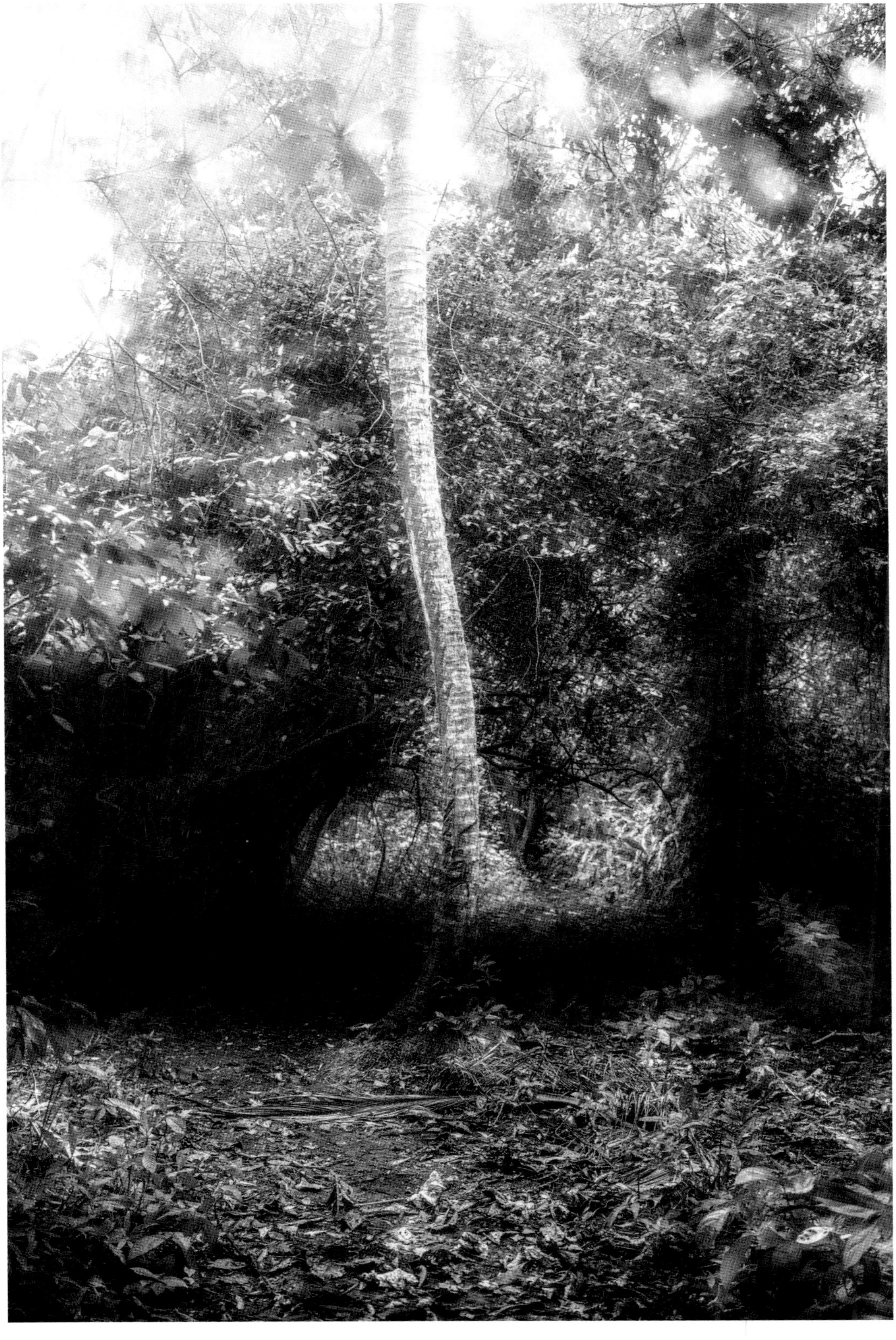

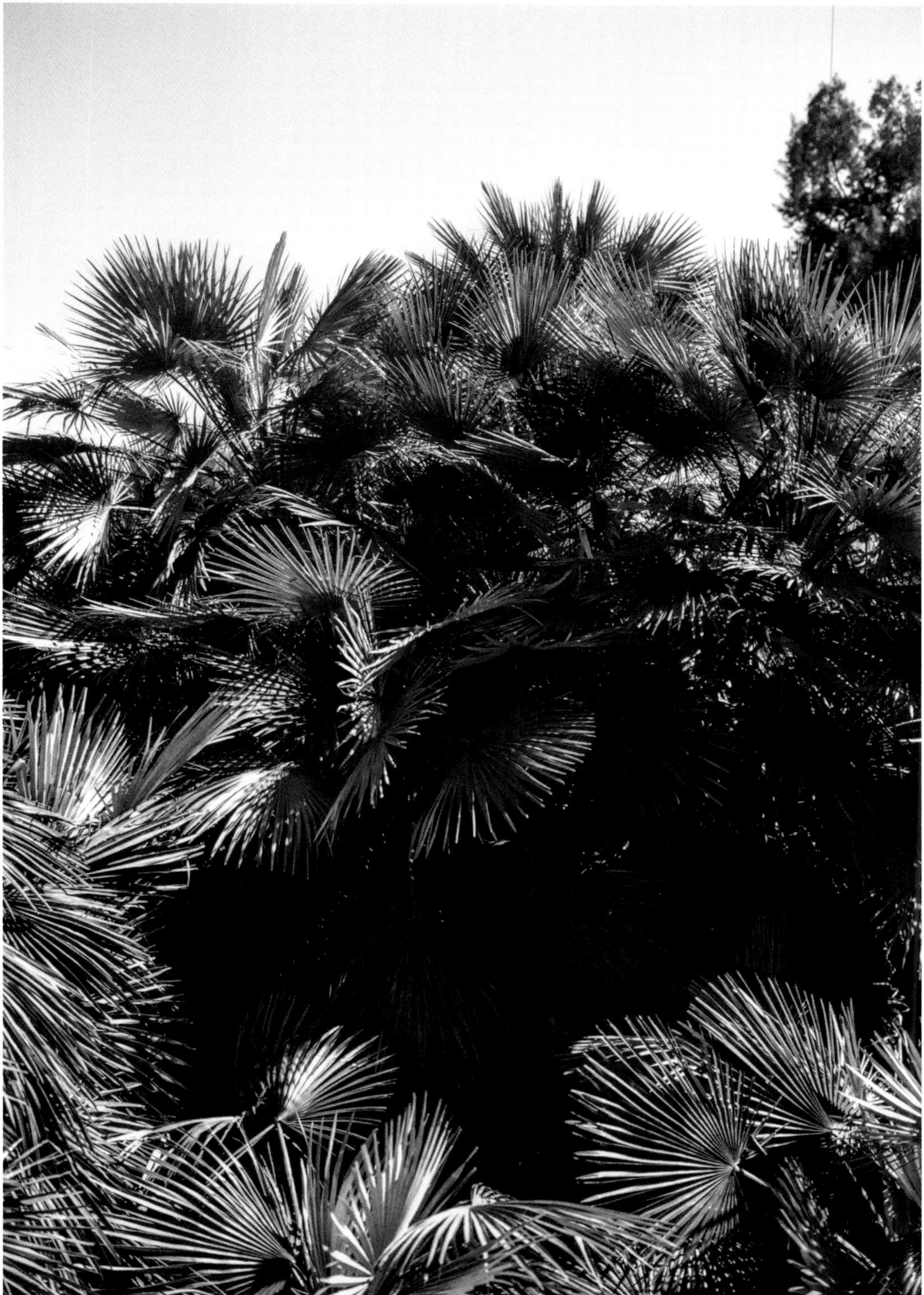

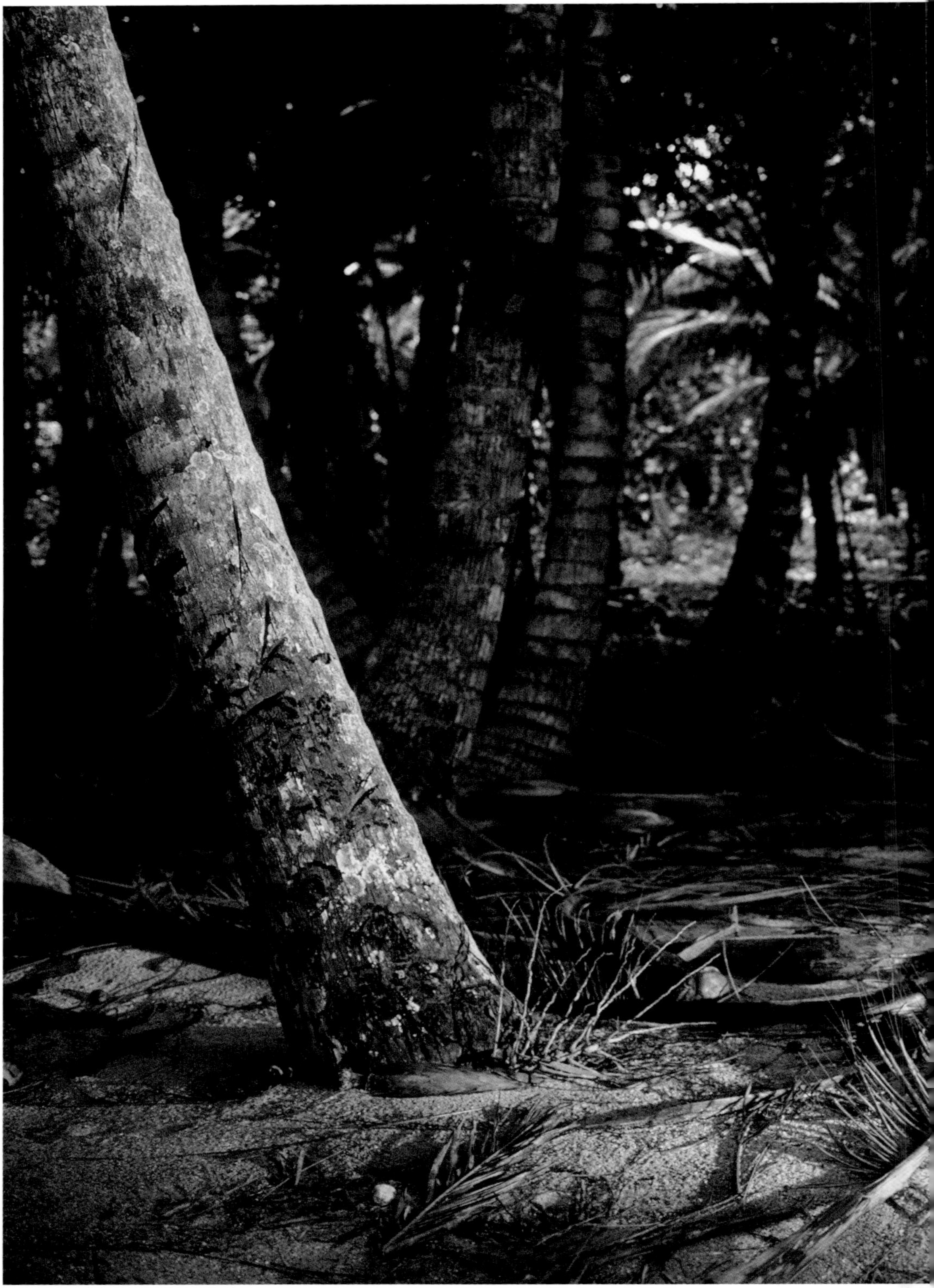

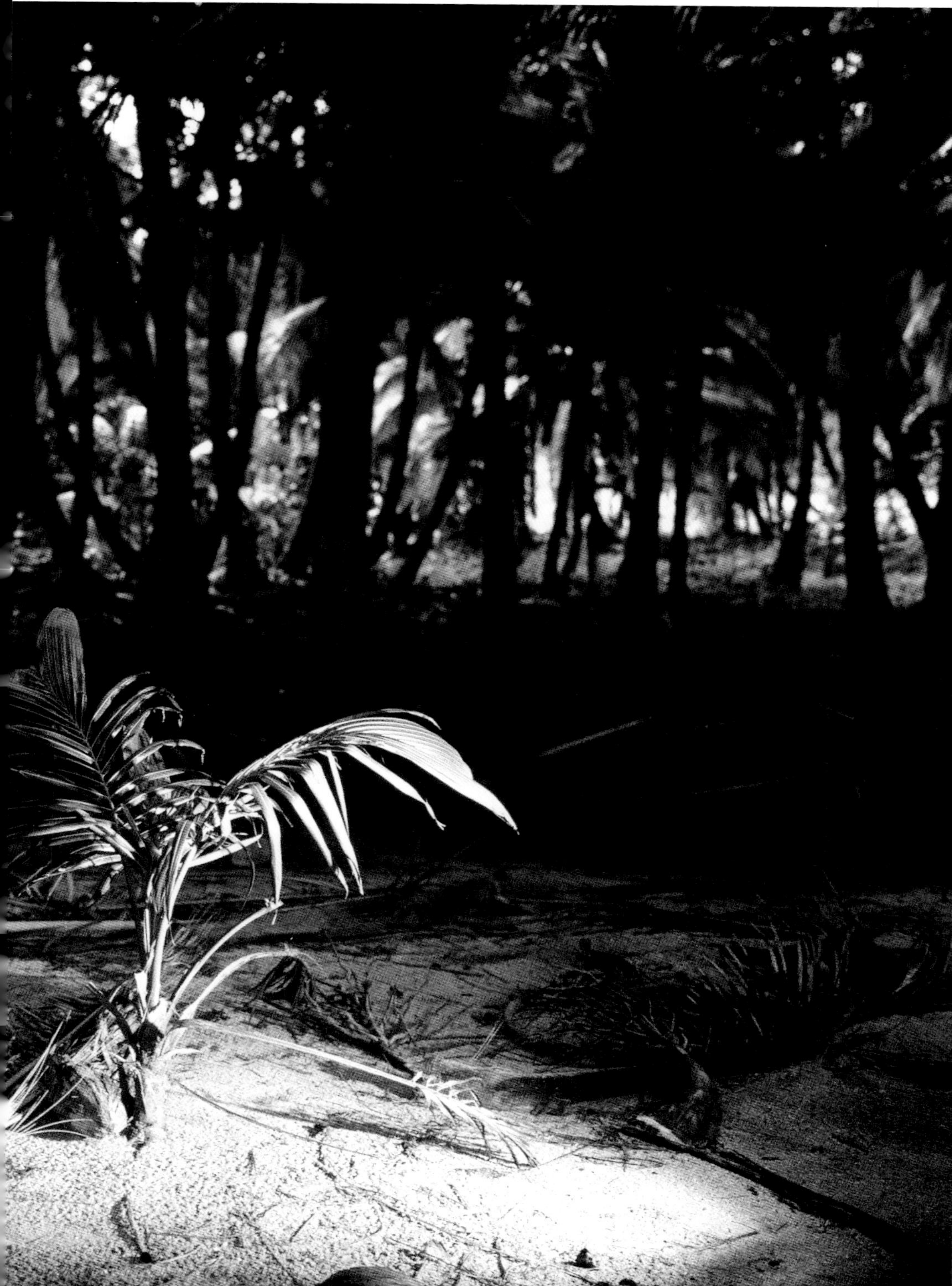

I arrived back at the compound after dark, clambering over the moonlit rocks along the indirect shoreline, now that the jungle paths had gone black. In my hut, I poured a nightcap of rum into a plastic cup and headed back outside to appreciate the pearlescent palms in their lunar glow. Dim light spilled from the dining area, expired on the lawn. I had no energy for conversation, but curiosity drew me, and I approached. Isabel sat at the head of the long table. Some kind of notebook or accounts log lay open in front of her, but she wasn't looking at it. As I slipped into the small gazebo nearby, Isabel began to speak, apparently to no one, eyes fixed on the closed doorway across from her: "Okay," she began. "One more story before bed..."

She paused, summoning the narrative, and then continued:

In most tides: an island. A lonely bluff, populated only with men. It is the job of these men to nurse and translate the stories. Each man is the same: tall, lean, tan. Each of the stories, suckled into being, is new. But they do not come easily.

Every story is born beautiful and incomprehensible on this almost-island: like a fresh-hewn locust log, pocked with small thorns, and speckled with fragrant white blossoms—flowers that fall off easily with a gesture of the hand, or a gentle shaking.

Sometimes, images are born in and with the stories. It is necessary for the men to wean these pictures from the words, and tease them from the logs. But it is not necessary to know when the log becomes a story, or when that story, which is also a baby, becomes a beautiful, longhaired boy.

When the stories are deciphered, they are harvested. Is that the right word? Prepared and exported to the world of people, they always attain great acclaim. But the strange sanctuary remains obscure.

Once the little boys—that is, the stories—reach a certain age, they become poisonous. There is nothing for it, and the tall men must dash their dangerous little heads against the big rock to still them.

They say it sounds like a ripe coconut dropping on wet sand.

Looking up from the story, Isabel stands, almost ceremonially, moves several steps to her right and bows to retrieve a sleeping child—dressed in full Superman costume—from the depths of the hammock that cradled him. She then kneels, and with one swift gesture lifts another smaller body from under the table. Thus burdened, she walks off into the dark jungle path, a flash of red cape trailing momentarily behind her.

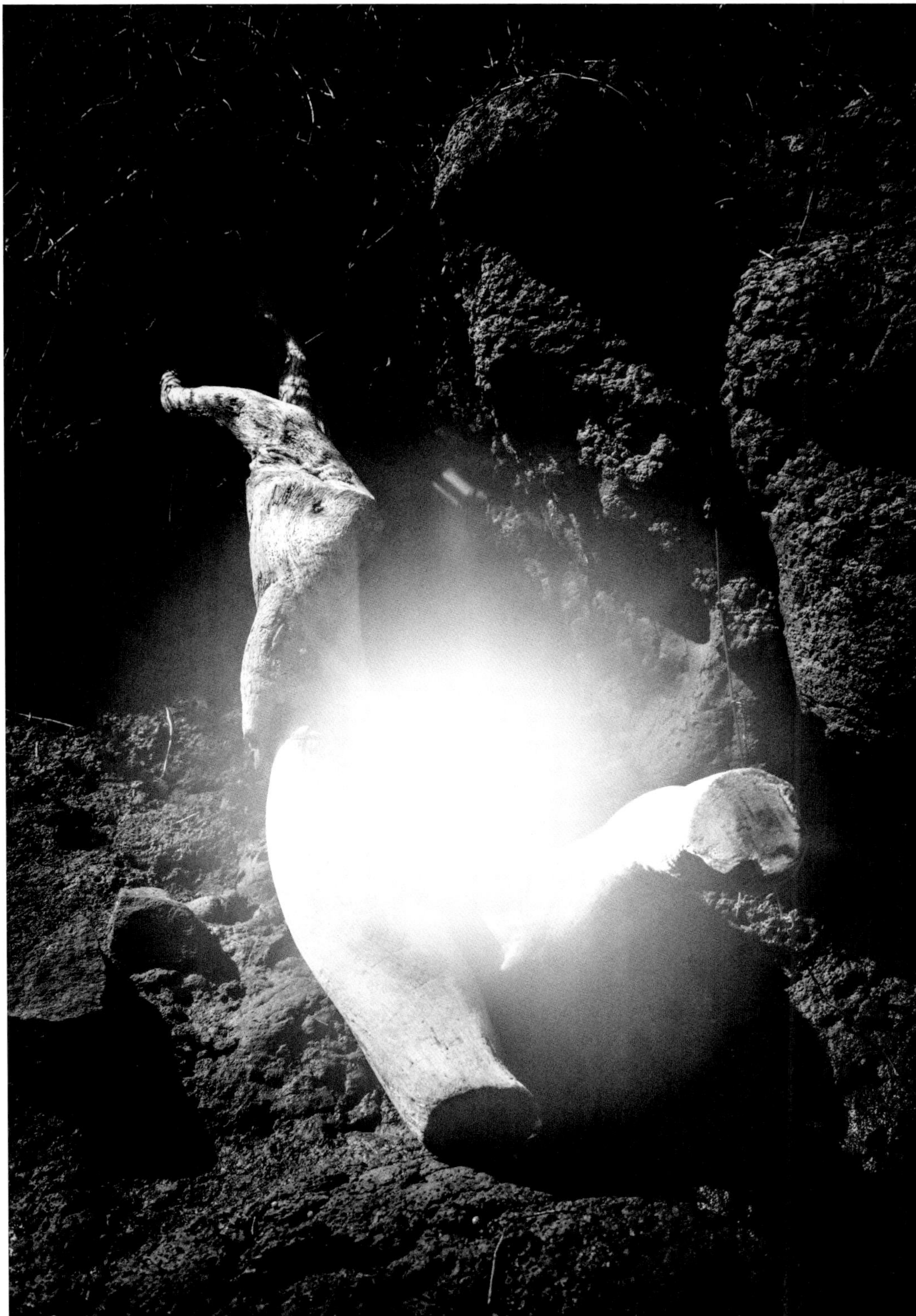

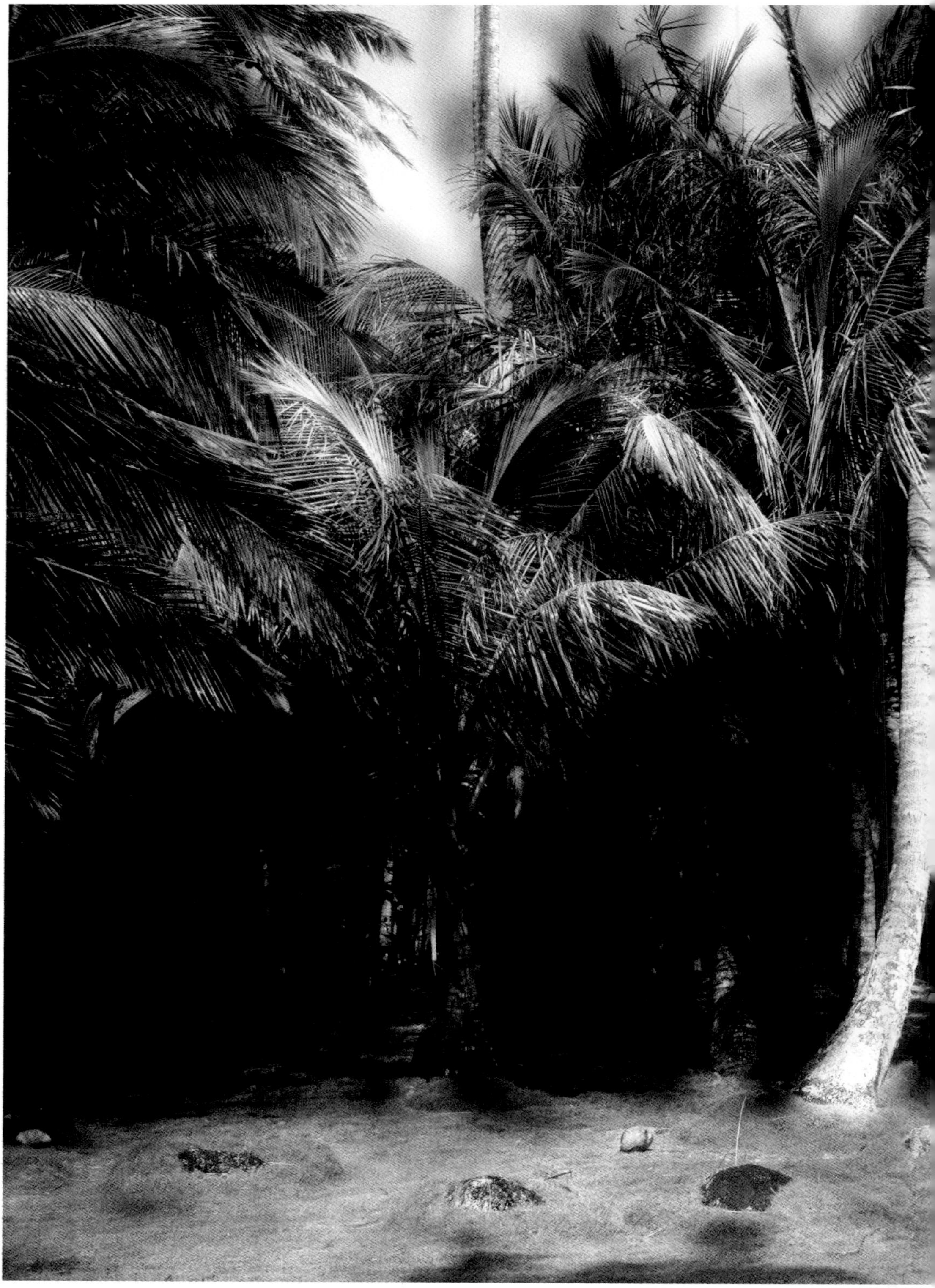

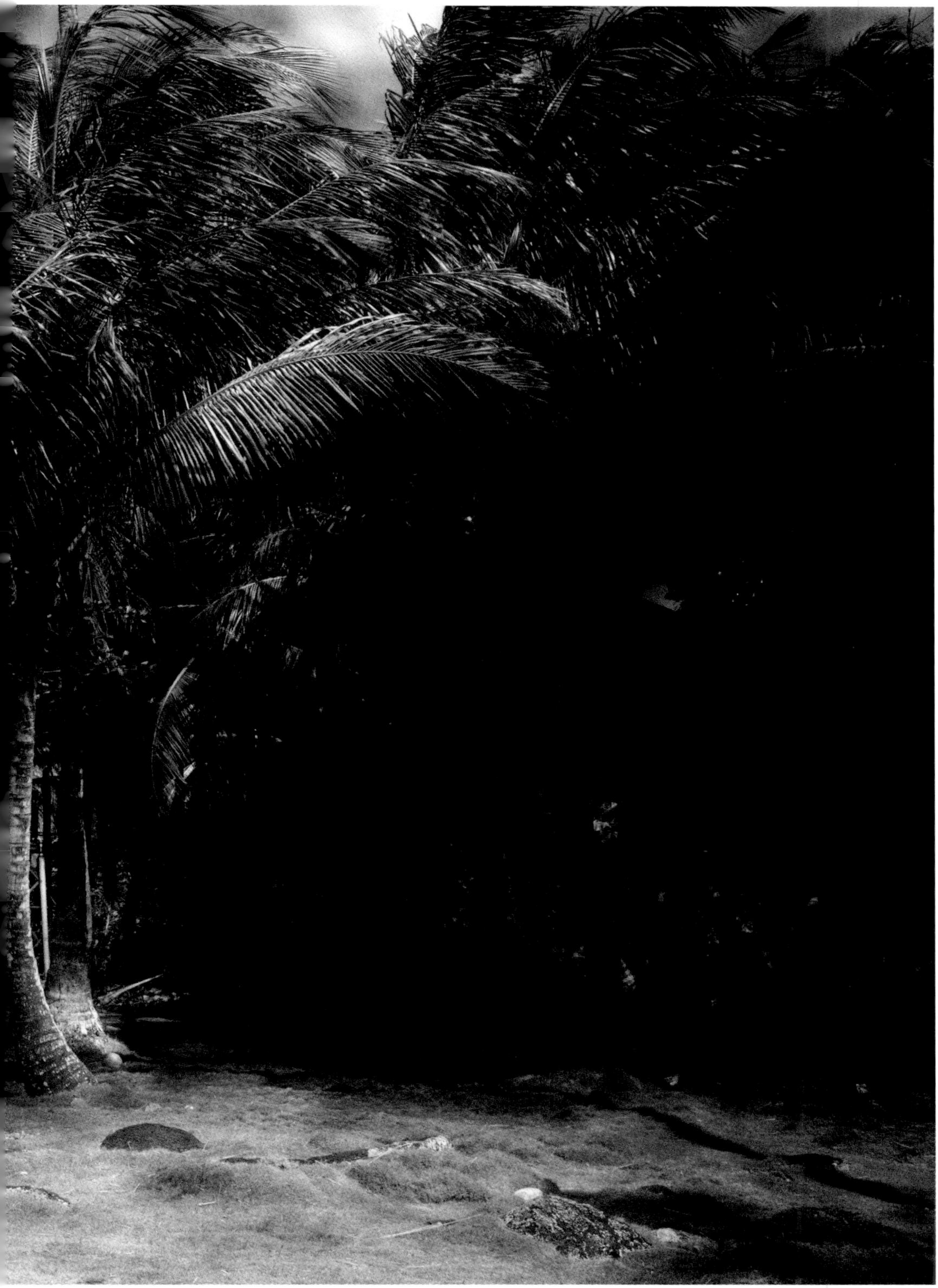

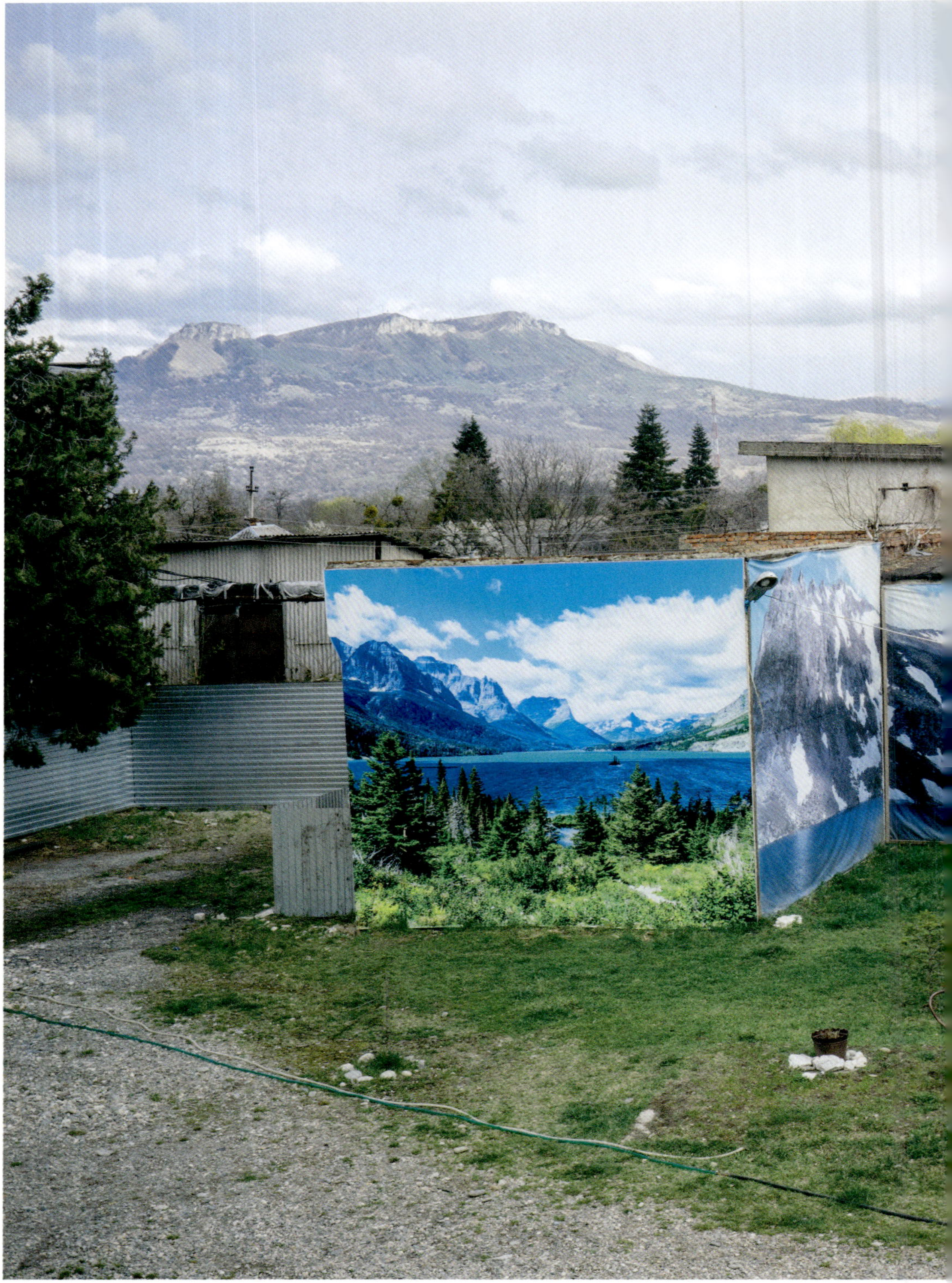

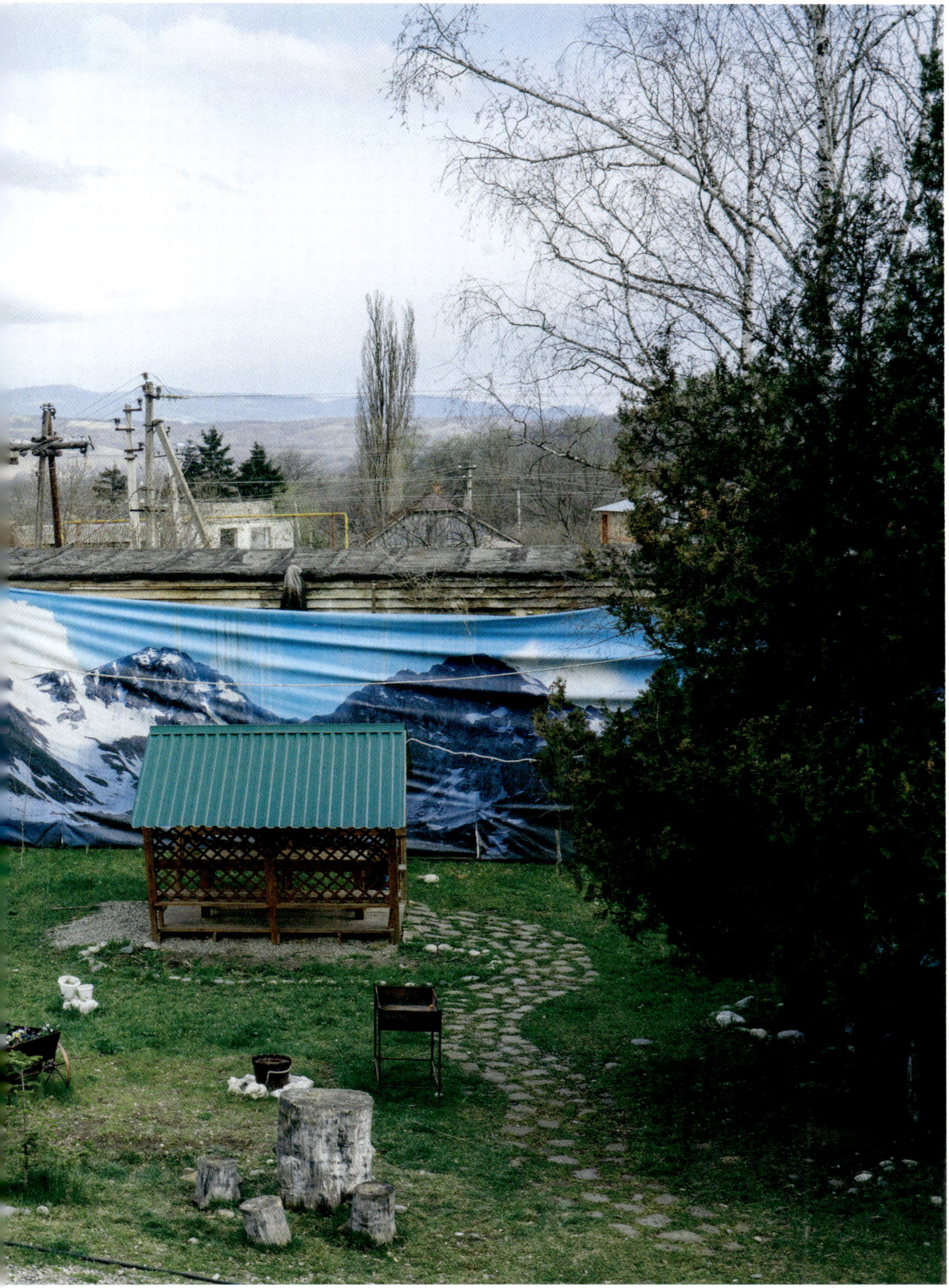

Chapter VI

It didn't begin here, at the edge of the village of G., on a dead-end dirt drive rising several hundred yards into darkness. Night had been settling in as we walked, but Petya still wanted to show me something. Something essential: the Stone of Love. We reached the end of the short road, marked by a few dark piles of earth, barely visible in the gloaming. There, across our path, darker than the deepening dusk, lay a freshly dug ditch.

I had arrived in G. earlier that afternoon. In an American film, the bus would have dropped me off, and then pulled out of the shot to reveal Petya standing in the sunshine on the other side, smiling to see me. But G. was the end of the line, and there was no sun. The bus had nowhere to go for a good long while, so the big dusky hulk just ground silent between the bus station and the road. I walked around the front and caught sight of Petya in the distance, loping up the town's one paved street to meet me. He was smiling.

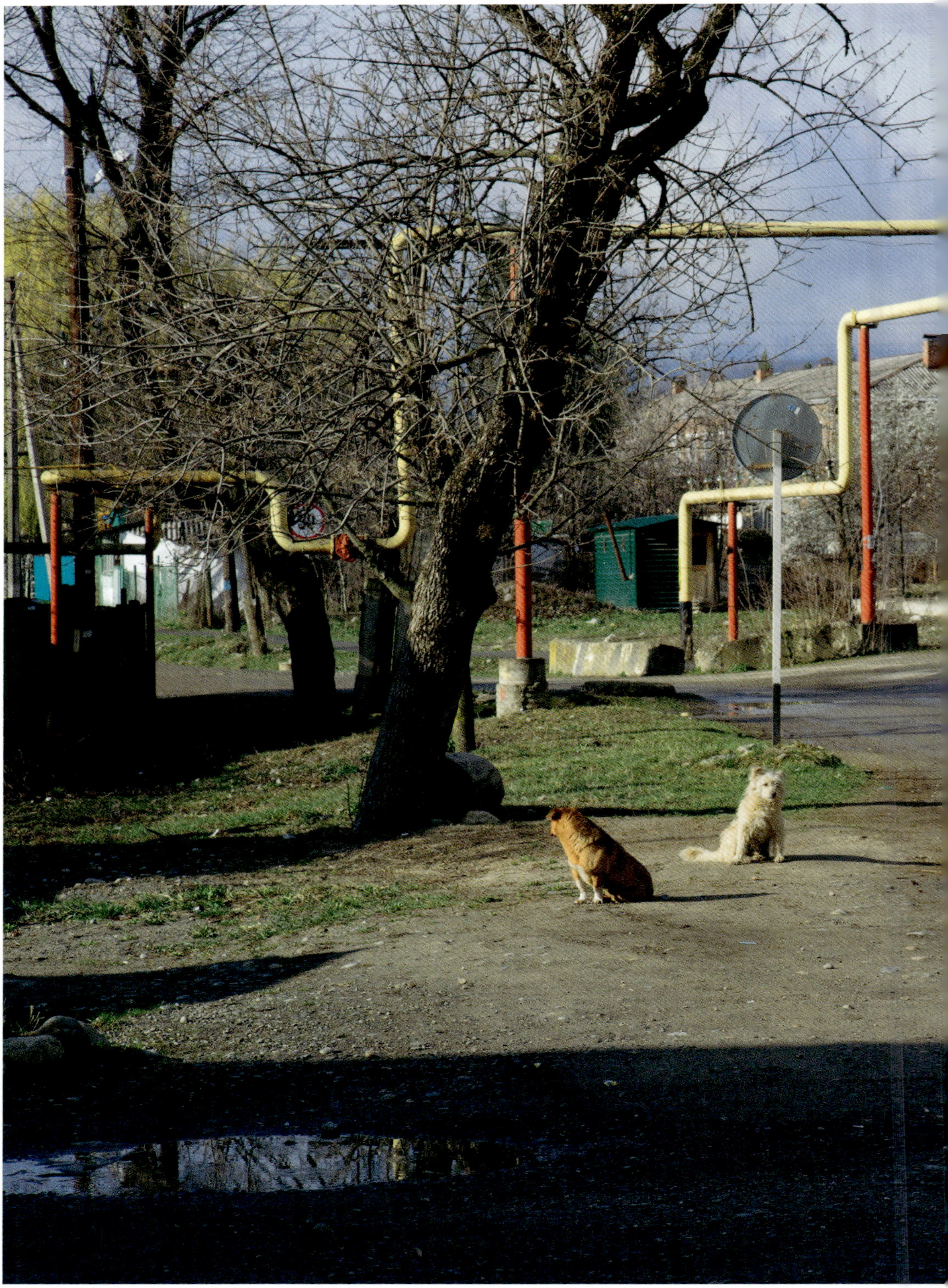

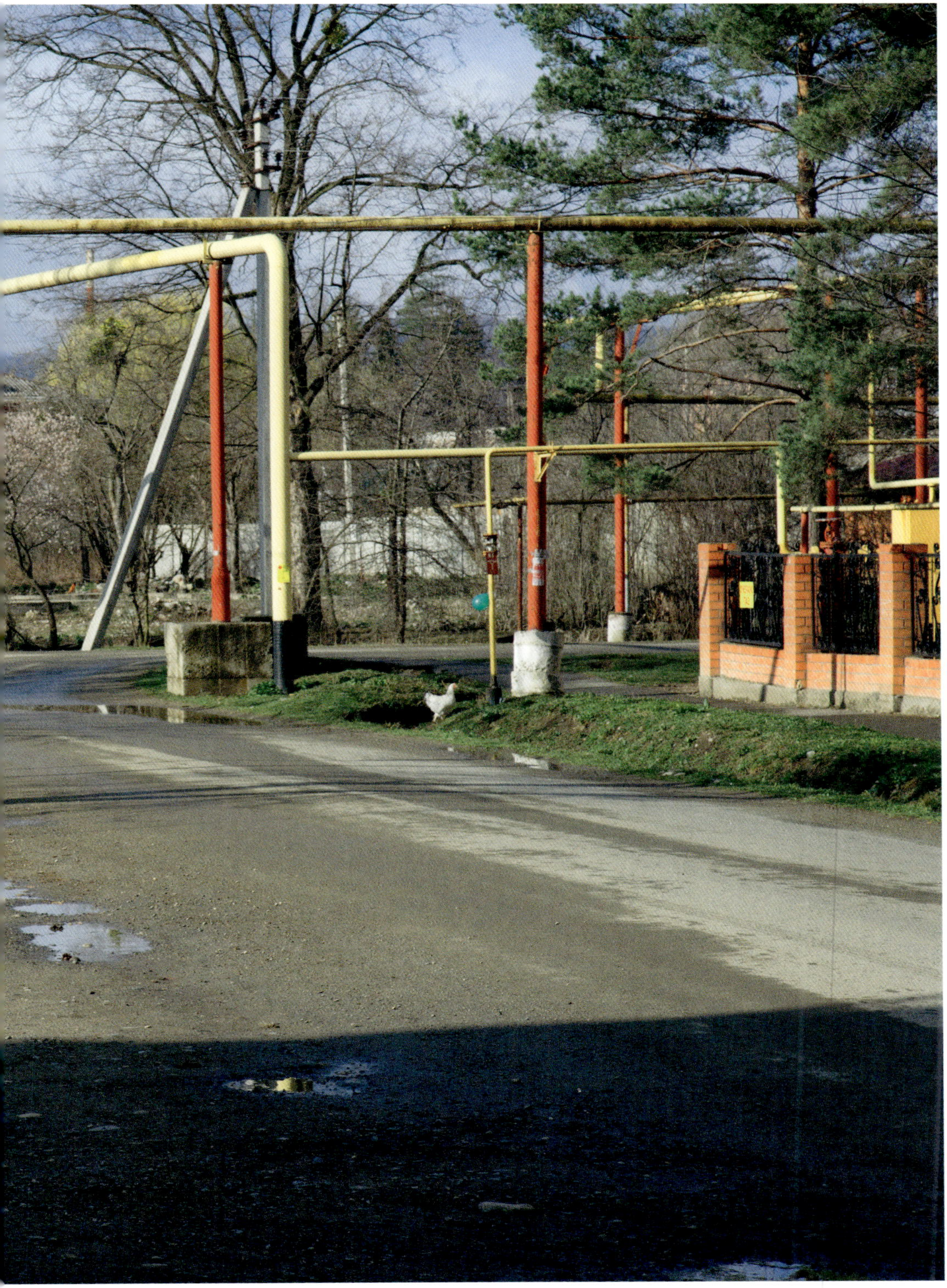

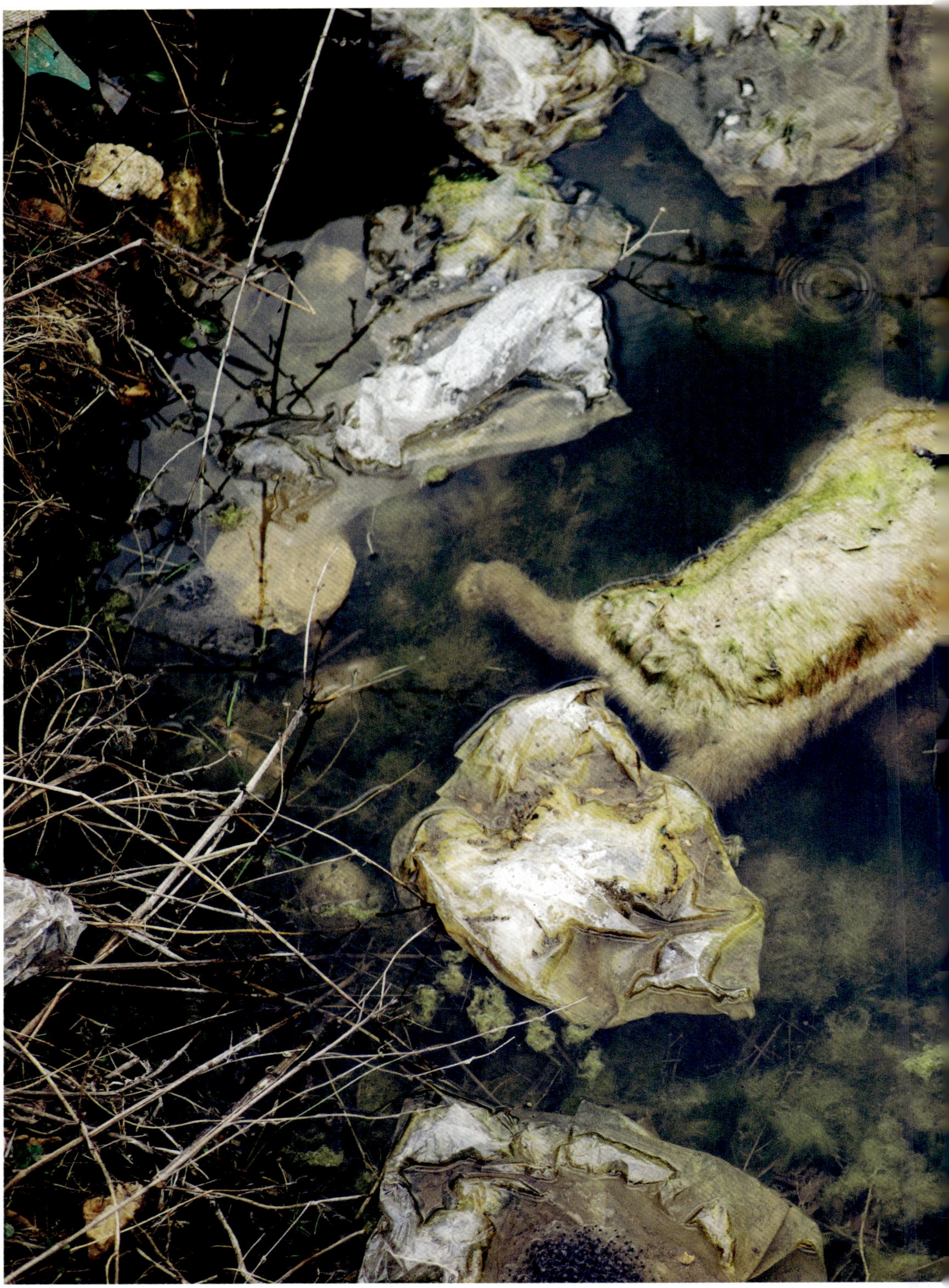

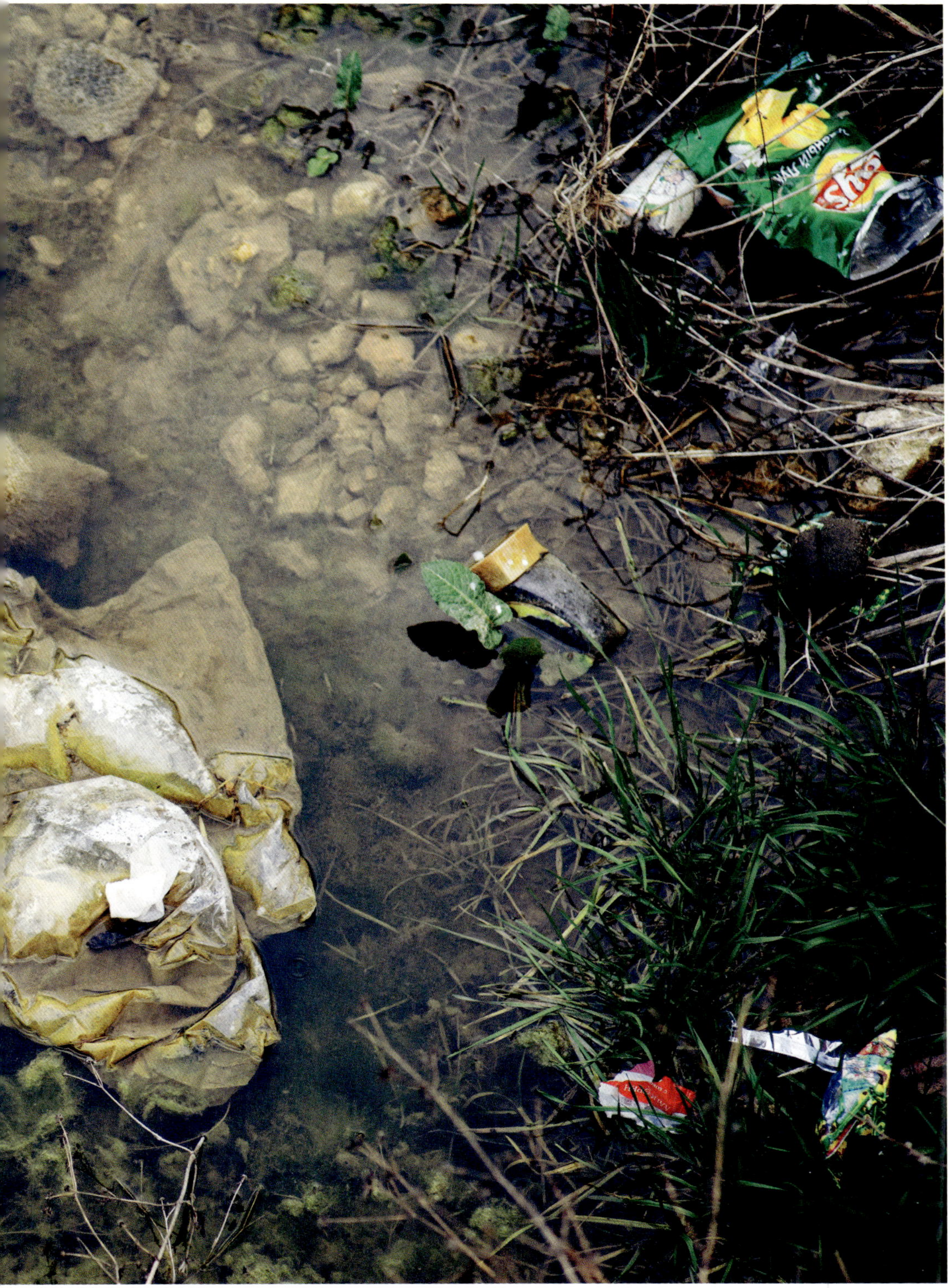

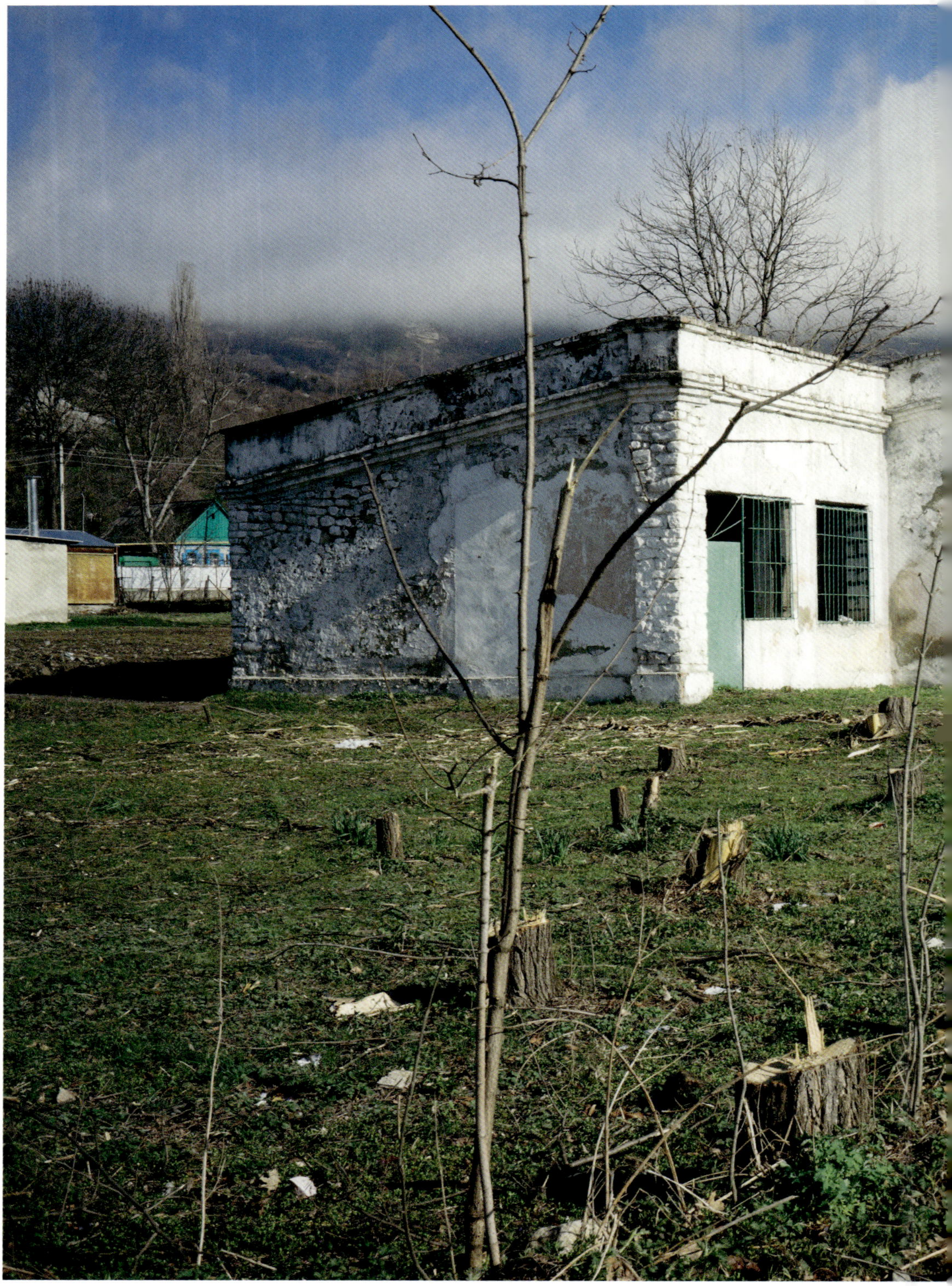

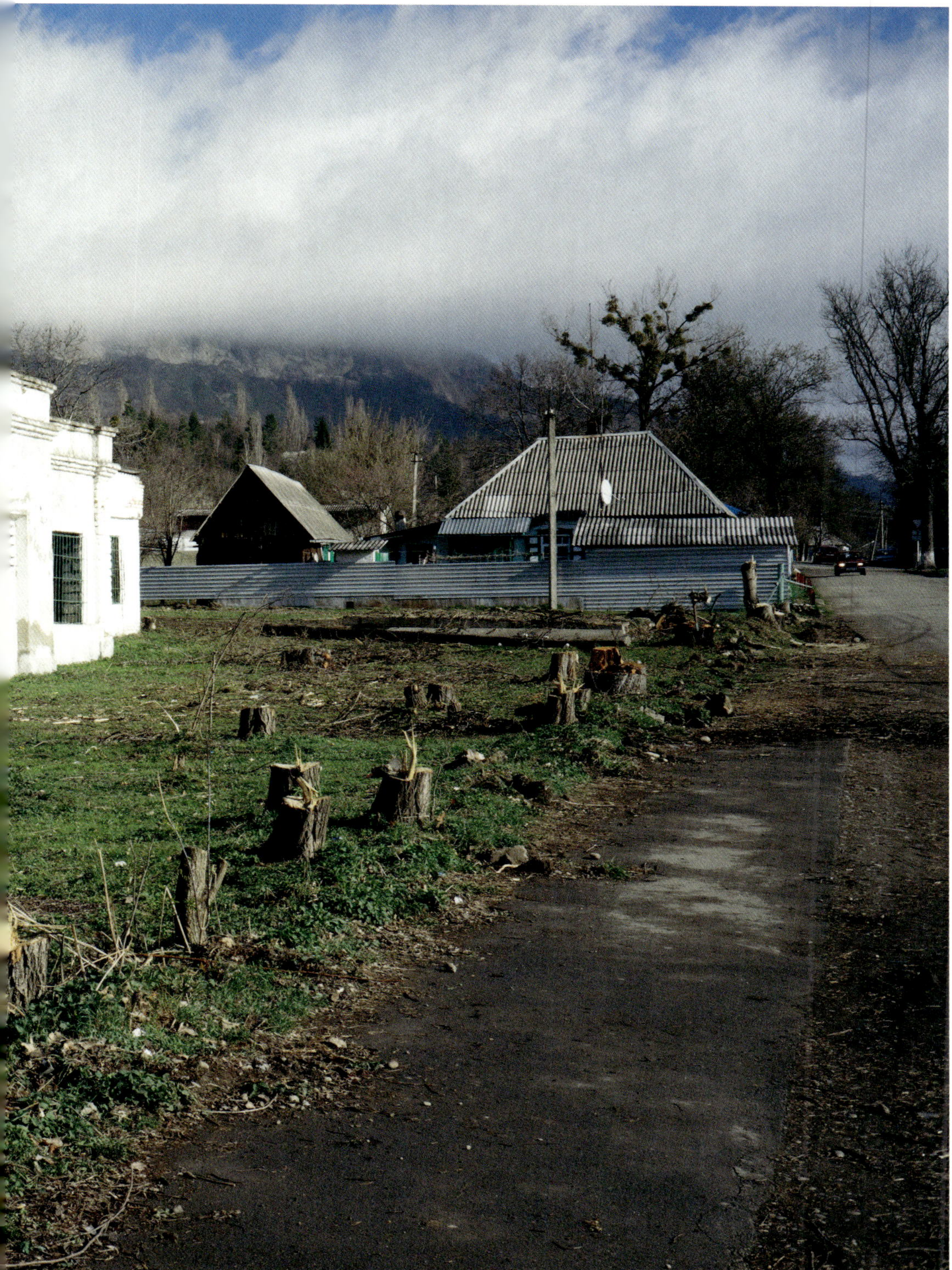

At the town hotel, which also seemed to serve as a boarding house for wayward teens, we tracked down the errant desk clerk, who set me up with a room while Petya went back to work for an hour at the factory next-door. We met again on the street, walking back past the bus station and down a dirt road that followed what had once, I learned, been a stream. A wide metal pipe acted as a small bridge, and we crossed over to a row of one-story wood and concrete homes, each hidden behind a high wall. Petya's yard housed a wire-faced chicken coop and a terrified white dog, in a long low cage, barking piercingly at my arrival. His mother, inside, was quieter but no less suspicious, and after a chilly introduction, she left us alone in the kitchen to eat.

The light, it seemed clear, was already setting into a slow fade when Petya pointed up at the high flat ridge to the west of the village and said: I want to take you there. It was an enchanting place to imagine oneself: walking for what might be miles along the high open path—rocky, table-flat, touched with green. Still, unless my sense of depth failed me, there was no way to reach the spot in daylight. But I had not come 11 hours on two local buses to meet a stranger, only to suddenly become practical. So I said yes.

We passed back over the filthy stream, across a road, and struck up into the rocky fields and wooded copses of the hillside. Patches of scorched pasture glowed golden and black in the cloud-filtered light of early evening. The landscape was abraded but beautiful. Petya veered left and right as we ascended, always looking for a walking trail towards the summit that eluded him. Sometimes we came to an impasse—a wall of thorns, a small ravine—that forced us into a sideways stroll, across the hills, before climbing some more. Above, in each leafless tree, hung enormous green balls of mistletoe, like Christmas decorations that had not been taken down. I had never imagined the possibility of this much mistletoe. We both kept our eyes open for the first flowers of spring: impossibly perfect pale yellow primroses, and small patches of common violets. Each time he came to the violets, Petya stopped, stooped and picked one, holding it to his nose in the hopes of sharing its fragrance with me. To his repeated disappointment, the blossoms held no scent.

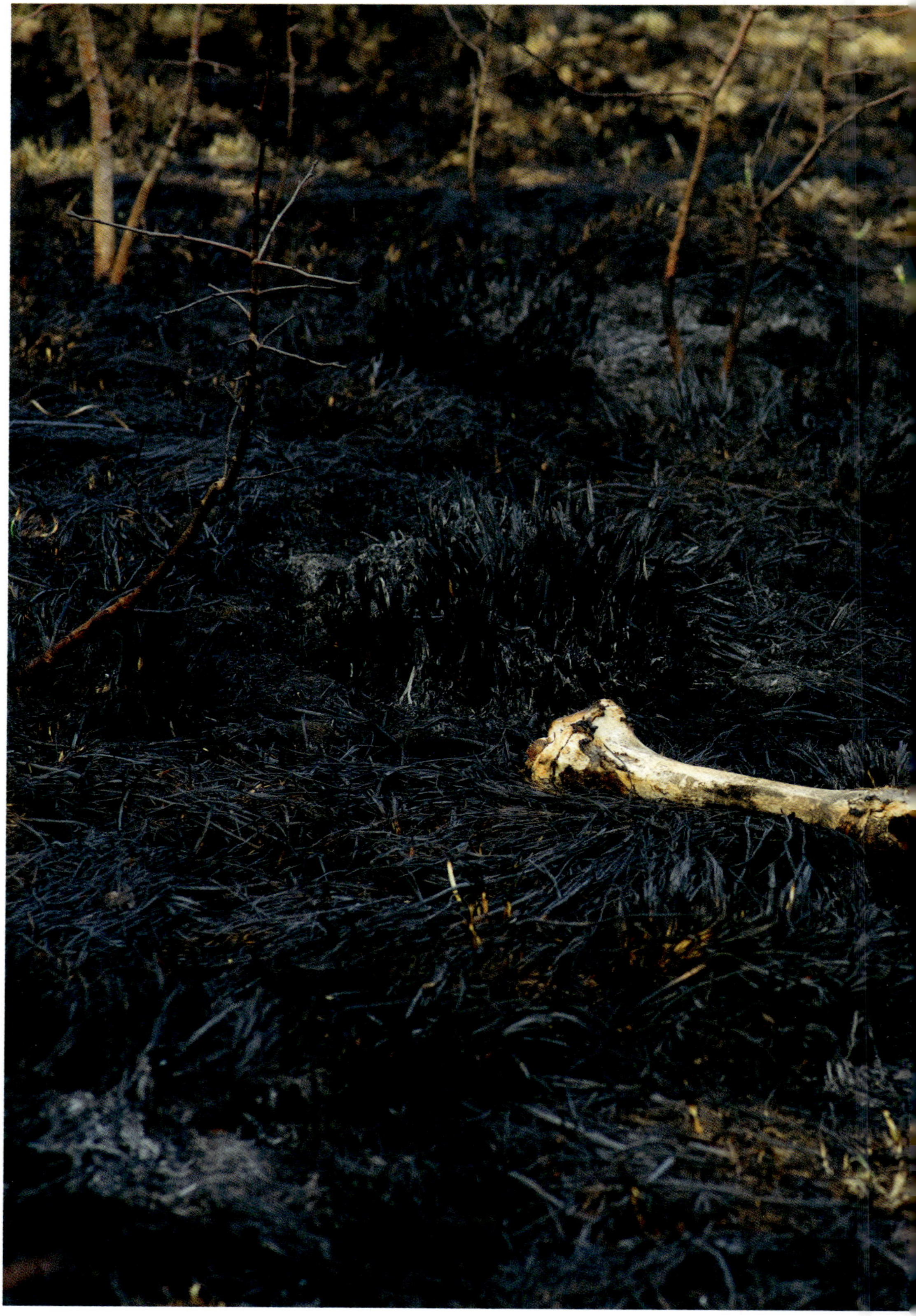

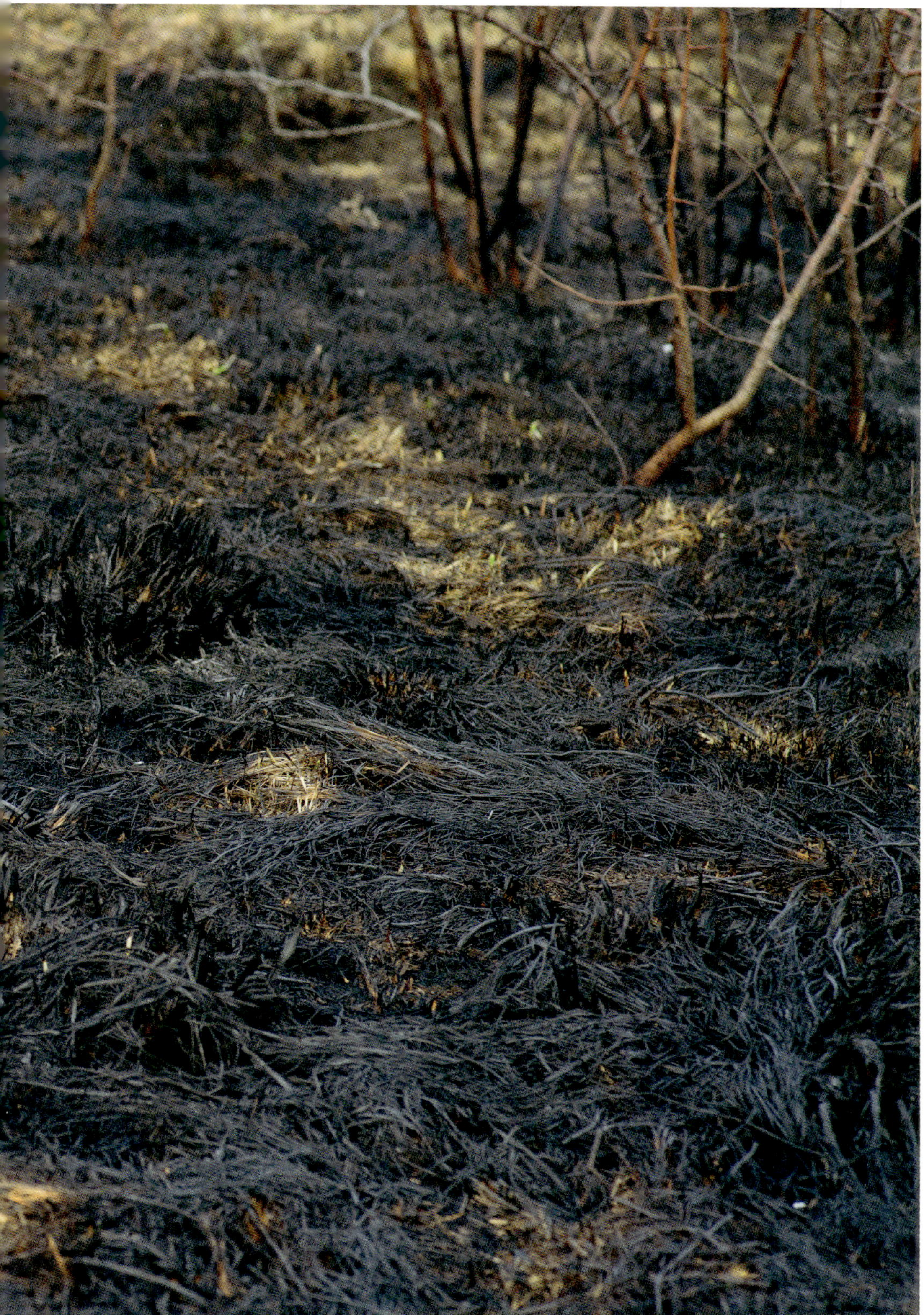

As we walked, we talked about nature, and family, but mostly about love. Petya had been in love twice in his life, and had declared himself done with romance. The first was an older man, a friend of the family, whom he had pursued and seduced while a teen-ager. The second, a man he had gone to R. to meet just last year. They had spent two passionate weekends together, before the other man decided he didn't want to see him. He was heartbroken. He was finished now, he said, smiling demurely between pursed lips, a scentless violet dangling limply between thumb and forefinger.

We stopped often as we talked, my gaze repeatedly scanning the distant vista of snow-capped mountains. But when my eyes caught Petya's he always seemed to be looking right at me, standing straight in front of me with the parallel lines of his puffy jacket measuring the level of the jagged horizon. His grey-blue eyes said: I am here and you are here, on this burnt hillside in Russia.

We reached an outcropping of rock in an open area of pasture, and stopped to assess the situation. Clearly, we were no longer going to the top, but I was leaving tomorrow morning, and Petya had to work. There was another rock, just outside the other end of the village, that he wanted me to see. The secret treasure of G., the Stone of Love. I said: we should go. But first, at this rock, we took portraits.

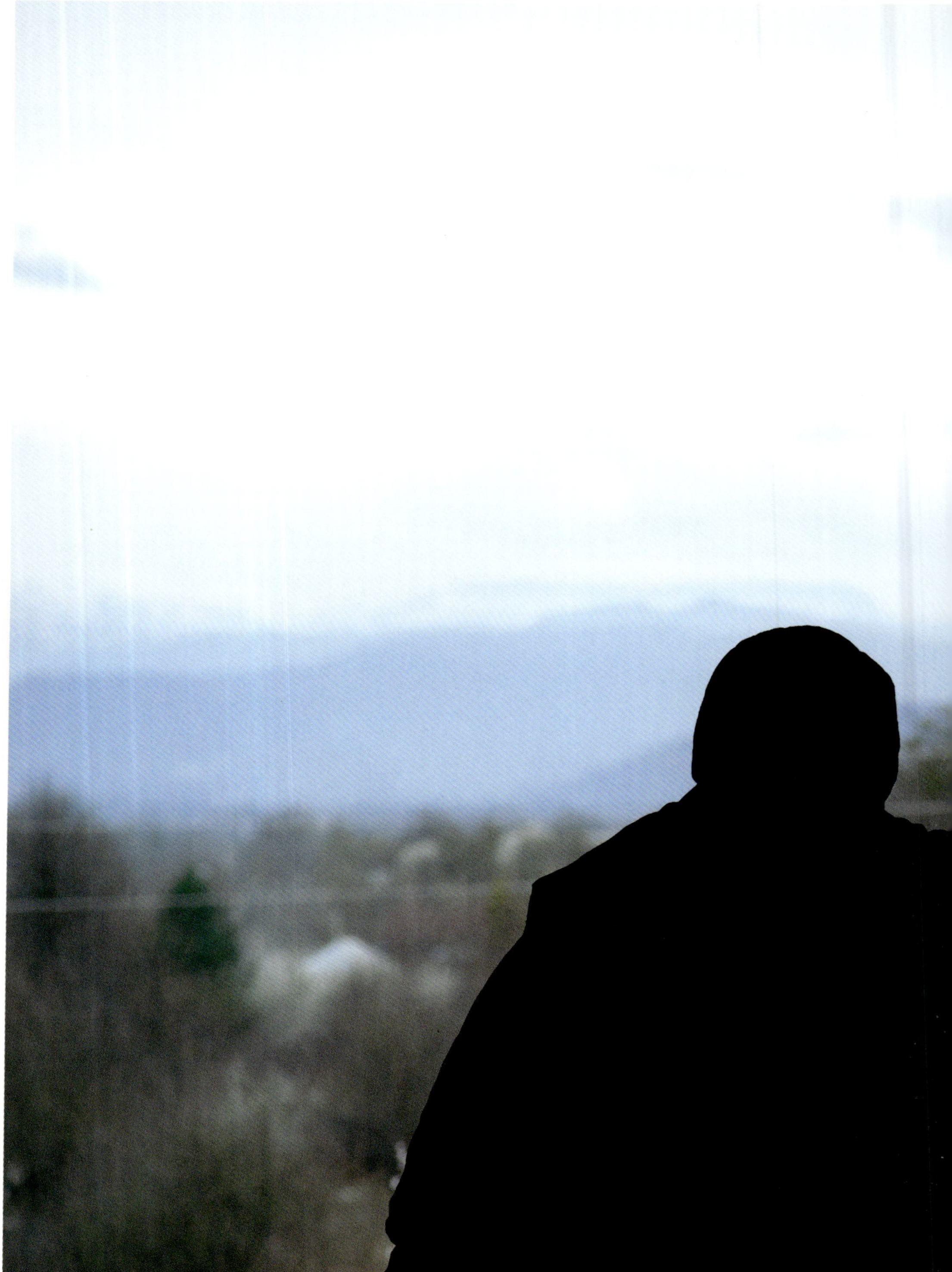

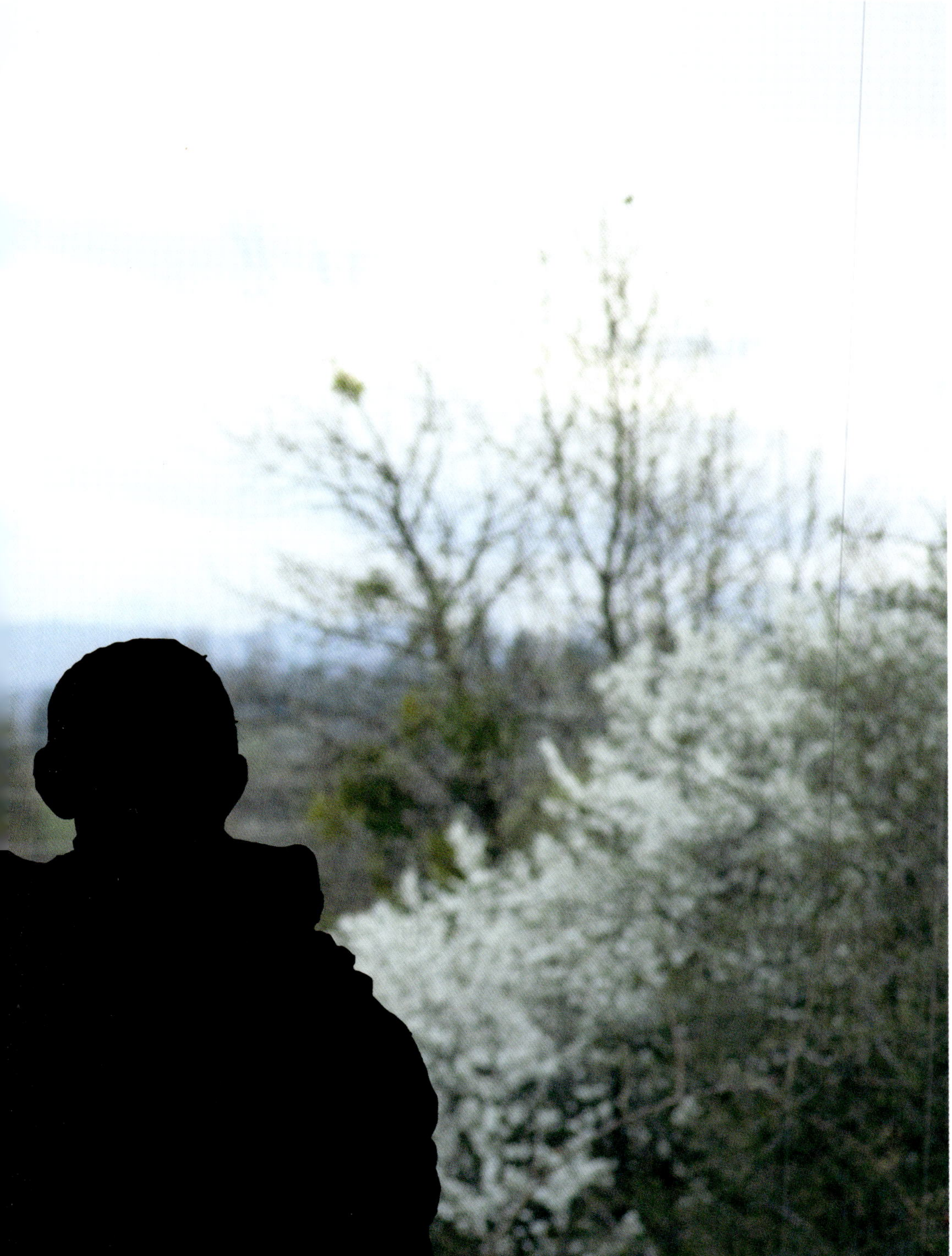

Dusk covered the village by the time we hurried past the bus station, continuing straight along the main road, as if chasing darkness rather than the fading light. Petya explained as we walked: the Stone of Love is a small, almost perfectly shaped natural stone pyramid, with a tiny plateau at its apex. It is G.'s one true wonder. He took his cousin to see the Stone of Love when she visited from Moscow. She liked it so much, he said, that she determined to take a piece home, attacking the pyramid with other, smaller rocks to break off a fragment for herself.

When we reached the ditch at the end of the road, the scents of dung and grass were more tangible than the fathomless black depth in front of us. We followed it for 40 or 50 feet, finally finding a narrow stretch to jump across. From there we traveled quickly and erratically upwards, darting around rocks and brambles and cow dung as we went, tacking repeatedly as Petya continually recalibrated his route. In this darkness, he didn't remember where the Stone of Love stood. Several times, as we were about to give up, he caught sight of some new feature—a jagged rock, a single bent tree—and thought we were nearly there.

At last we stopped on an open slope, the unevenly flowing town below us, barely broadcasting the faintest light to our faces. We agreed it was time to give up. And just then, Petya saw a large rock, not fifty feet from us. We hurried over, but it was certainly no pyramid. I leaned against it anyway. He put his hands on my shoulders, and there against the wrong rock, Petya kissed me.

We walked back into town, conspiratorially, our shoulders brushing in the unlit street. Petya called his mother to assure her the foreigner hadn't harmed him, and he was on his way home. In front of my hotel, he wished me goodnight with a shake of the hand.

As he turned to leave, I remembered to ask my question:

"Petya, why do they call it the Stone of Love?"

"Because," he answered, solemnly, his hands framing a small square in the air between us, "there is only room for two lovers atop the pyramid."

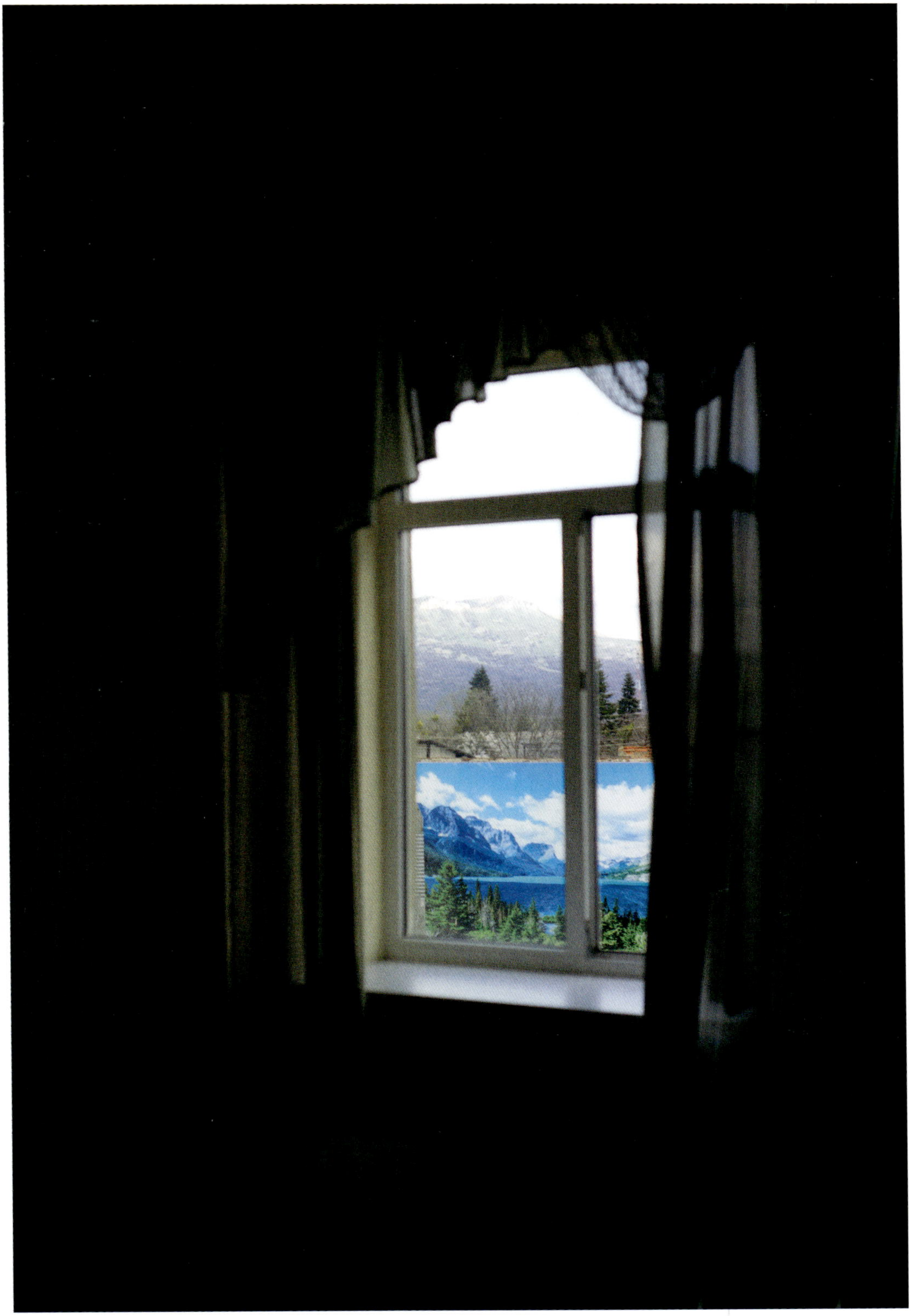

The next morning at noon I am on the day's only bus out of G., already idling in front of the station. Petya stands outside with the sun to his back, squinting at the light flashing off the window between us. He makes the 'call me' signal with his left hand, flicks his right hand up to check the time—diamond light flashing from the chrome of his watch—points off towards the factory building to his left, nods, winks and turns down the road, back to work, his proud gay ass, wide like the Russian steppe, wagging down the broad dirt avenue in all of its acid-washed glory. A silent, ineradicable swagger that says, defiantly, on alternating beats: Yes! and Love! and Never!

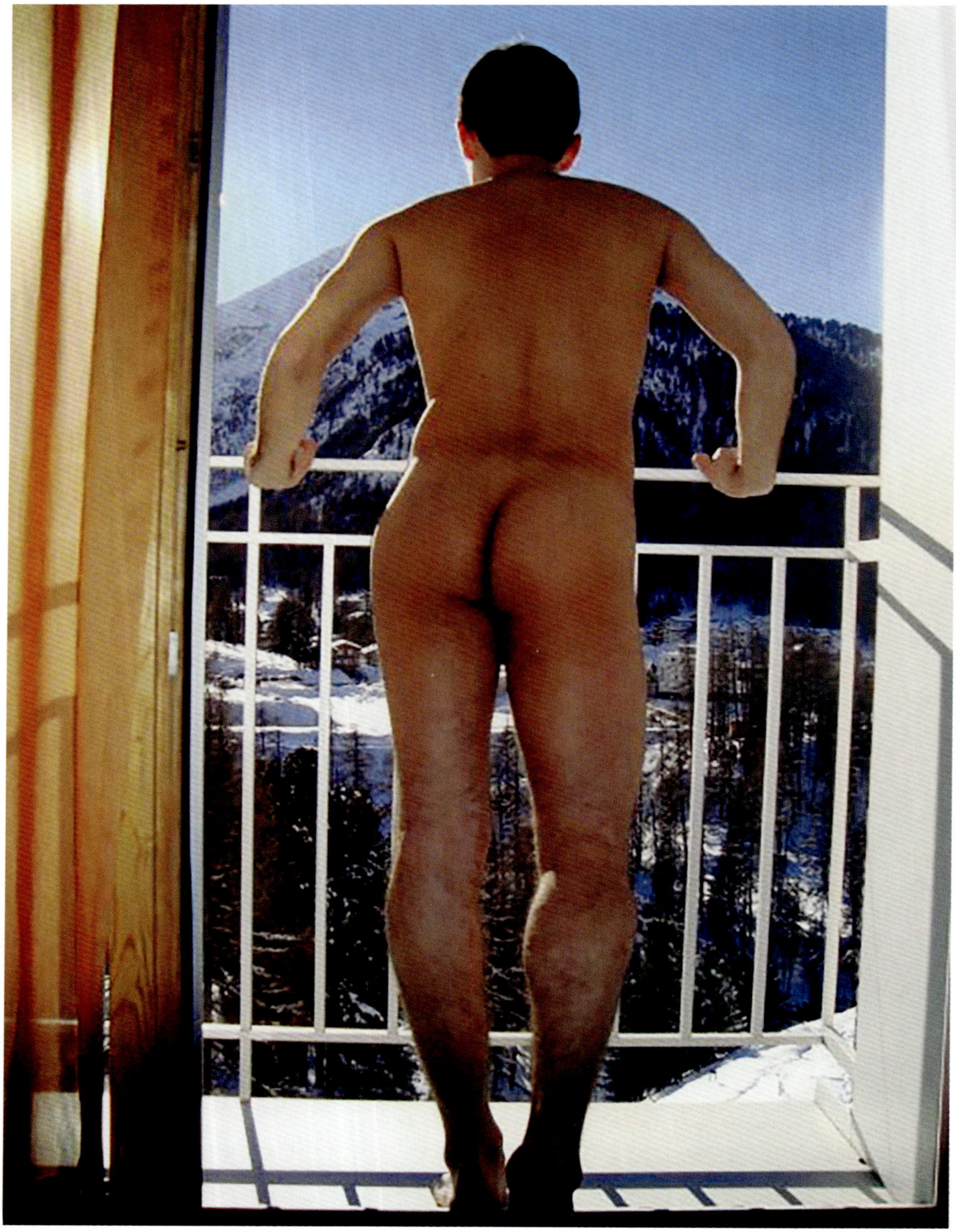

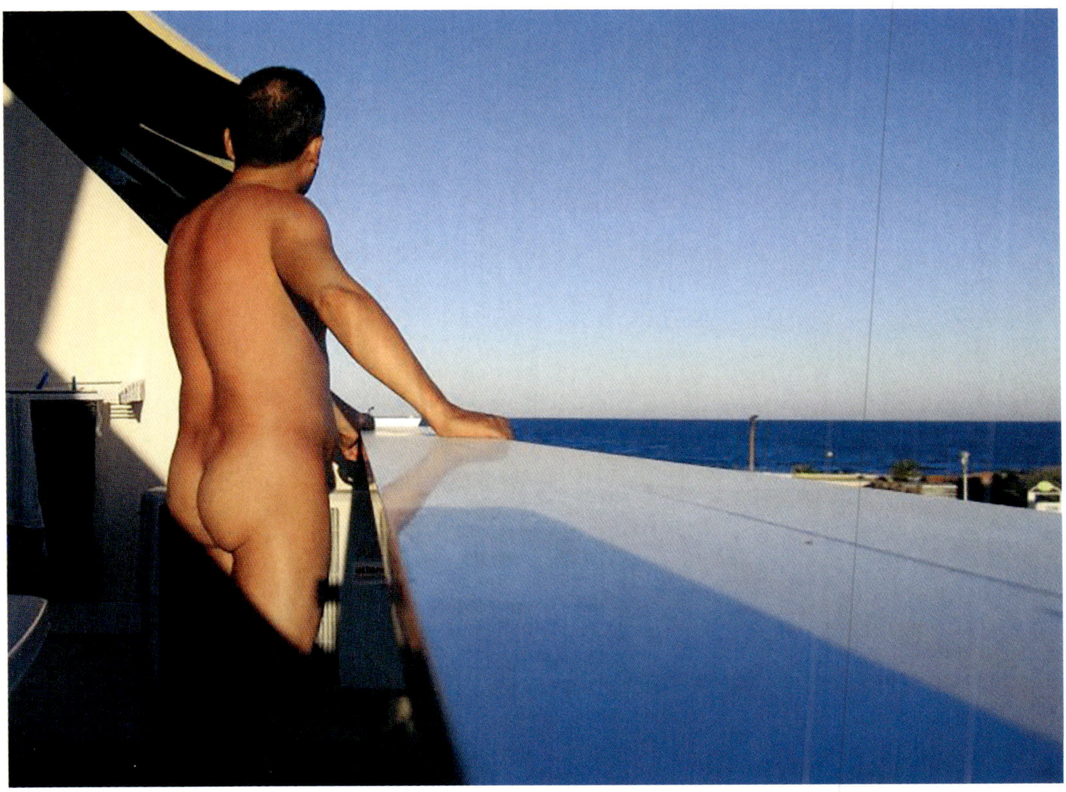

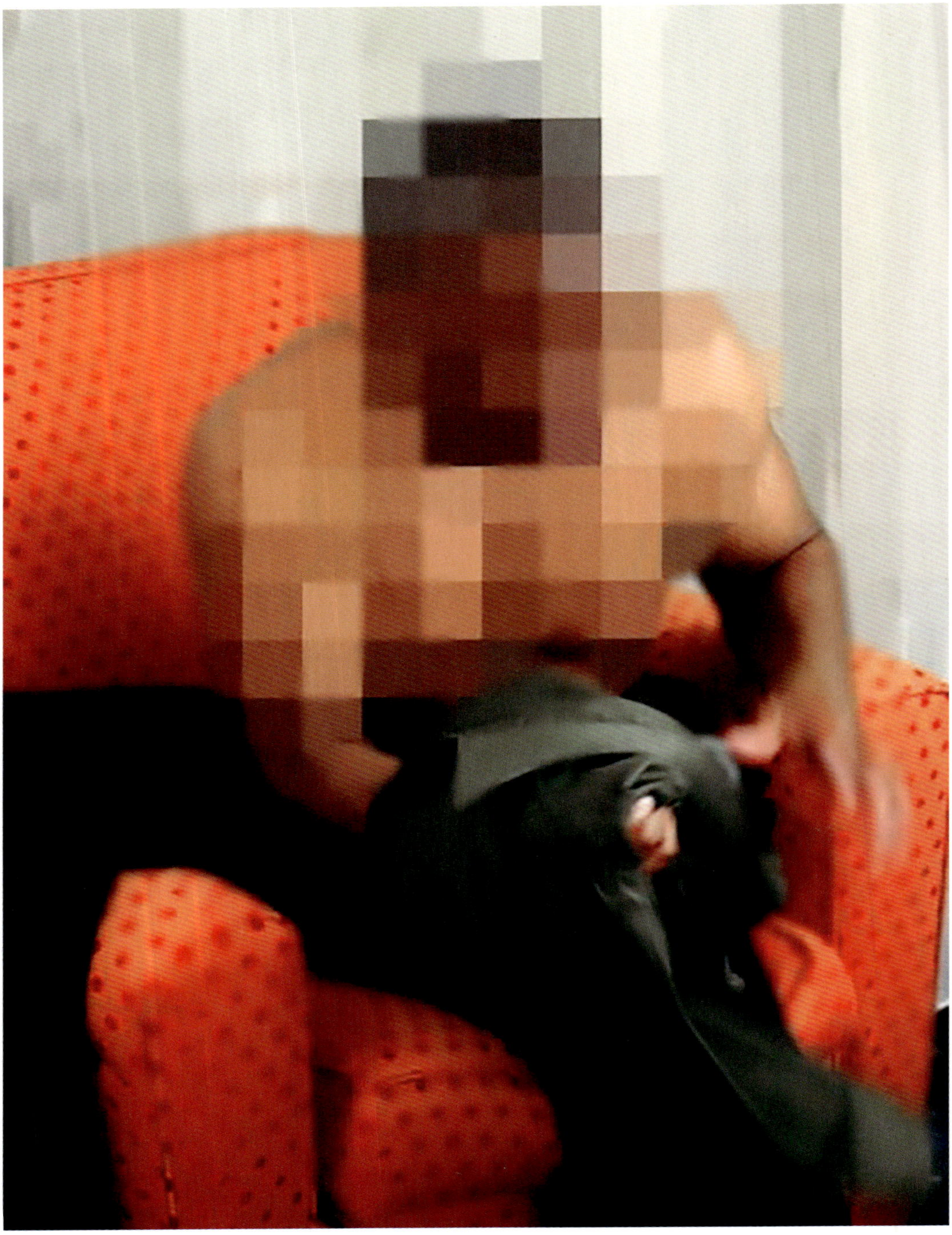

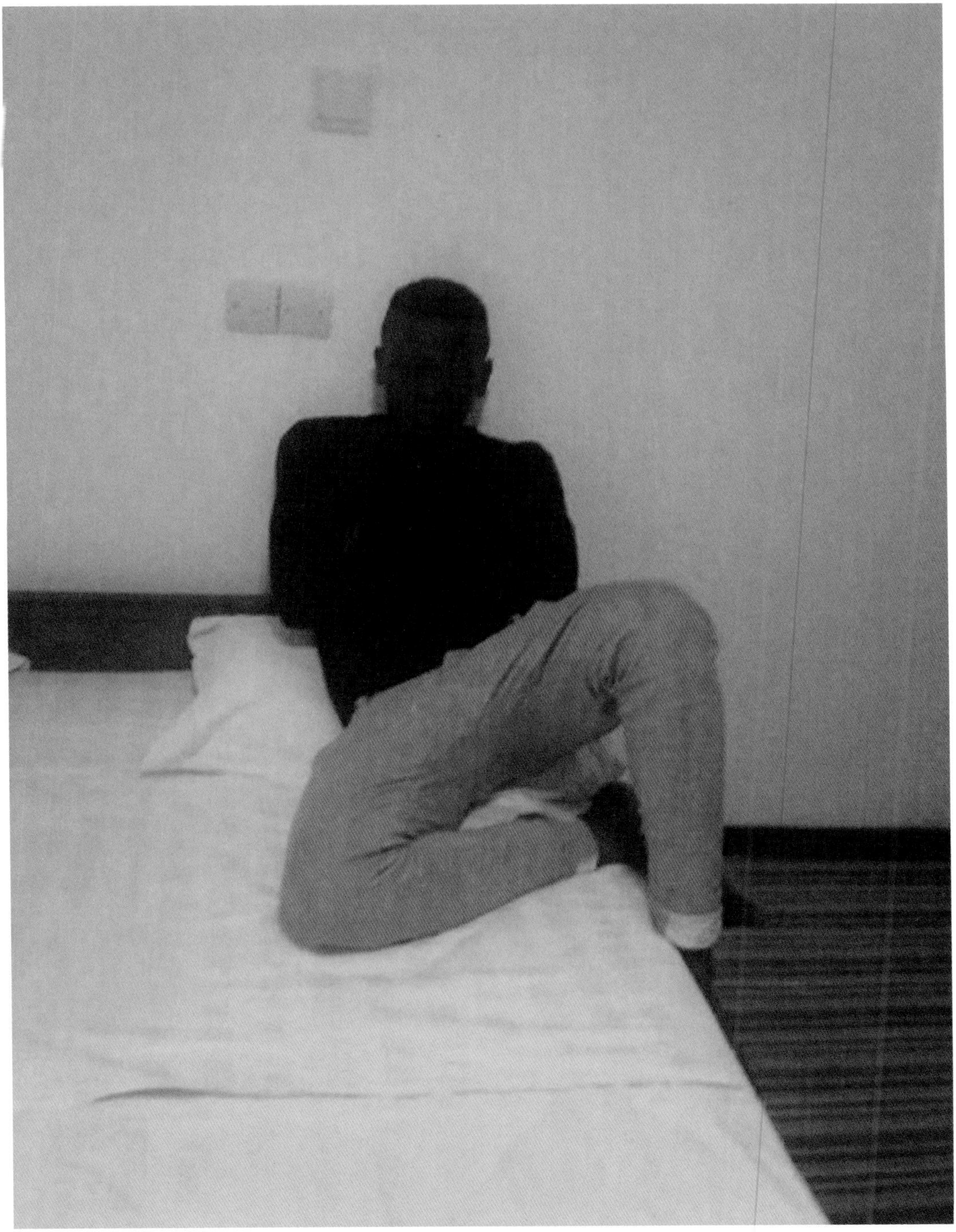

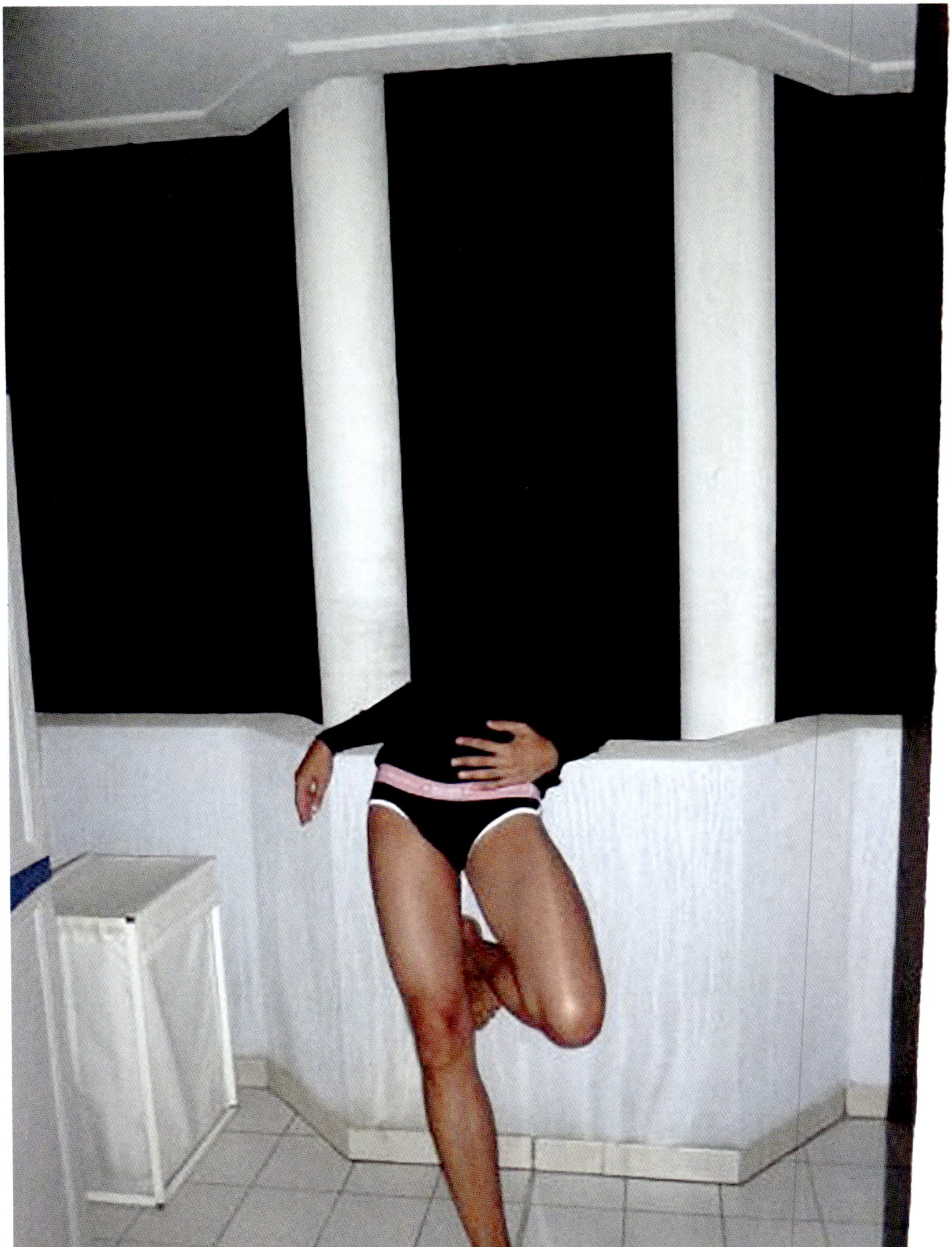

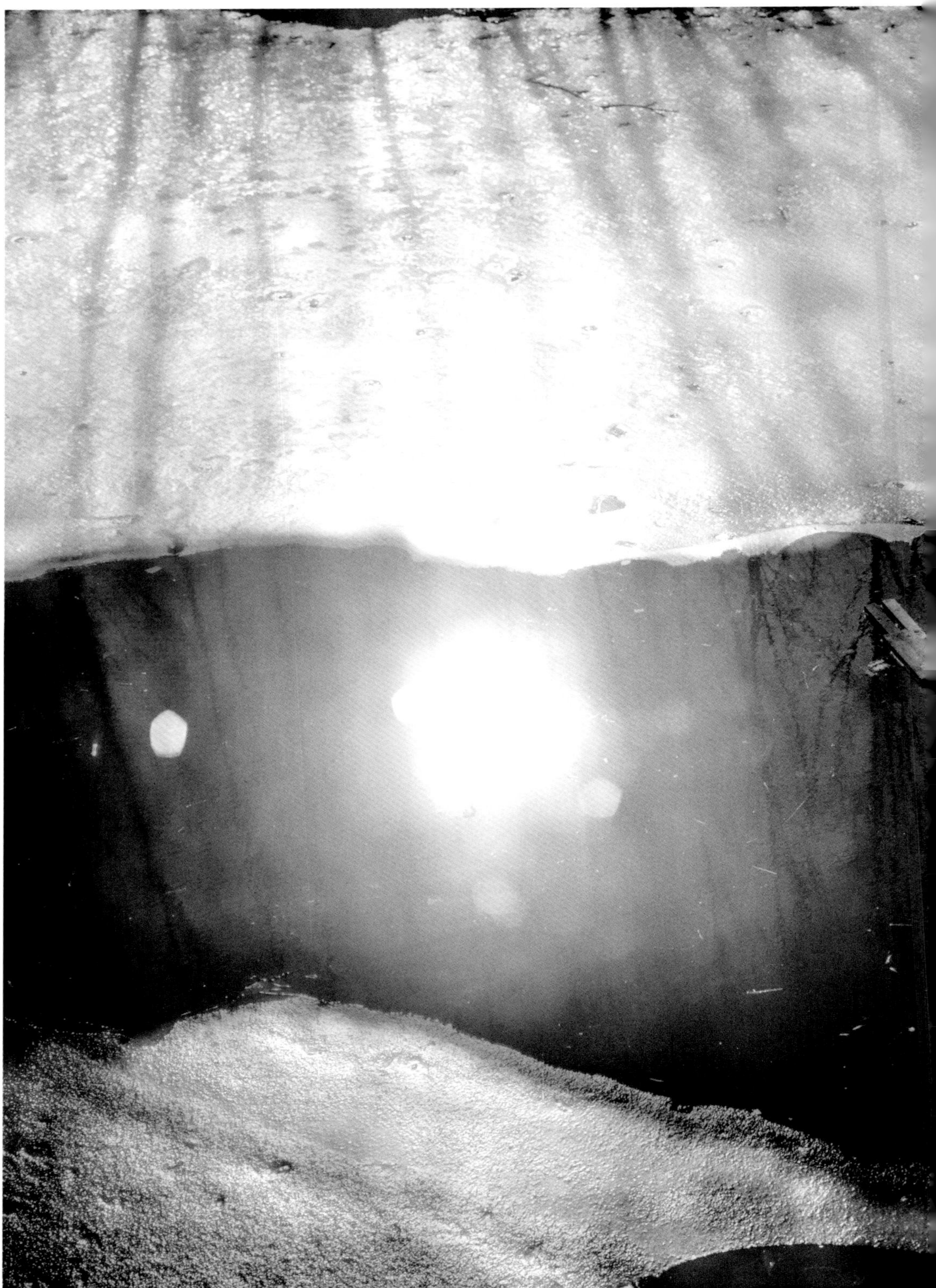

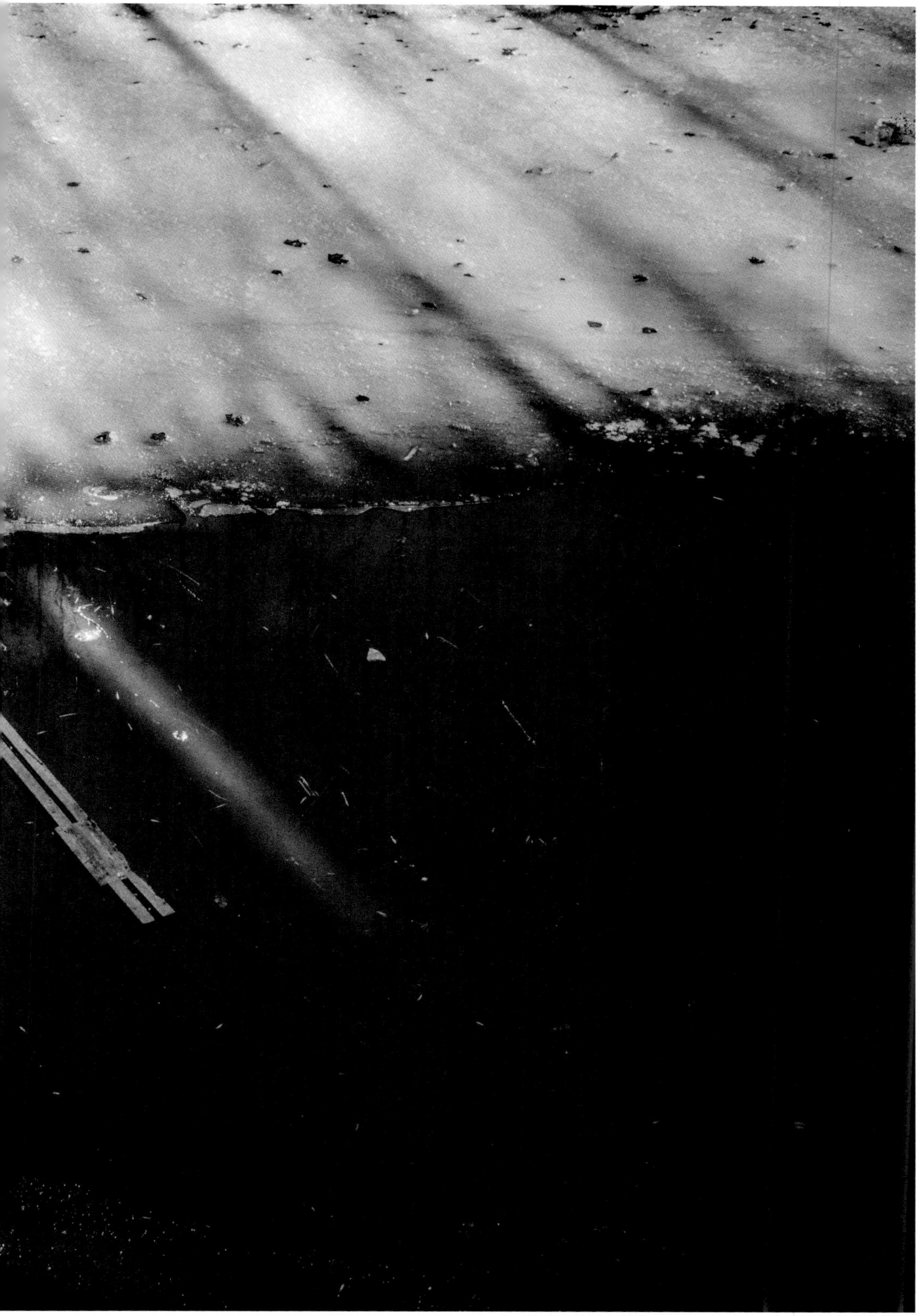

Chapter VII

March 28

I have never been to Lake Ladoga, and I don't know anything about it. But here, in Saint Petersburg, when the broad spans of the Neva thaw, this upstream abstraction rises to the lips of the city's citizenry. Ladoga, where the ice comes from. It surges downstream in vast broken hulks through the newly lucid river, enormous jagged pieces suggesting the unspeakable scale of its shattered source. Ladoga's ice sweeps through the gleaming city, heralding, as death does, life's return. Each tiny melting island flashes signals of recurrence and no-return—the suggestion of freedom lodged in the cold impenetrable wreckage of Lake Ladoga's thaw. The ice passes beneath us on the bridges, evidence against narrative, implacable and mute.

What lies within Ladoga's ice? I cannot know. But it feels like the idea of other consciousness, crystallized, impenetrable, its insular presence glimpsed but never known. The ice feels like Russia, in which the isolated seizure of the other is always on display, and rarely penetrated; a constant pageant of internalization, the grand parade of icebergs. That's what I have come here for—to see the ice. But this spring, as if to thwart me, there isn't any. Nobody can say why with any certainty. It was a mild winter, and the ice was never formed? It is a late thaw; the ice is yet to come? Nobody can say, and still they say: Ladoga.

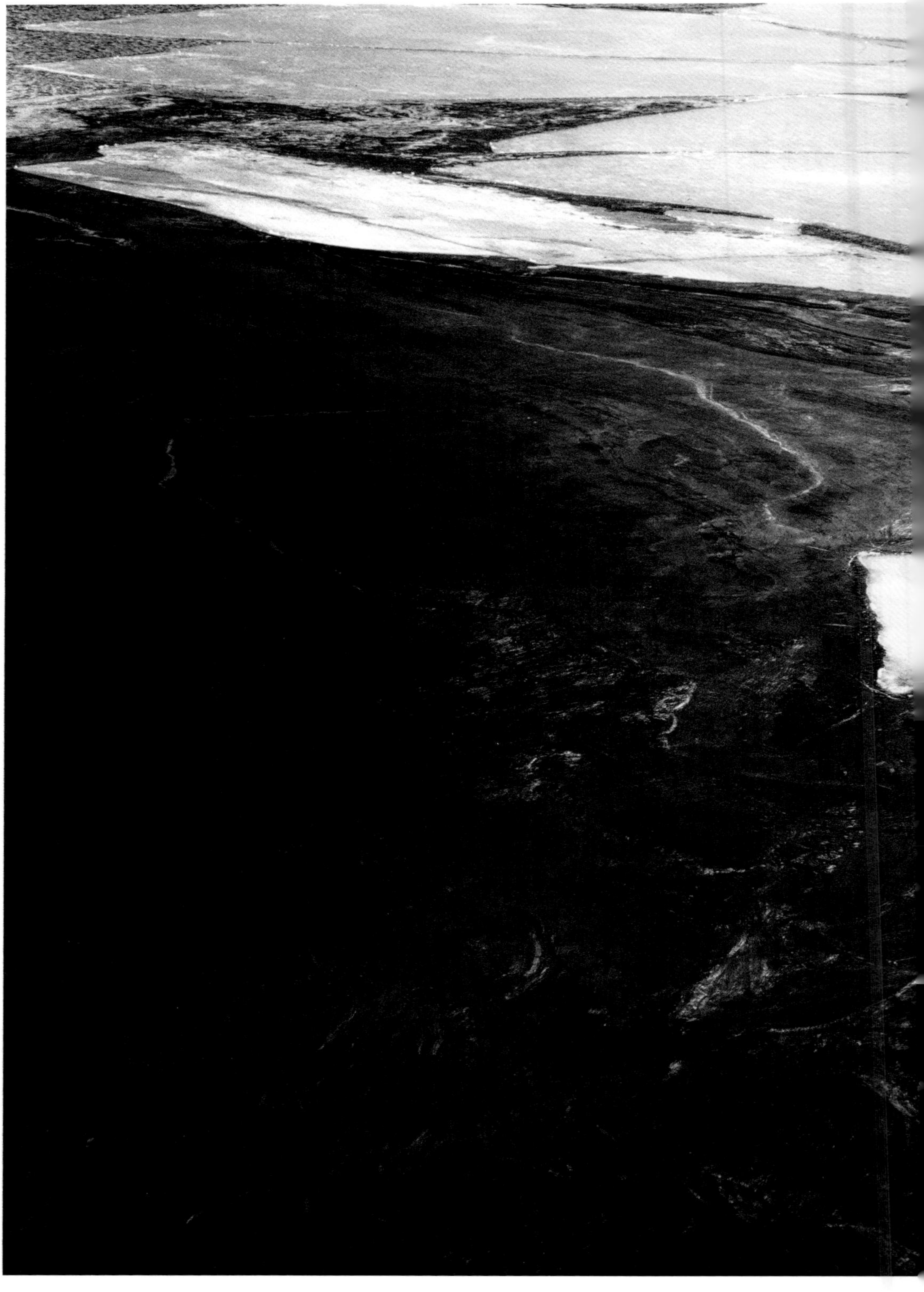

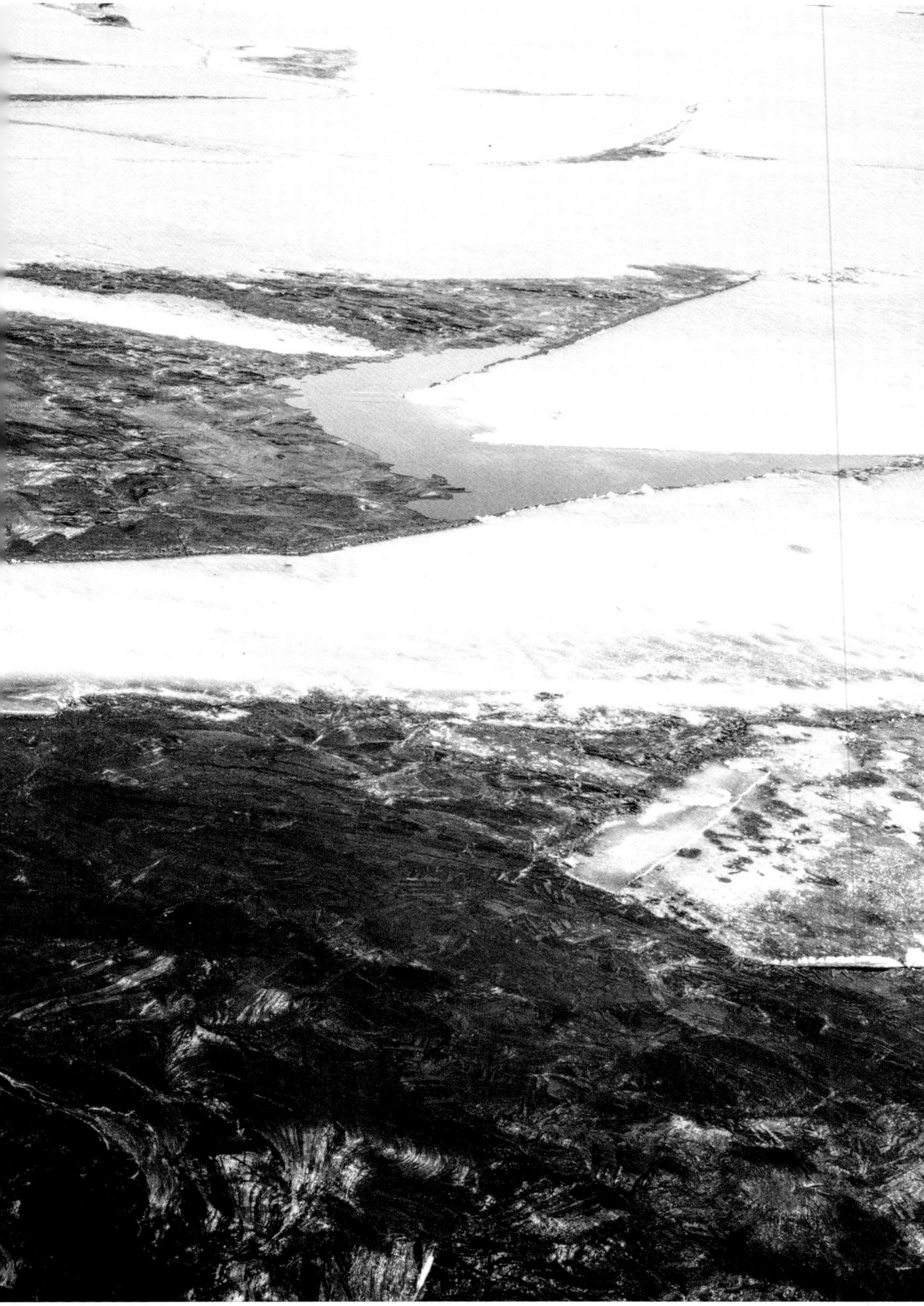

March 30

Two long causeways connect it to Petersburg—the old one from the east, the new one from the south. On a map, Kronstadt appears as if set in place by a jeweler's pincers, balanced in the blue aperture of the Bay of Finland. The island is fixed, like a rare insect, lest it drift closer, or float away. I don't know which causeway will carry me, but I will be there tonight. I am told this island has ice.

March 31

I've never had a firm hold on poetry. It remains for me, in its strangeness of construction, in its un-grippably slick purpose, as if under ice.

Today, according to plan, I began my stay on Kronstadt with Mikhail Kuzmin's 1929 verse, *The Trout Breaks the Ice*. It's a queer object - a romance shaped around rupture rather than union, suffused with omens and visions. And keeping the beat of this thwarted love song, one desperate fish, banging against the ice from below.

The Trout Breaks the Ice. Even the title is slippery like a fish. The verb's aspect suggests, with the merciless cruelty of the Russian tongue, that the breaking of the ice is not complete, or permanent. But to translate it with grammatical exactitude—The Trout is Breaking Through the Ice—implies the reportage of process, a play-by-play. And that is not right either. Kuzmin's ice is breaking, has broken, and may need to be broken again. This ice is always with us.

This morning I reread all twelve chapters before a long walk along the canals. These waterways, like the poem, follow a logic that eludes me. Large and irregular sheets of ice seem to drift one way and then another, not attending to the daily rhythm of tides but reversing across minutes, repeatedly, as I walk; caged waters, pacing between their narrow walls. Sometimes, for forty feet or more, there is no ice, and a fathomless black unfolds, impenetrable as oil, reflecting darkly the late winter sky. I cannot imagine the depth of this water, and have the sudden vertiginous feeling that I am under and not above it. Its surface is the opposite of a window, and the enemy of mirrors.

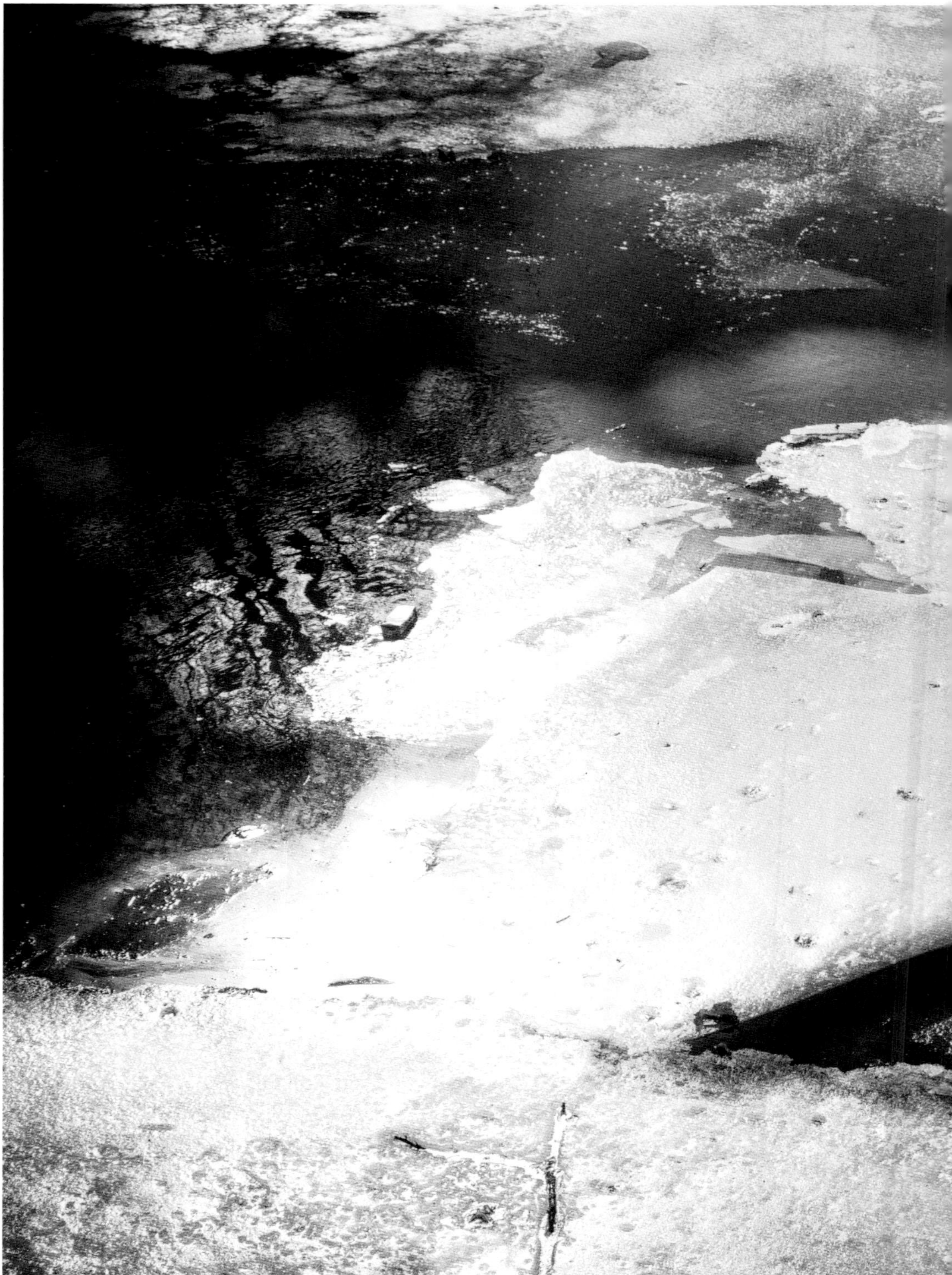

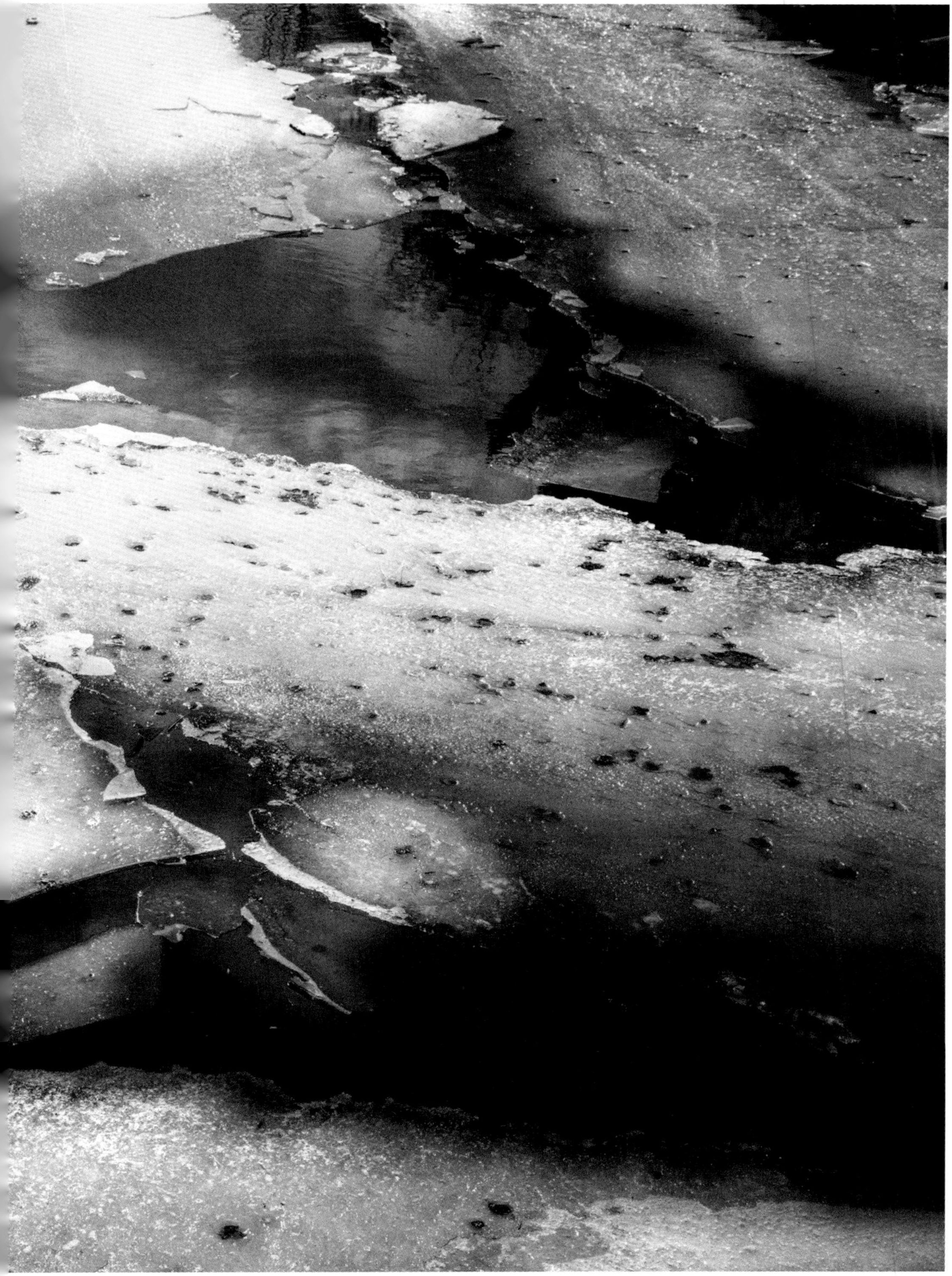

April 3

Rifling through last year's text fragments, I come across a file named "unintegrated." Perhaps, I conjecture, it will explain why I've come so far to look for ice:

Every year, I introduce my students to Freud's mechanisms of the unconscious—repression, displacement, anxiety—and their implications for photography. These are bright, curious youths, staring at me earnestly, carefully transcribing the arguments. They understand my words, and yet every year, it becomes clearer to me that these ideas mean less. I'm starting to think I shouldn't teach them anymore. Perhaps, I think, it's impossible to really feel Freud if you were born into a world entirely liberated from the deflective opacities of Modernism, or Liberace. These 21-year-olds are smarter, more serious, and more lustrously confident than those before them, but perhaps they grew up in a culture of such relentless affirmation and support that repression, while theoretically possible (perhaps in some less fortunate world?), does not resonate with their vaccinated lives.

I should probably view this development as just that, but instead, I always feel heartbroken, for them or myself I cannot tell. I persevere, telling myself it's worth it to reach and change just one—the one with a history, a pain, unsettled in the world even just a little. Today, again, I introduced André Breton and his radical embrace of the unconscious as a superior source of experience. This year, against a sharp impulse that tells me "No!" I read aloud my favorite lines from the First Surrealist Manifesto: "From childhood memories, and from a few others, there emanates a sentiment of being unintegrated, and then later of having gone astray, which I hold to be the most fertile that exists."

Their faces, upturned, attentive, do not change. They wait for me to tell them why this matters, or what it could mean … and I am suddenly overwhelmed with solitude. I want them all to leave me alone with the dust of those words on my lips—unintegrated, astray—to write them on the blackboard in large letters, lit by fluorescence, brilliant and fugitive for me alone. I think: these days, there are online resources for the "unintegrated" and student clubs for those "astray". There's an App for the alienated, with customizable options and voice recognition. Suddenly, while they wait, I want to tell them something small and painful from my past—anything to summon the ghost of that figment called alone. I am mourning their loss of the consolation called solitude.

Reading this again, here, with Kuzmin warbling in my mind and the grey ice shifting against itself in the canals outside, I start to feel I have come to Kronstadt in resistance to that exact loss.

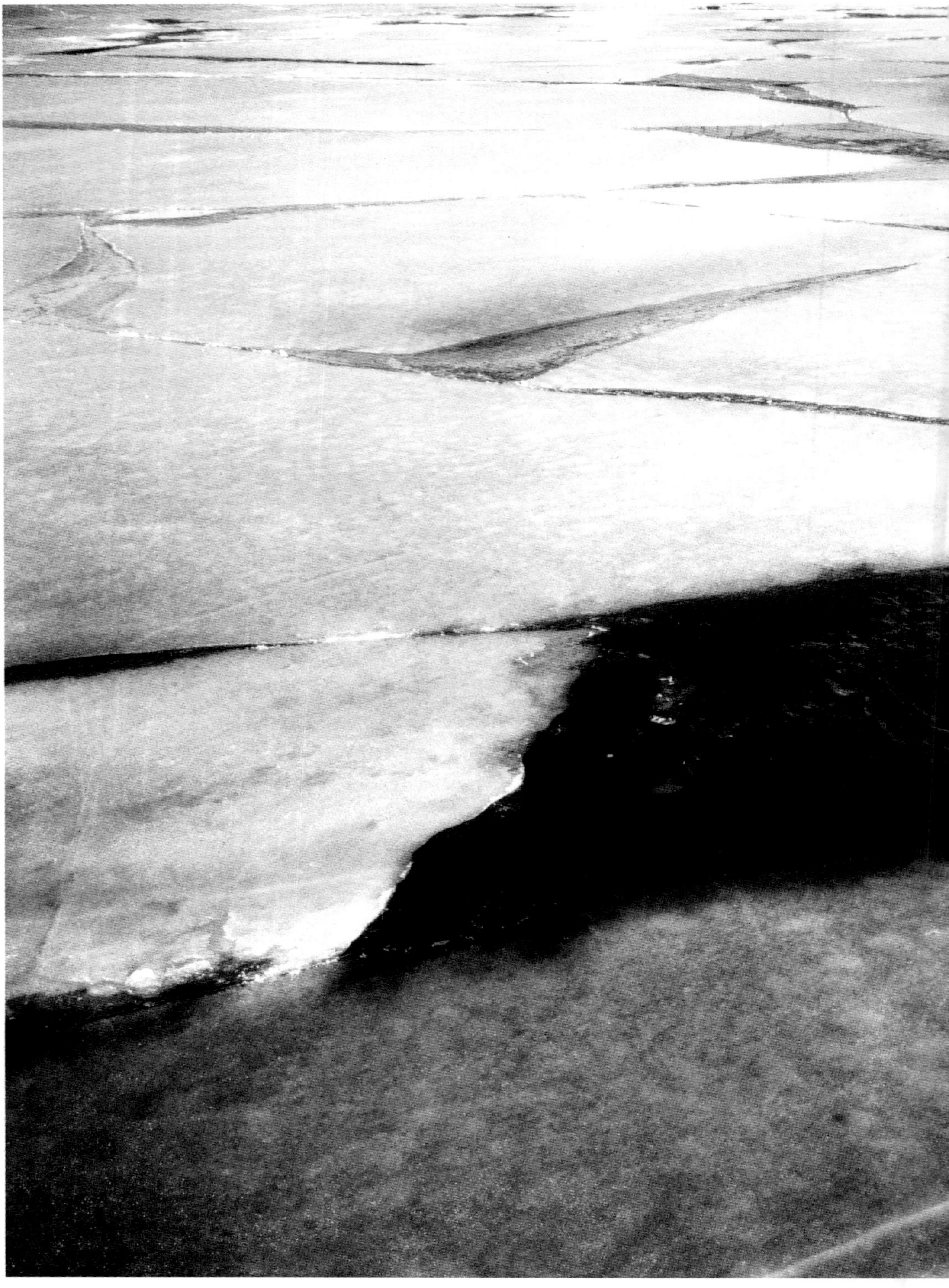

April 4

Kuzmin's 1907 novel, *Wings*, is celebrated as one of the first openly gay works in western literature. But to see this queerness, a modern reader must squint, blurring the narrative into a soft gay diffusion of atmosphere, with the divine a gauzy cover for the possibility of the profane. In the book's concluding moment, having decided to travel to Italy with his philo-spiritual mentor, our liberated hero throws open a pair of shuttered windows, filling his world with light. But what comes in and what goes out? What new vista is framed before him? We see nothing. A wise reader is supposed to know, perhaps, that mentor means lover, Italy means fucking, and open windows mean fuck me, but we can't be sure. He throws open the windows and I am left wondering: do they ever even kiss? It is an astonishingly bold obliqueness of expression, but allusion without object is not freedom. What does it mean to have a window without a door?

Somehow, despite all of this deferral and evasion, I find *Wings* oddly seductive. Perhaps because the hot stuff remains hidden, ever the protagonist's (and the author's) fantastic secret. Everyone's sexual awakening, to one degree or another, happens largely alone, with oneself. But gay eroticism is particularly allied with solitude for obvious reasons: the fear of discovery, the lack of a map, or an available object. Attraction born as a secret, even to other parts of consciousness. And secrets are most delicious, and safest, when consumed alone, and kept for oneself. That is their consolation. There's a part of that link, between the erotic, the alone and the imaginary, that still turns me on. Solitude makes me hard, and sad.

April 6

I awoke in the night with a pain in my chest. It wouldn't go away. Still, I drifted back to sleep. There, I met a man. He said, "come with me." We scrolled from one cartoon sketched village to the next, in sideways relief, for reasons unknown to me. We were looking for something. Along the way, he handed me some red and he said, "take this." I did, and it helped with the pain in my chest, which was still there. The man had amazing powers of red, and others followed him, or gathered in his orbit too. At one point we came across a vast wall of pills, each with the letter S in the center. To neutralize them, he sent out an invisible wave that reached all the pills at once, turning each S red. Immediately, all of the pills were neutralized. Finally, though the pain in peoples' chests was better, but not gone, everyone said, "enough with this red." The man went off to a village, and threw himself down on the ground, and that is how the water of that village began to flow red.

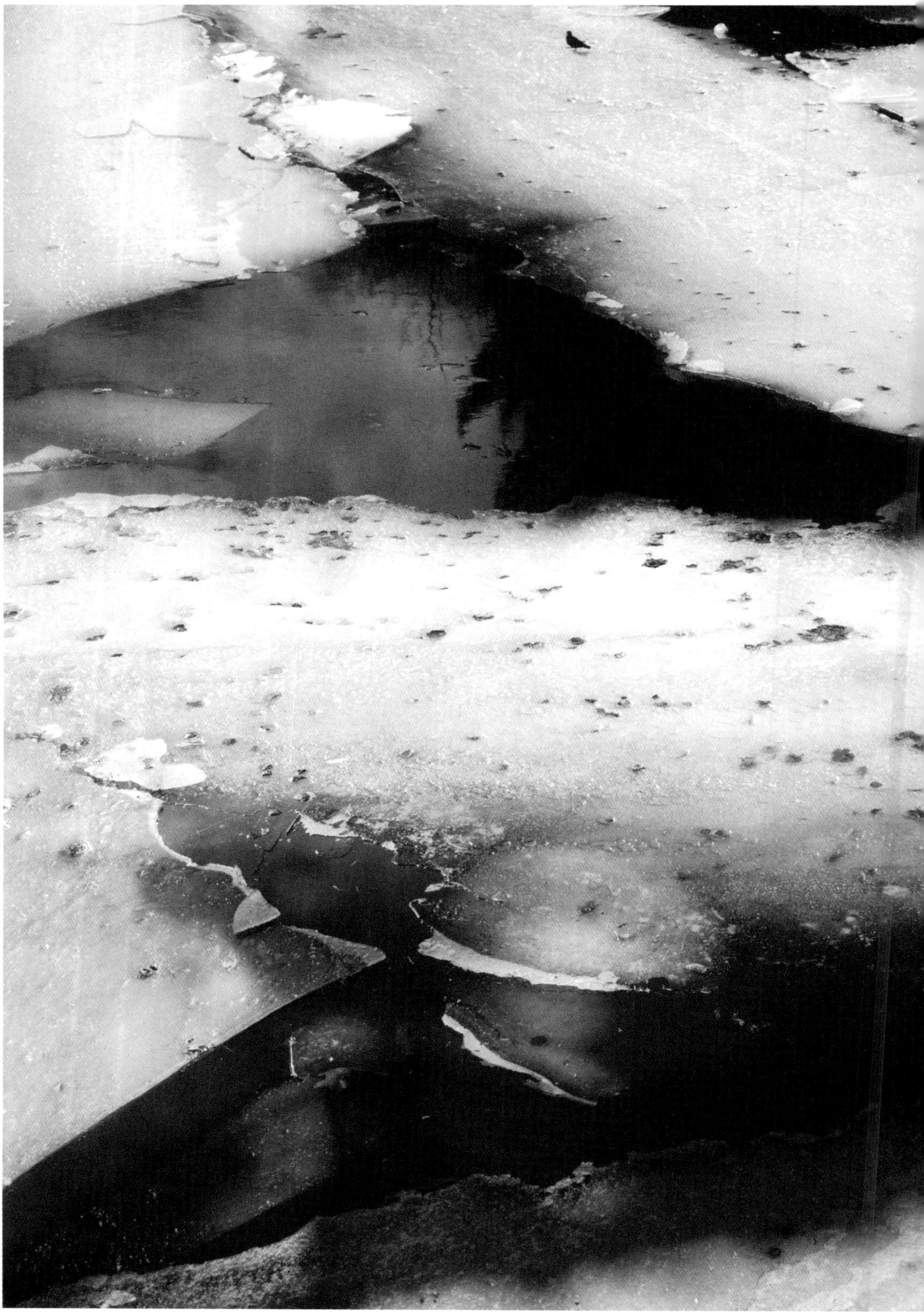

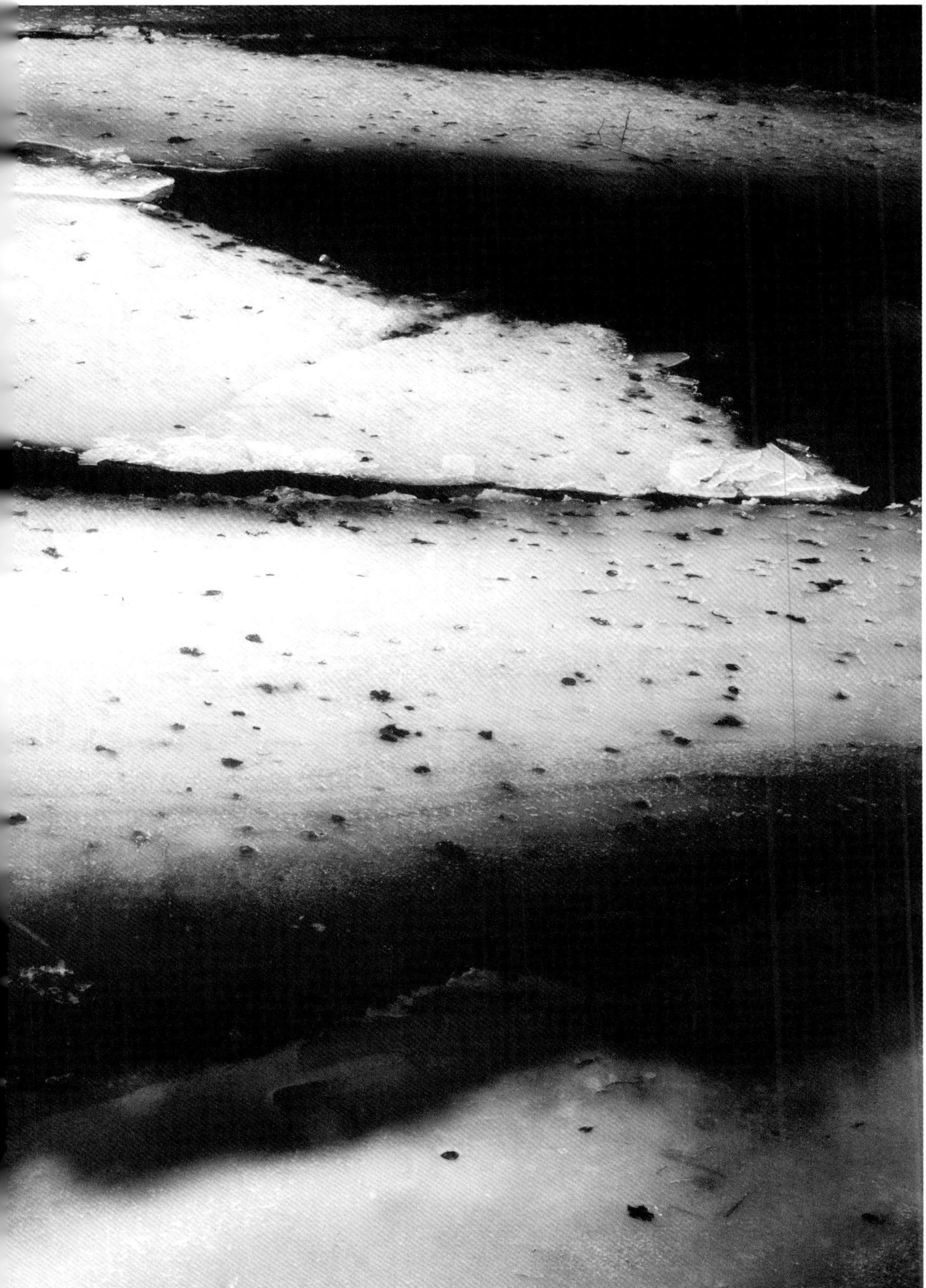

April 8

Walking home from the bus, along an icy canal at dusk, I stopped to watch a chubby bundled boy of 8 or 9, lingering behind his ambling babushka. He was busily rearranging a collection of small sticks and blocks of ice in the spaces between the wrought iron fence-posts along the canal. It seemed some obscure form of occult checkers, with inscrutable narrative implications.

His isolation and absorption, not to mention his heft, reminded me that the social lives of boys are unpredictable, and cruel. Some years of my childhood there were friends, and games, and afternoon adventures, and other times, there were none of these things. One year, nobody turned up to my birthday party. Such deserts arrived sometimes in winter, when there was always and everywhere ice, and I would walk to elementary school alone.

To give meter to this solitude, I played a daily game. Leaving the house, I would select a modest piece of ice from the sidewalk, small enough to move briskly with a kick, but large enough to bear the erosive forces of cement and brick. The goal: to kick this little block all the way to school, without ever breaking my stride to make contact. There was no turning back, and no stopping. But the rules, to avoid loss, were sometimes bent to include half-stepping and swerving, or even substituting an almost indistinguishable icy double. In the case of outside intruders—other kids on the sidewalk, an unsympathetic crossing guard—a quick into-the-pocket airlift might occur. But if the ice strayed too far from my path, sunk into snow-banks or dissolved into dust, I had no choice but to move on alone, without my ice. If my block made it to school with me, I would hide it in a snow-bank, around the corner from the school gates, so I could later kick it home. The ice was not a friend, but a thread of solitude through the labyrinth of childhood.

April 10

Last night, I dreamt that I came across a deep seam of internet-posted monologues—all of them instructions for wanderers in search of love. It was never clear if they were intended as advice or report, knowledge, or desperate imagining. And though this stream was one exclusively of words, all I could see was a sequence of pictures: shallow focus, with blurred foregrounds, luminous, muddled, in black and white.

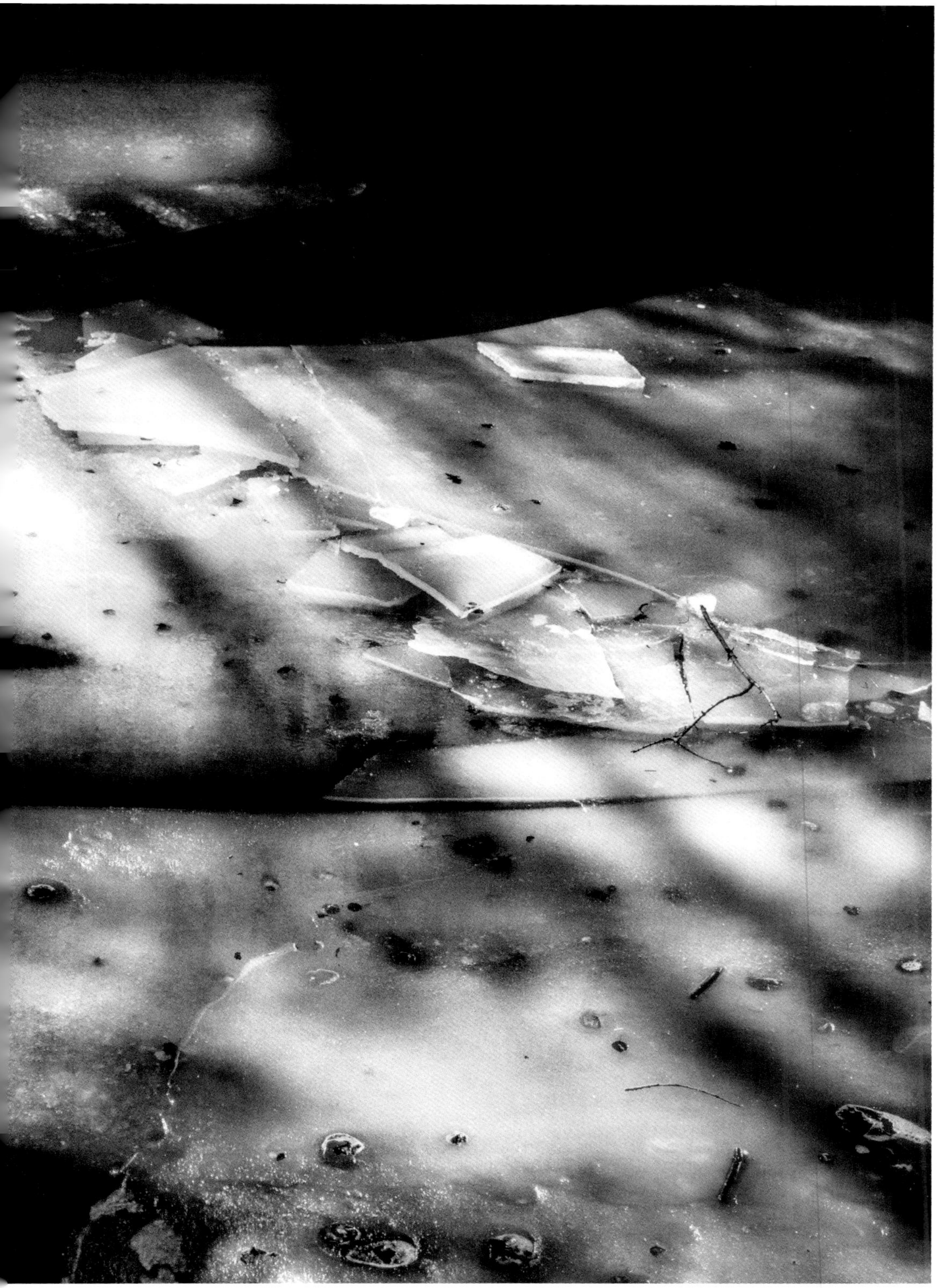

April 12

Metaphors cloud the atmosphere like smoke,
But vanish in a ring below the ceiling.
—Kuzmin

Still there is The Trout [who] Breaks the Ice. I am relentlessly determined to find my place in this poem, though I have failed to do so for over twenty years. So far, I've learned only that everything is offered, and just as soon removed: intimacy is always present, and never captured; love is always real, and a mirage. Everything is offered, and just as soon reversed: the ice is inside us, then we are inside the ice. The fish's struggle, light flashing off its muscular iridescence, telegraphs freedom, eros, anguish, hope, life and death. The trout's telegram is unreadable.

This poem has two endings to its central romance, offered up one after the other as alternate outcomes: terrible rupture and joyous reunion. The happy ending (lovers, re-united, by a hapless agent of fate) is as unreal as the tragic one. Resolution is not credible in a world in which *that* love is impossible. And so desire loses the clarity of its object, becomes merely a voltage, expressed as a measure of unstable currents—of the heart's erratic output—but never as itself. The result is not success or failure; it is rather the opening of windows, a fracturing of ice, a sparkling eye, a fish in a tank, an automobile vanishing around a bend. Metaphor as its own terminus: the purest of solitudes.

April 13

I am in a vast Russian gay club that seems to have no space for drinking, socializing or performance—only endless dark hallways flanked by the dressing rooms of Russian drag queens, each one readying and rehearsing alone. I move down these hallways as if on a long tracking shot, glimpsing cracks of light and shards of movement and sound behind closed doors, but never really seeing in. I am trying to decide which drag queen to choose for my final show.

Eventually, I find myself in front of an open doorway into a brightly lit dressing room. She sits with her back to me facing the mirror, luxuriant blond hair below the neck, elegant arms stretched out along the dressing table, studying herself before she performs ... except her mirror does not reflect. I am right behind her, but neither of us is there. I find that my limbs are stretched out just like hers. She raises her left arm and I do identically, as if we are mirroring each other's movements, standing before a glass that absorbs but can't refract.

April 16

For Kuzmin, consummation cannot be unyoked from repression. Deferral, the tension of what's held in, on the point of breaking but never shattered, is the invisible coin of desire. The Trout Breaks the Ice. The conundrum is in his title. The trout's watery prison is not to be escaped—it is his only habitable environment. Kuzmin, icy monocle in suspect eye, stands immured on the limpid brink of freedom one hundred years ago. That brink, in Russia, is still right there. The poet writes: *Walls of unshatterable glass imprison me.*

Contemplating Kuzmin, looking at the cracking and melting ice, I come to think of poets—and the poem—as what lies beneath. Under the ice: a place not of helpless entrapment but of safety, isolation, privacy; the chilly certainty of self.

We know it is a poem, because we can read it through the ice. We know it is a poem because we cannot reach it. In this sense, the banality is sometimes true: *we are all poets* or *we are all poems*. When anyone, each of us, is the one who lies beneath.

April 17

This morning I reread Kuzmin's short essay-manifesto "On Beautiful Clarity," regarded by literary historians as "influential". It's hard to read that significance from here. Peel away the qualifications and generalizations and the reader is left with nothing but the collegiate dictum: write clearly! But I visited it again, and was bored again, because the title keeps dappling my mind, flashing the electrically lit image of a rotating diamond through my brain. On, beautiful clarity! Last night, I realized: I know that diamond!

The world whose pressures formed me was full of surfaces, obdurate and blank, endlessly refracting me to someplace else. They did not give or hold; I was never invited in. A number of years ago, my boyfriend shot a video of an enormous diamond, spinning in lonely perpetuity at the Liberace Museum in Las Vegas. He showed it to me on the tiny screen of his mid-2000s digital pocket camera, so that it appeared as if held inside the small box of that device. The museum has since closed, but the image of the glittering jewel, trapped behind a glass vitrine in endless rotation, haunts me still. I imagine the great man, alone in his over-lit dressing room, draped in white ermine. He rocks his gem gently, setting the facets alight with refraction—bright rays entering his cool blue eyes. Blinded by the cold diamond purity of form, he might wonder: if I am brilliant enough, will I simply disappear?

April 20

Last night, again, that dream:

Circling wires, surrounding unknown territory. A wasteland, windswept, elevated and vast. The wind, howling, offers me its mouth, and I put my cock in. My eyes are wide open. The landscape I see is bereft and super-saturated: ochres, indigos, charcoals and rust. The wind offers me its mouth. I am face-fucking the wind.

I woke up still seeing the landscape, and thinking of Gustave Le Gray's achingly melancholy 1850s photographs from the Forest of Fontainbleau. The artist spent whole days alone in those exquisitely landscaped woods, silent among old oaks and enormous boulders. Later, he made salted paper contact prints under direct sun. Le Gray tinted these images in an experimental range of tones, arriving at forests the colors of "gray-blue, neutral, violet blue, black-blue, black, bistered black, bister, sepia, yellowish sepia, dead-leaf yellow, greenish gray." His journals tell us so.

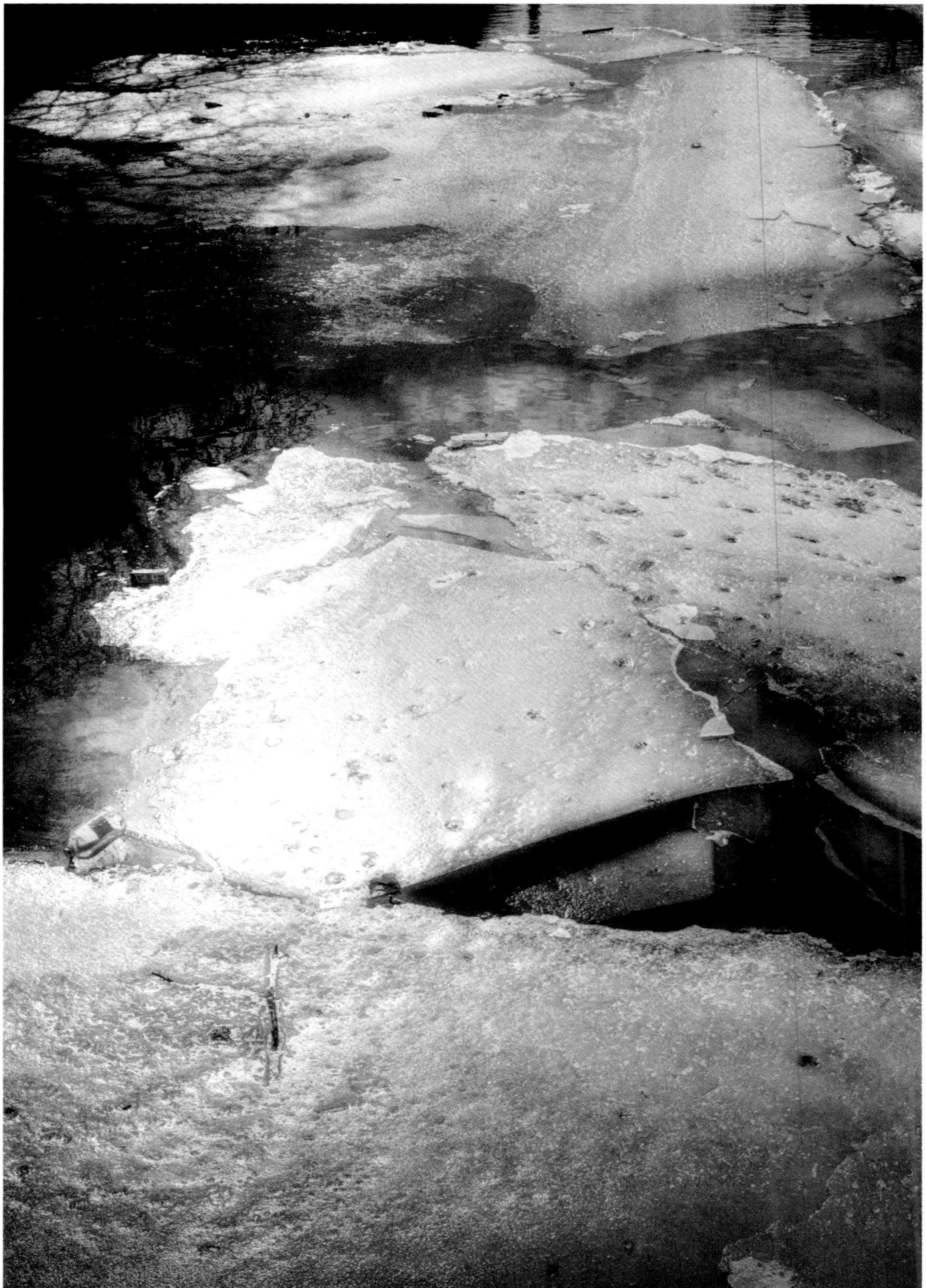

April 21

In a letter to a friend, Kuzmin wrote:

I like cats and peacocks.

I like pearls, garnets, opals, and such semiprecious stones as "bull's-eye," "moonstone," "cat's-eye." I love silver and unalloyed bronze and amber. I like roses, mimosa, narcissus and stock, I don't like lilies of the valley, violets and forget-me-nots. I don't like vegetation without flowers. I like to sleep in the nude under fur.

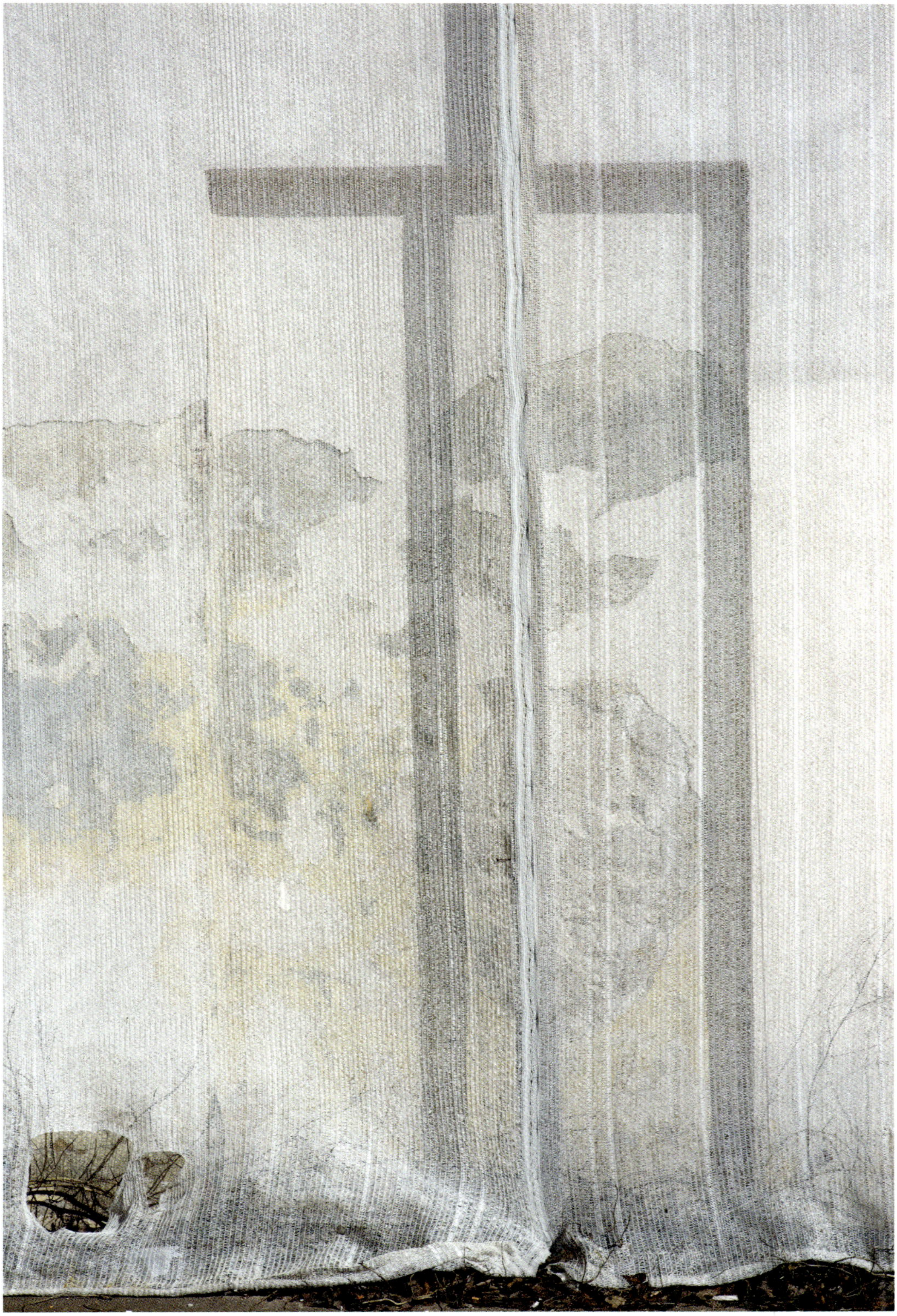

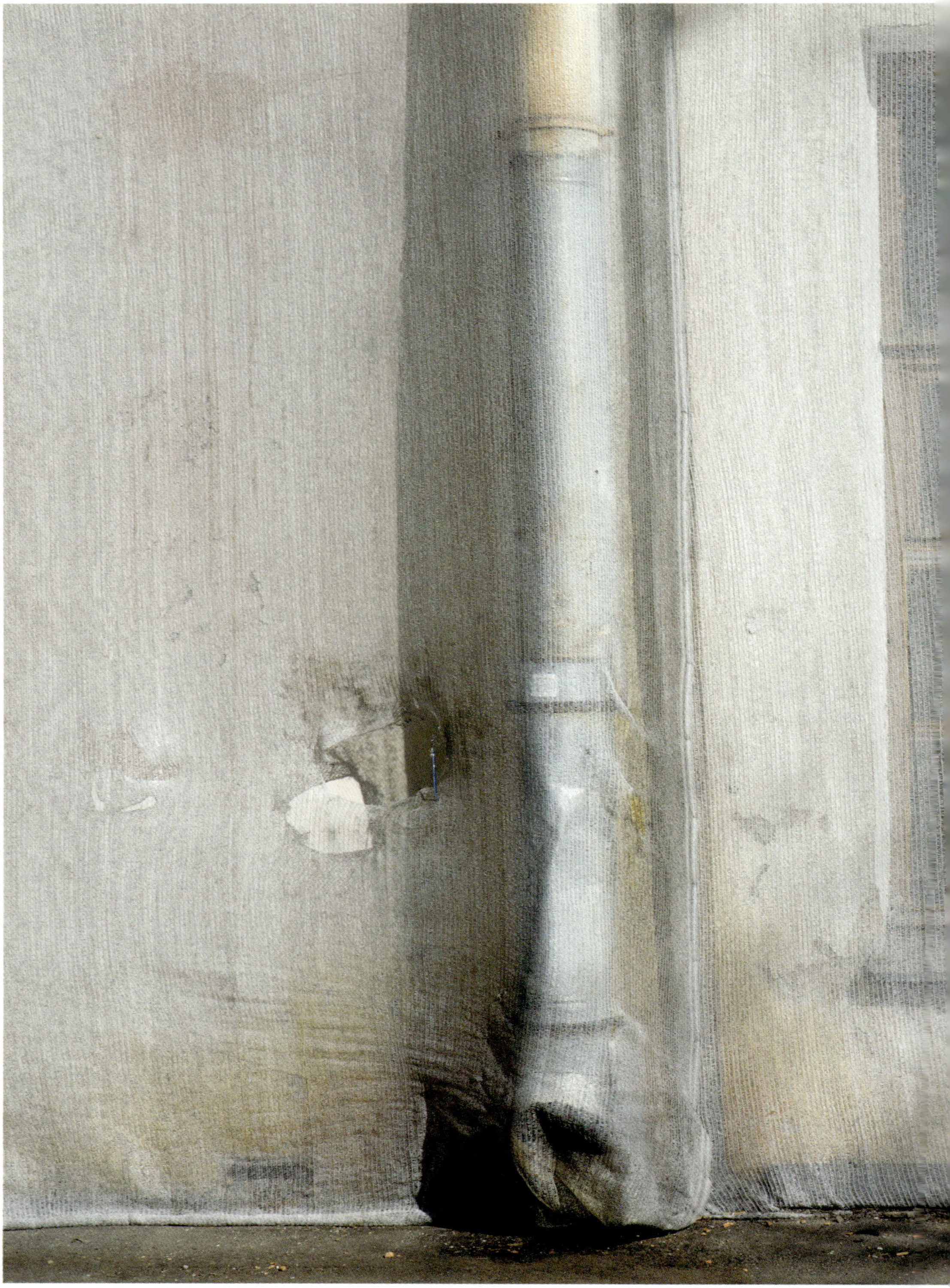

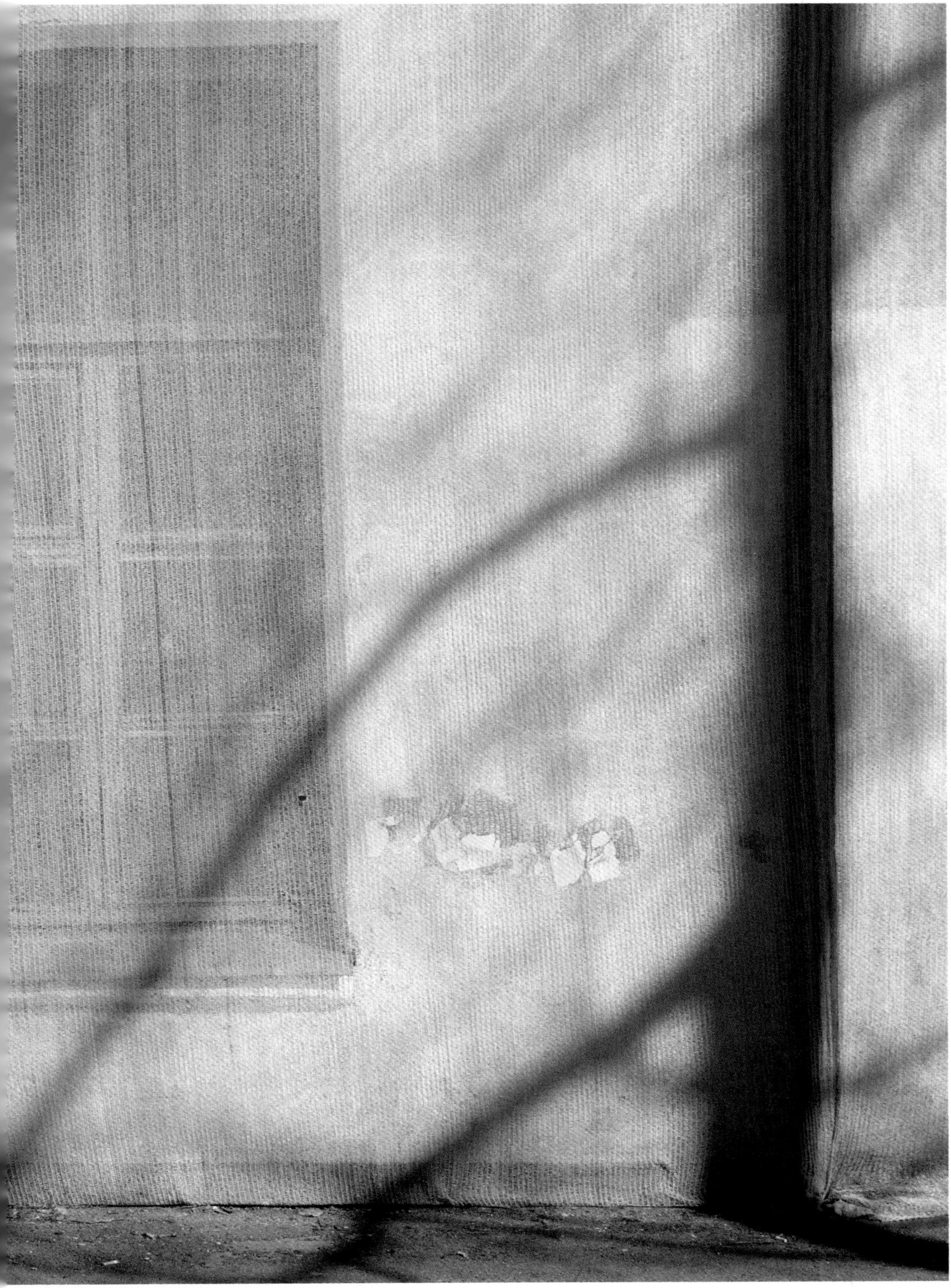

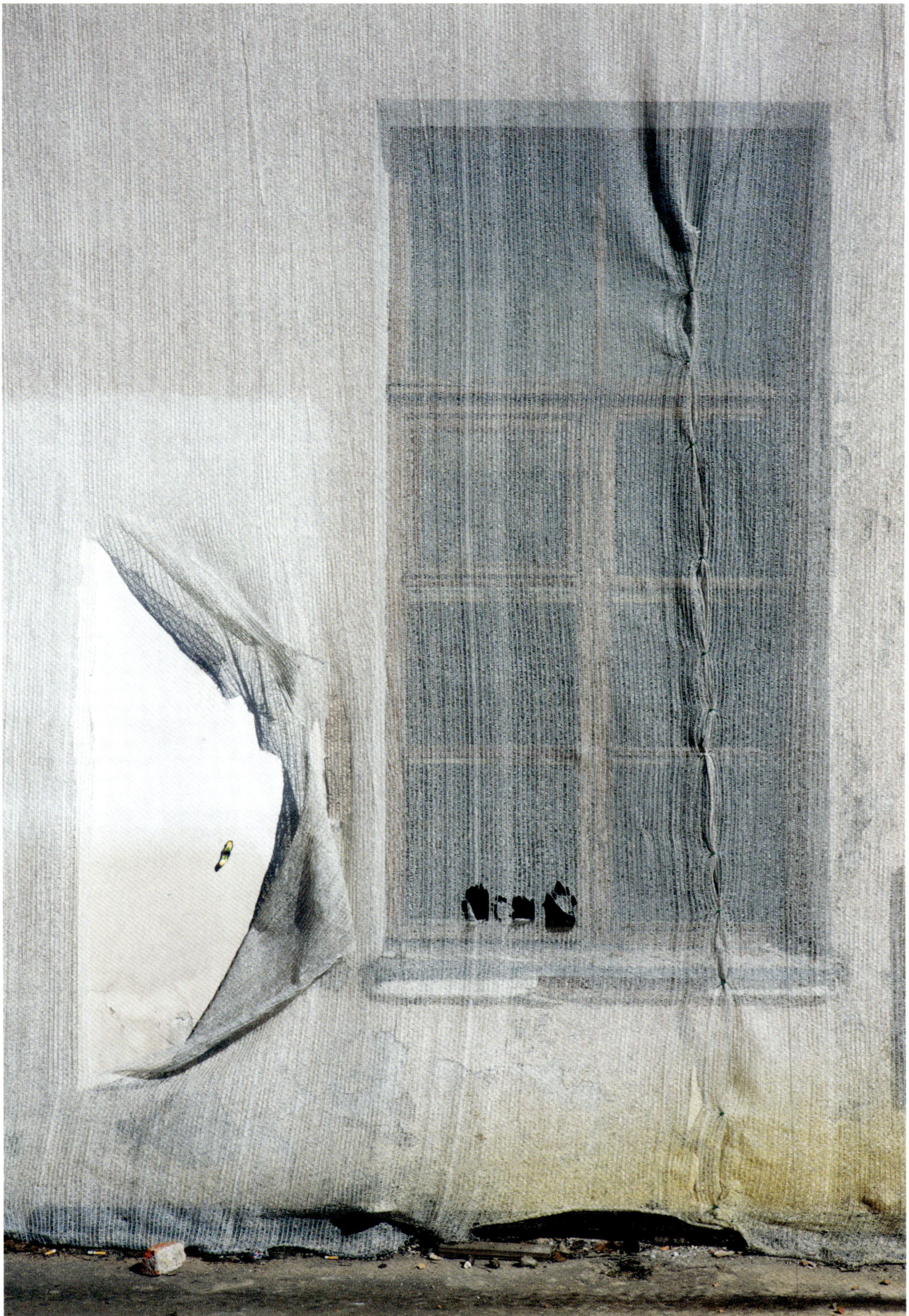

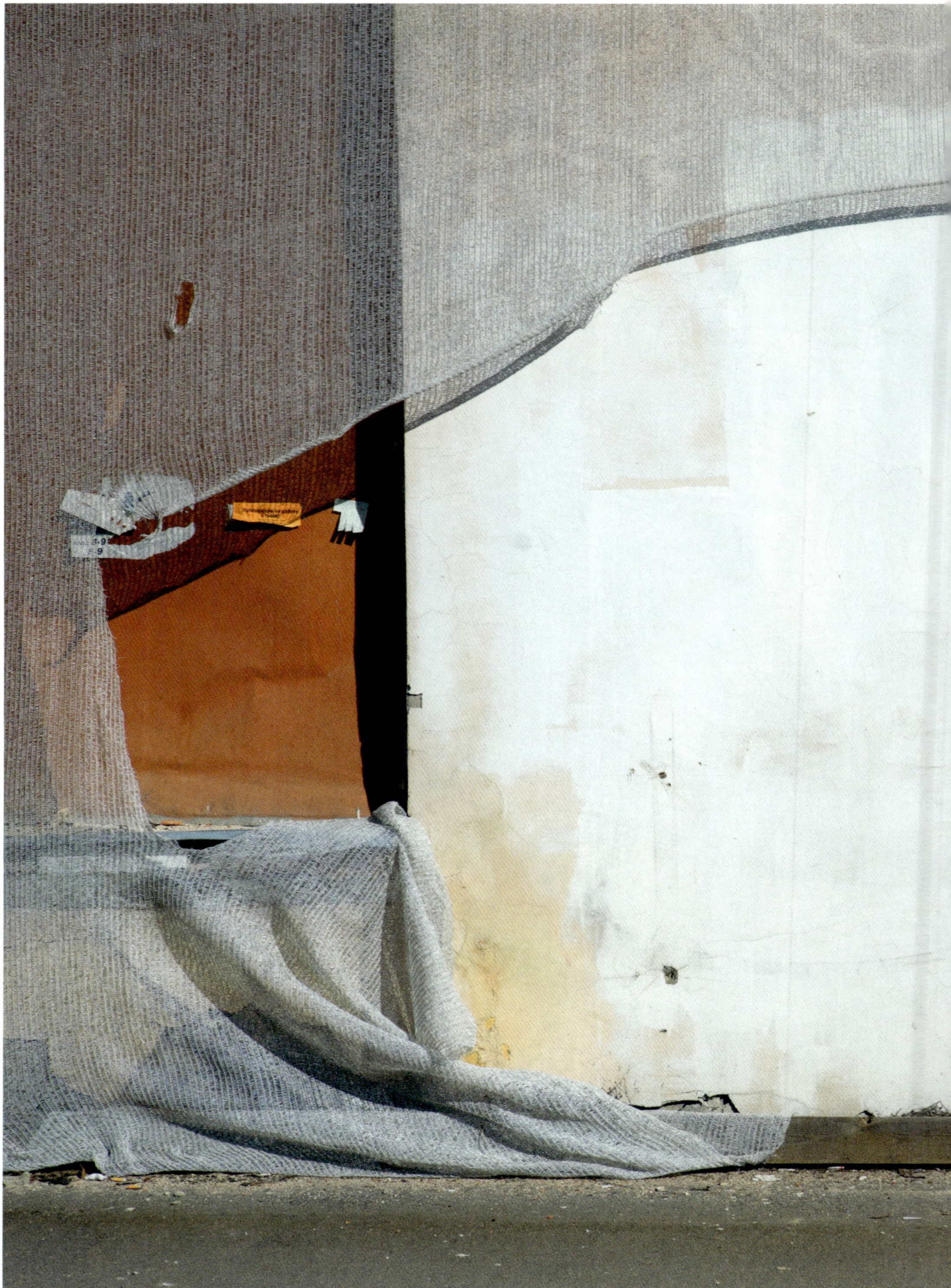

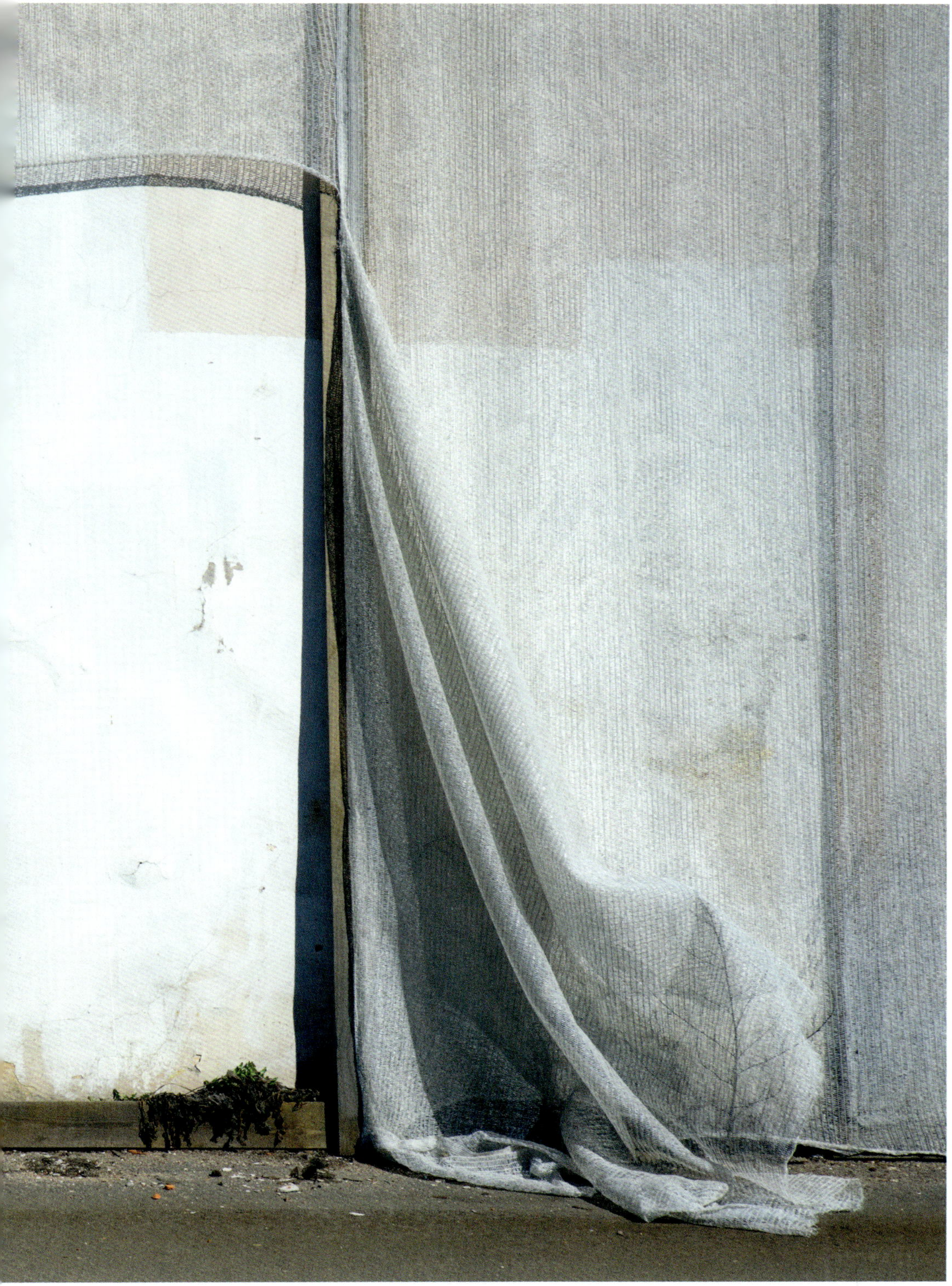

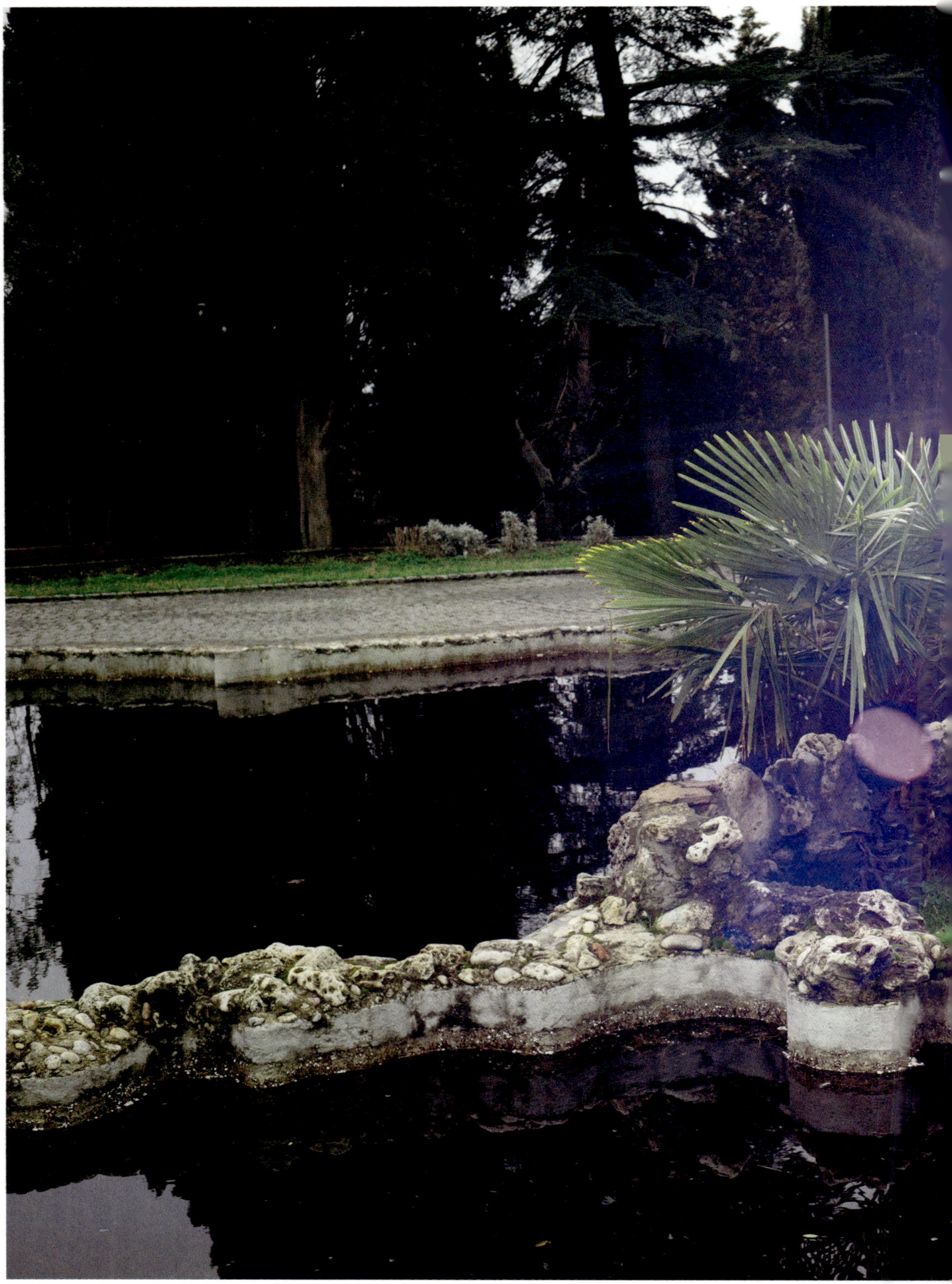

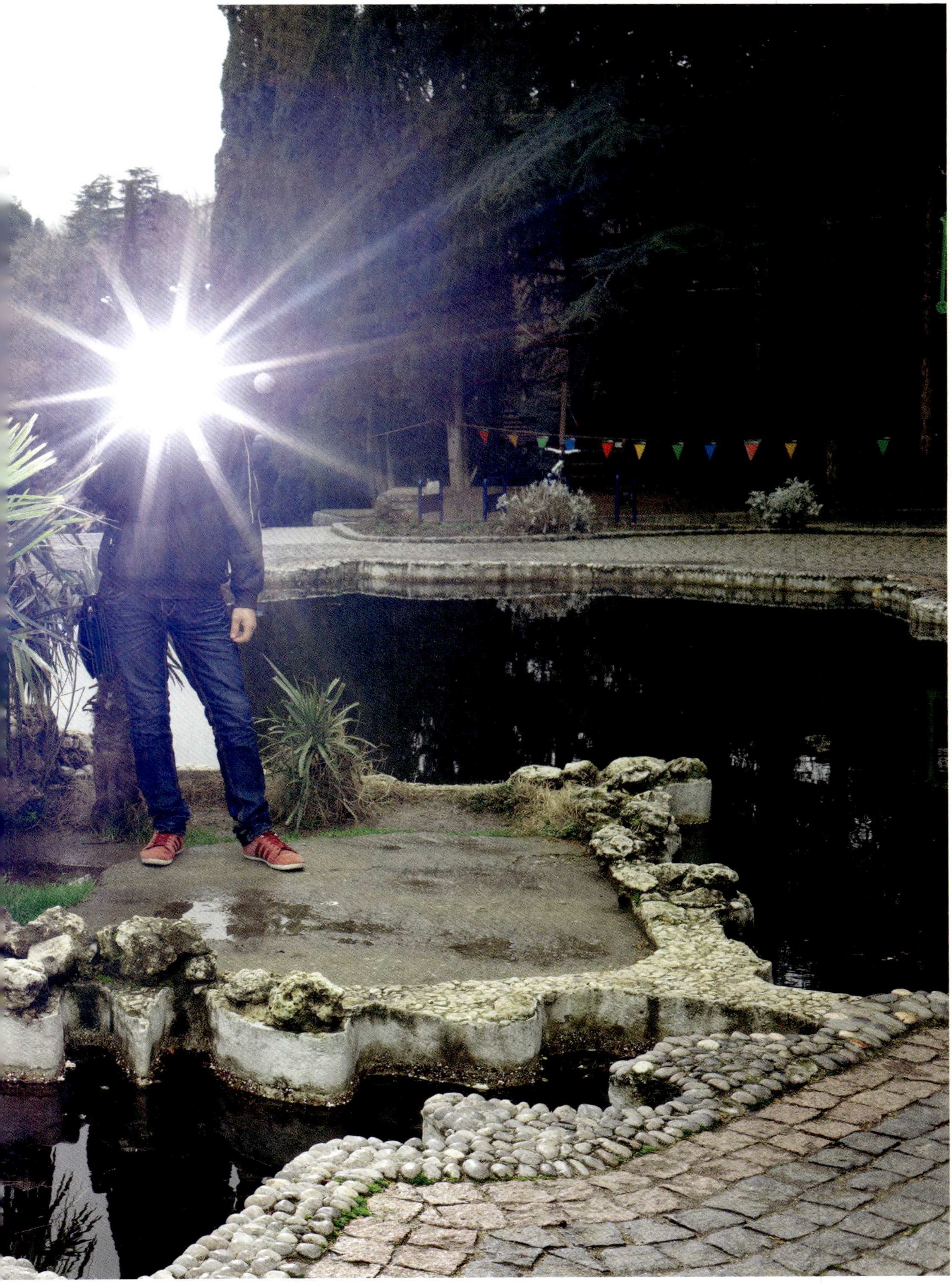

Chapter VIII

I encountered two kinds of heart on this almost-island.

The first appeared only fleetingly, glimpsed from moving vehicles, in desolate places, at the margins of towns, high up on walls, often at a bend in the road, so that the moment I saw them, they disappeared. Such hearts were spray-painted in the persistently hopeful yellow and blue of the Ukrainian flag, with urgent haste, over rough and mottled surfaces, so that the bright colors bled and discolored within.

The second kind of heart was found hiding in the sturdy little gay chests of certain men. Both kinds of heart appeared bruised.

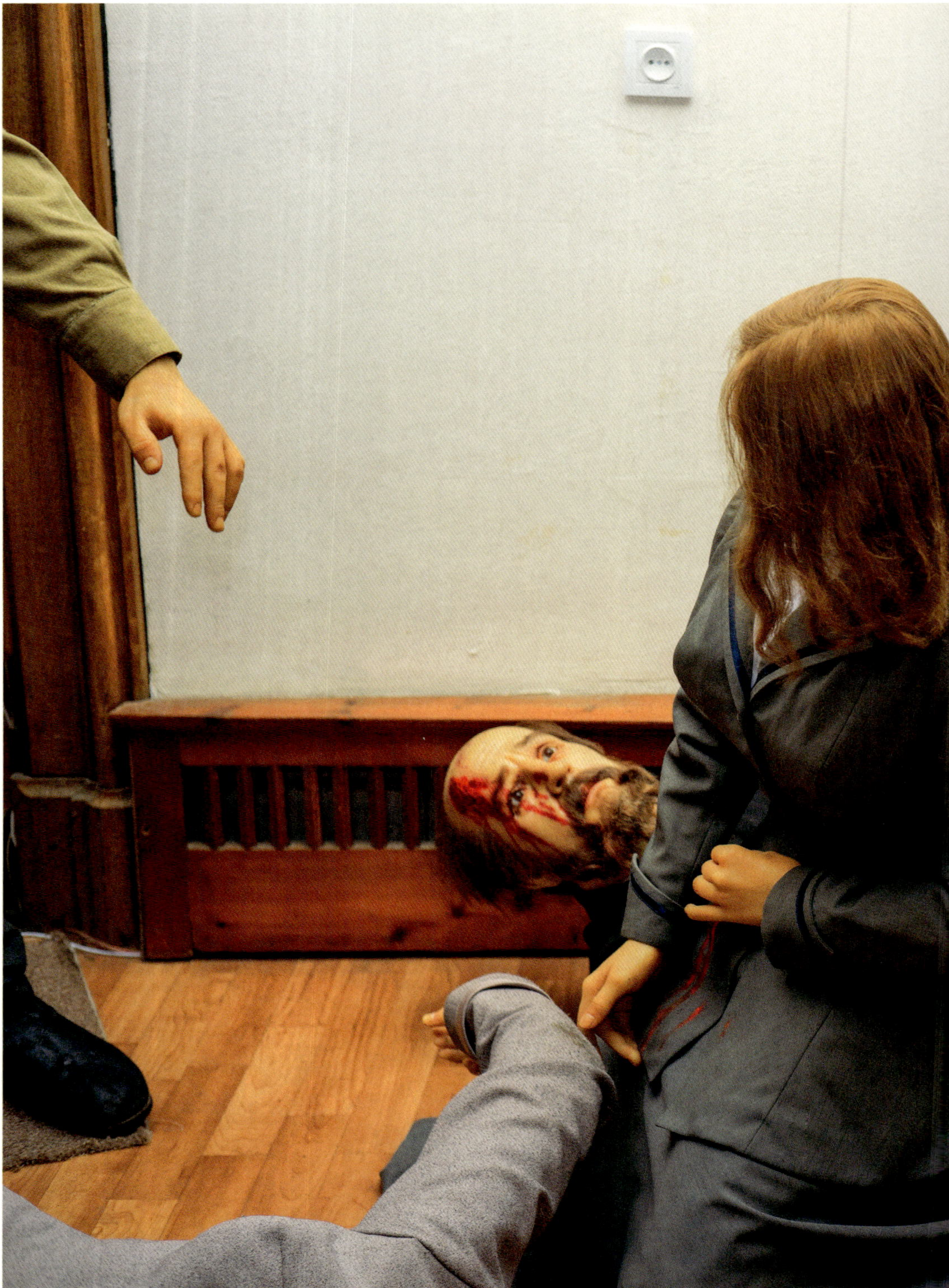

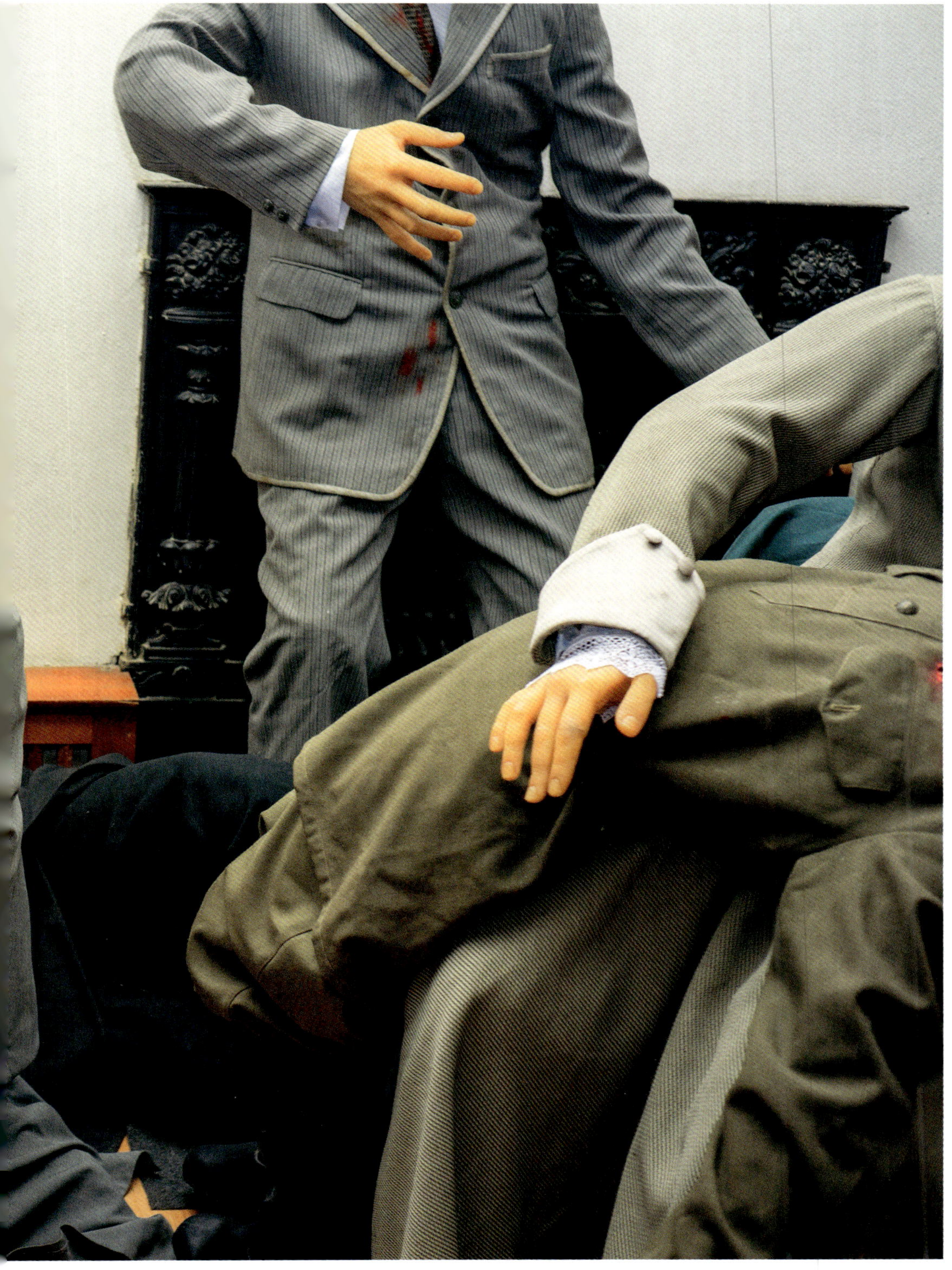

When you are not allowed to love, what, then, will you find to love? The answer to such questions can be geographical. For example, Ukraine is one vast expanse of low steppe, exposed for millennia to the surge of invaders and the flight of refugees, to flooding river basins, sweeping famines, and the migrations of language and species. Except for the sharp jag of the eastern Carpathians, on its European edge, Ukraine's plains are ever-undefended, allowing arctic winds to sweep south, unhindered, from Russia and the Baltics to its southernmost extremes. But just before the southeastern tip of its one dangling peninsula, jutting up from the Black Sea in bleached stone spikes, rise mountains. This steep range cuts a brief line before dissolving back into the low swells of coastal plains. Along these 50 miles of alpine wall, those merciless arctic winds, unhindered for thousands of miles, come to an abrupt stop. There, beyond this fortuitous defense, a tiny subtropical microclimate clings to the Eurasian landmass. Only here, on this folding strip of coast from Alushta to Foros, do certain tender species survive.

When you are not allowed to love, you will do so all the same. Behind the protective curtain of these coastal peaks, Seva the gardener loves palm trees. Not only as an expert or a specialist, but as a lover: physically, spiritually, with a sense of shared sentiments and mingled fates. Like him, the plants are not native to this land, but can survive. This is a tenuous subtropics, sometimes breached by icy winds, caught "halfway between the Russian arctic and the tropics," as Seva says, with the perspective of an amorous god. The fruiting coconut palm would never endure, and the tallest, most iconic of palms—*Washingtonia* or *Jubaea*—are prone to frost-snapped fronds, fighting back to life every spring, and ultimately expiring. Those that thrive, like the humble *Trachycarpus*, grow slowly, and stay low. They do not seek attention; they merely persevere. Most palms, he assures me, have a solitary habit.

Walking with Seva, the world is reorganized. All over town, palms are suddenly hiding in plain sight: in the gardens of four-star hotels, behind benches in pedestrian arcades, in the forecourts of various churches. They stand discretely in the shadows of statuary. Other trees loom over and around them; they blend in. But Seva sees them all, and he stops to photograph each one. His camera phone, rendered mute by the expulsion of Ukrainian telecoms, lives on as an archive of horticultural persistence.

As we stroll along the waterfront, towards the distant heated monkey cages astride the Intourist hotel, Seva draws innumerable palms out of their unassuming shadows. He touches their fronds, demonstrates their characteristic traits and assesses their progress or failure. None of them belongs to him, but they are all his. And in a gently Ukrainian-accented Latin, he speaks their names before taking their portraits.

In the capital, where he had once been a student, giant banners celebrating the Russian annexation hung across intersections: *A year of Crimean Spring. Our choice; our victory!* But many of Seva's closest friends and coworkers had fled north onto the mainland in the wake of the change, and he now found himself on the wrong side of a militarized border from his family. Neither his choice nor his victory, but the sickening sound of a door, barely ajar, swinging shut.

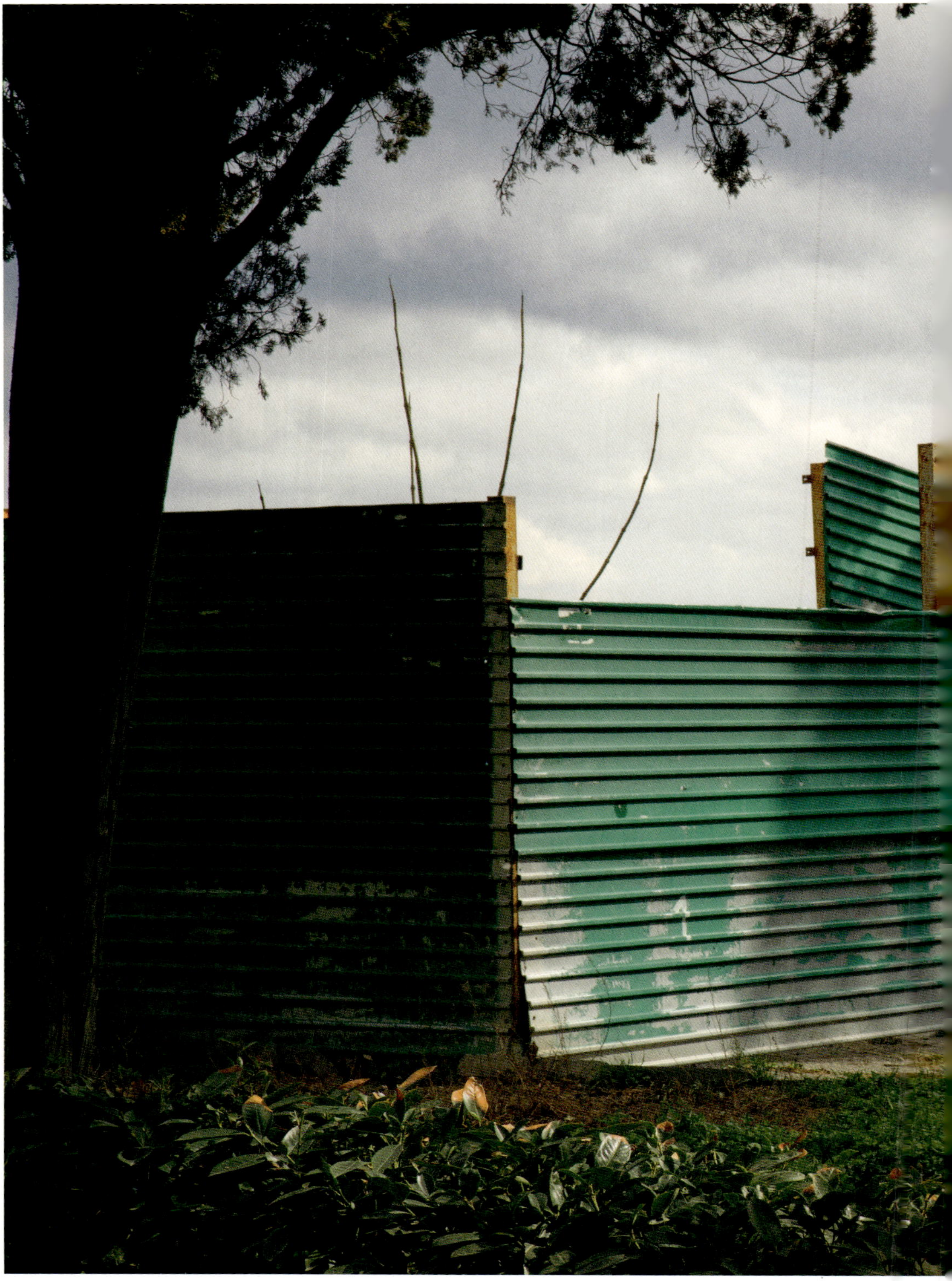

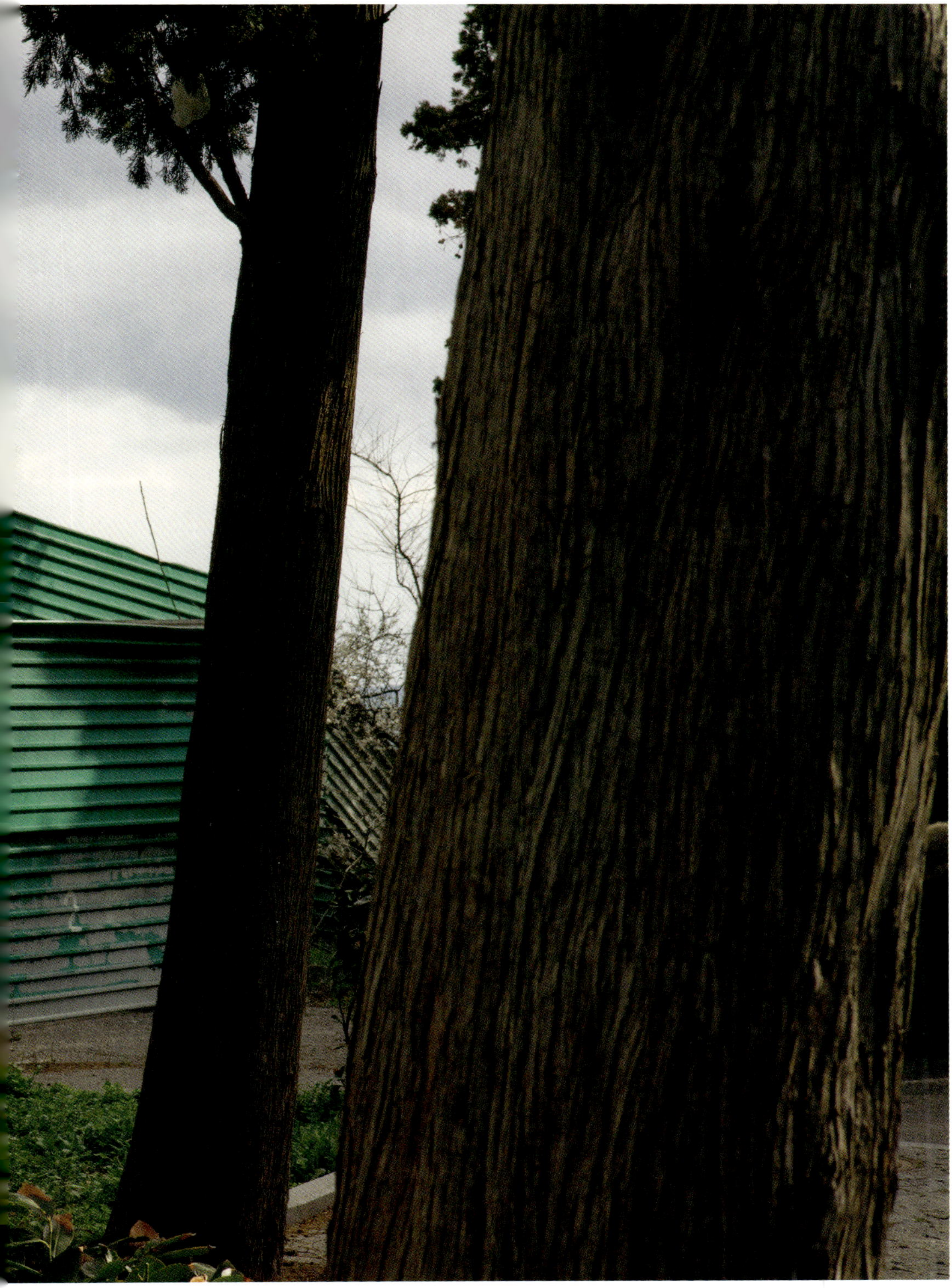

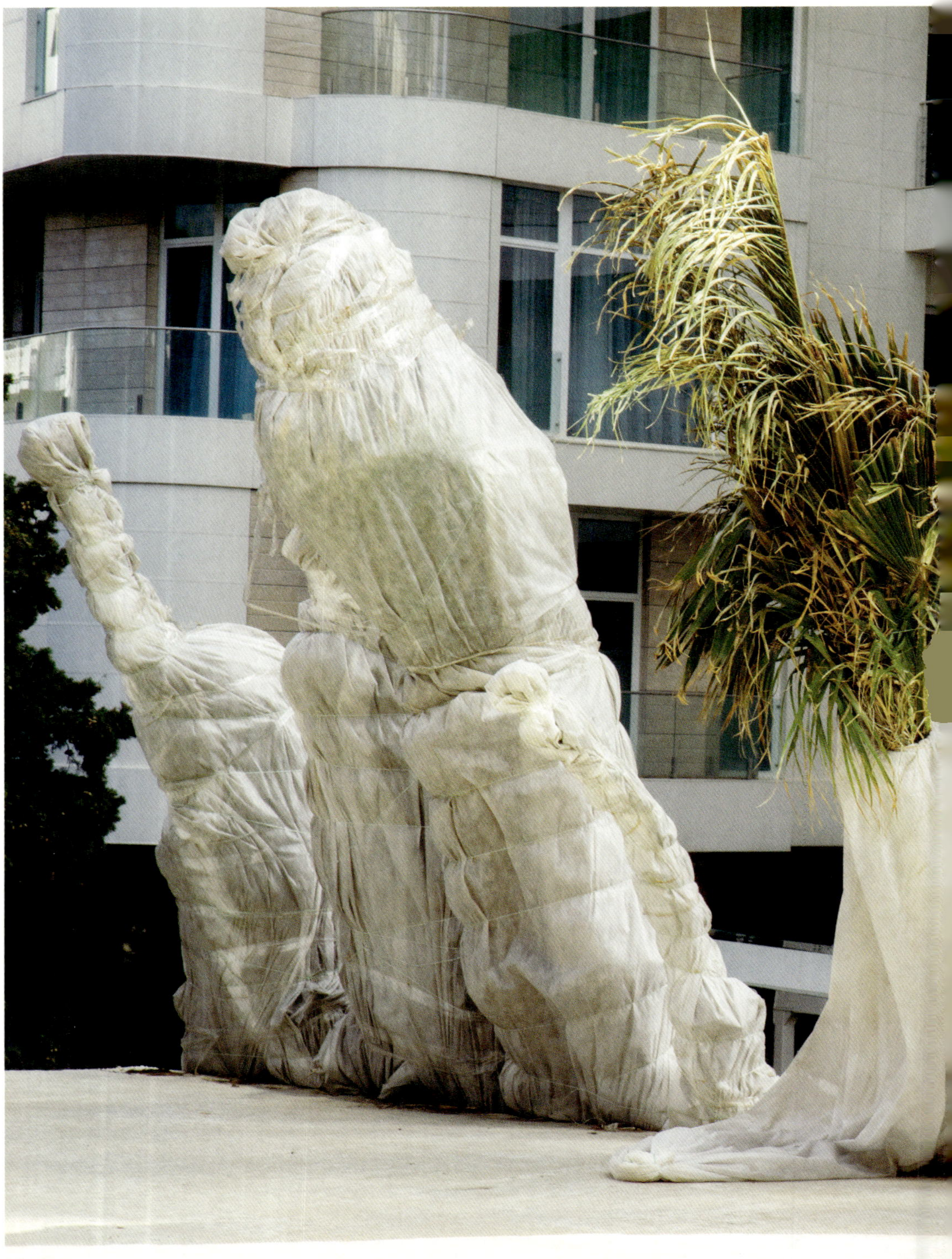

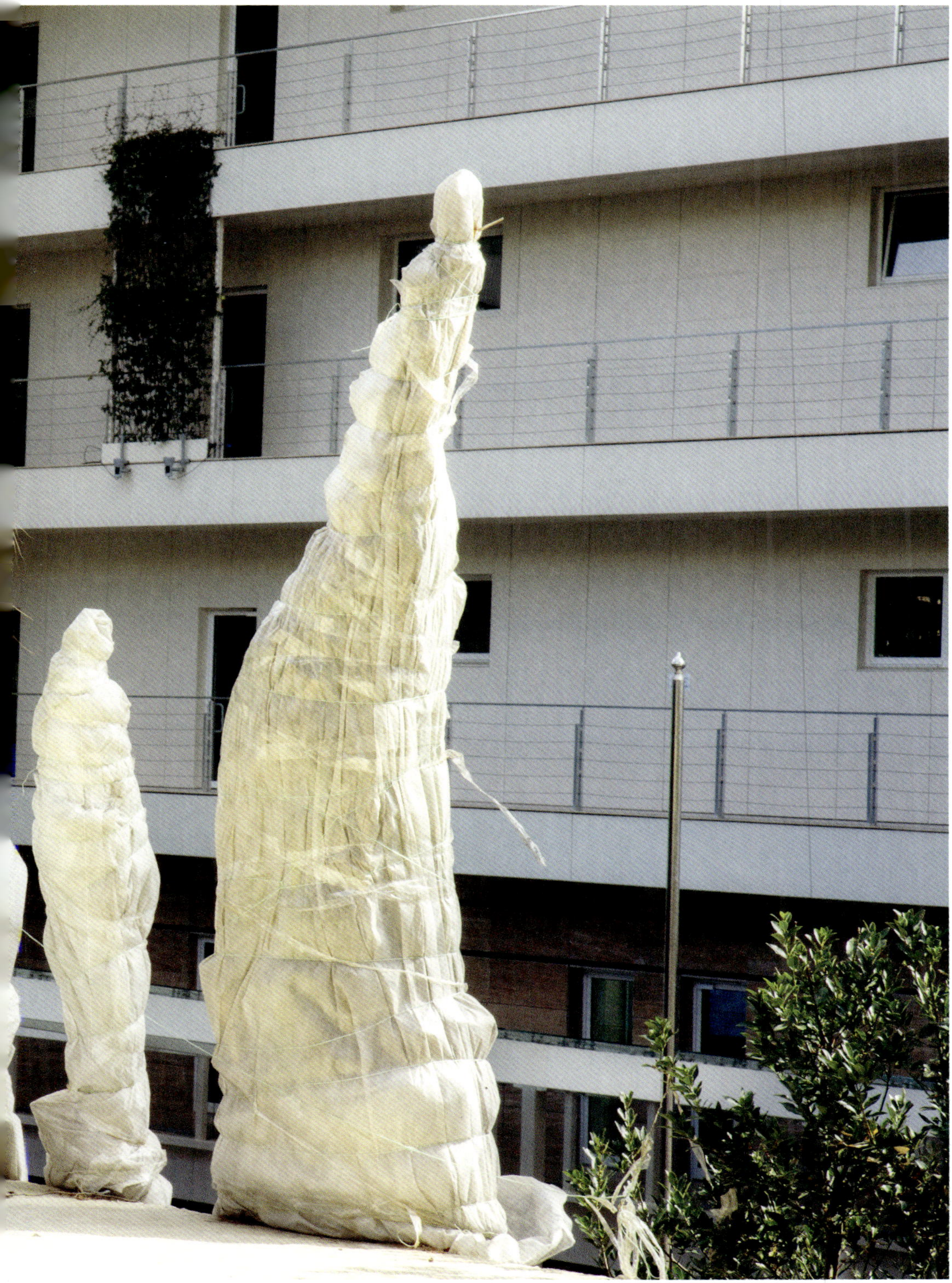

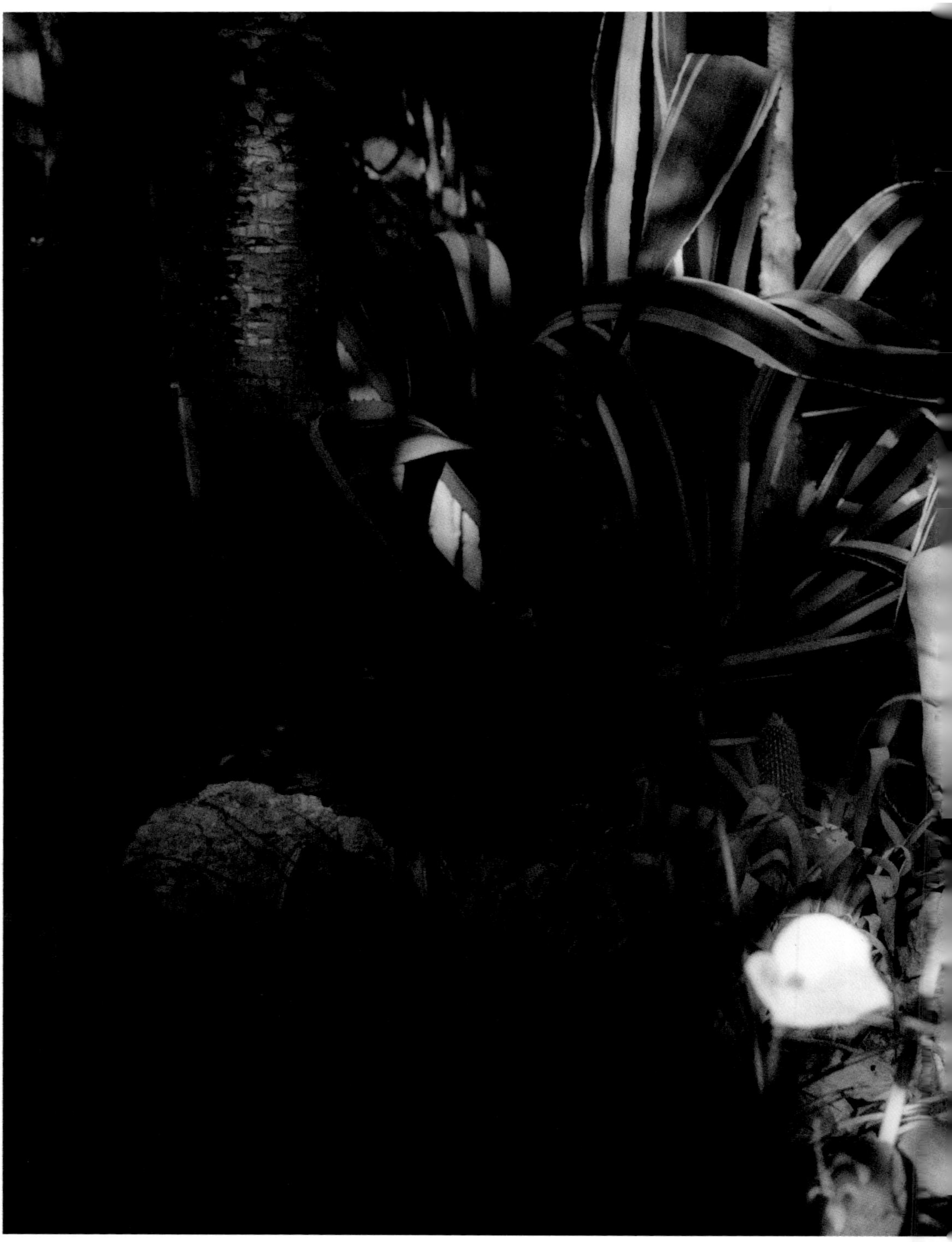

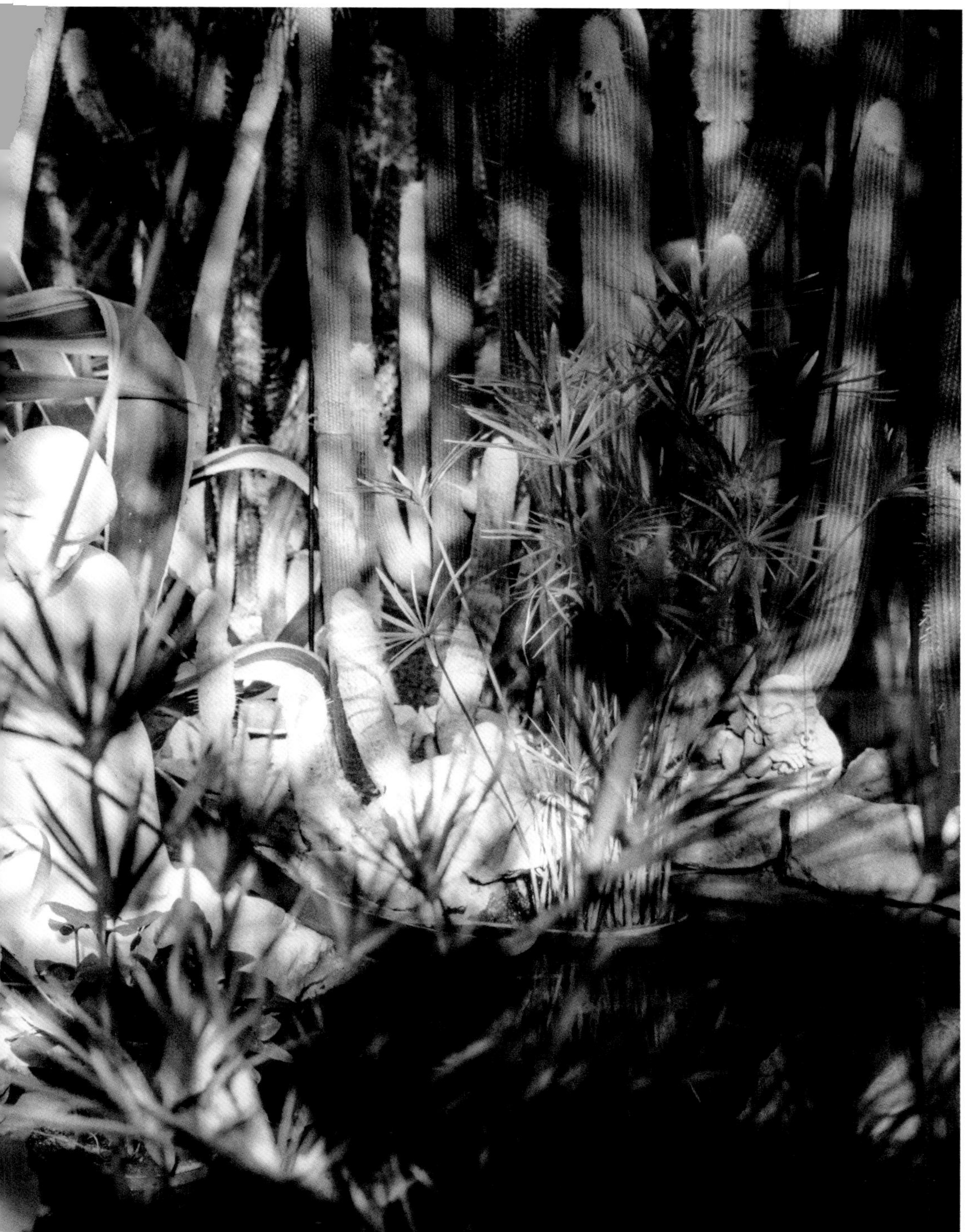

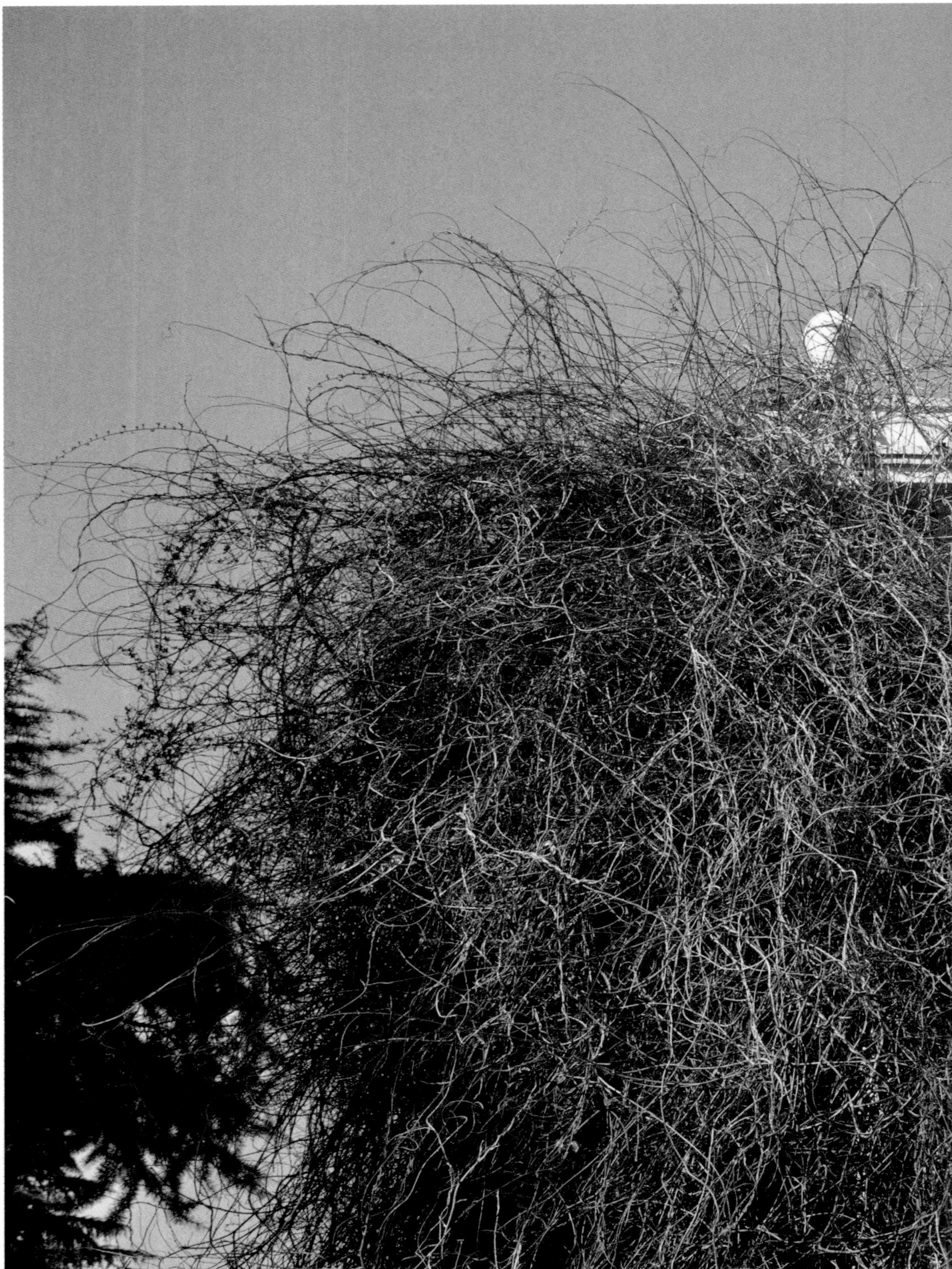

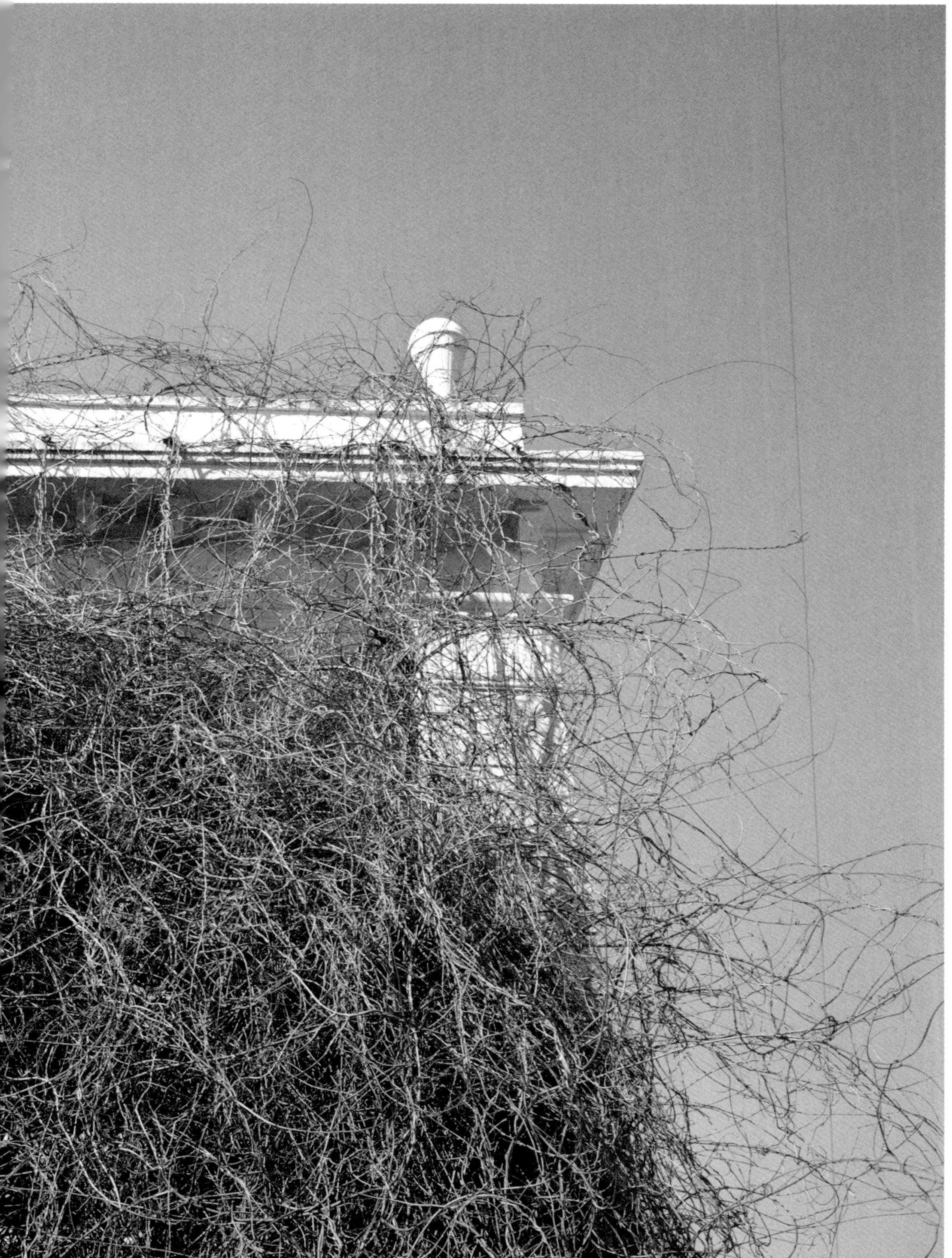

Seva steps out into the purple dusk, a small dark shape emerging from the dimming white cement of a late-Soviet façade. His body, an indigo drop, dissolves into the saturated blues and greens like watercolor. These gardens slope downward, south, opening out towards the sea—brilliant nearer the horizon, where the sun still strikes beyond the shadow of the high peaks.

The gardens are too dense and varied to take in as an image. They seem a series of journeys and impressions, marked by fragrance and decay, viewed only in passage. But Seva drops into them, following their pathways like vessels, a drifting corpuscle. Trails radiate outward and down from the cool white building, terraced through olive and cypress groves, past aromatic hedges of box and bay. In the gardens, change is silent and growth invisible, transition cyclical, movement minimal. Seva walks these memorized pathways, crushing certain leaves in his small hands, pausing to inhale the scent of the injury.

This garden has a heart of palms—a terraced plaza at its core. Extending eastward, lined with *Trachycarpus*, a narrow promenade culminates in a locked gate. Seva turns westward, slipping off the paved path to visit one stunted palm, its tender foliage still bound up tightly for winter. He struggles with a key to cut through the plastic cord. Ragged fronds drop downward in defeated release. He rests with the specimen for a moment, his backside brushing the encroaching bamboo. Delicate fingers run through disordered fronds, as if to comb them. Seva lets go, sinking into a sloping, dense grove of older palms. The day's warm air flees upward past his feet; a band of evening chill descends. These trees' trunks wear course protective thatch, like rough homemade sweaters. He touches two of them, one with each extended hand, and pauses between. A pearl of moisture appears on one cheek. It is likely a stray drop of rain. The grove is too dim to tell.

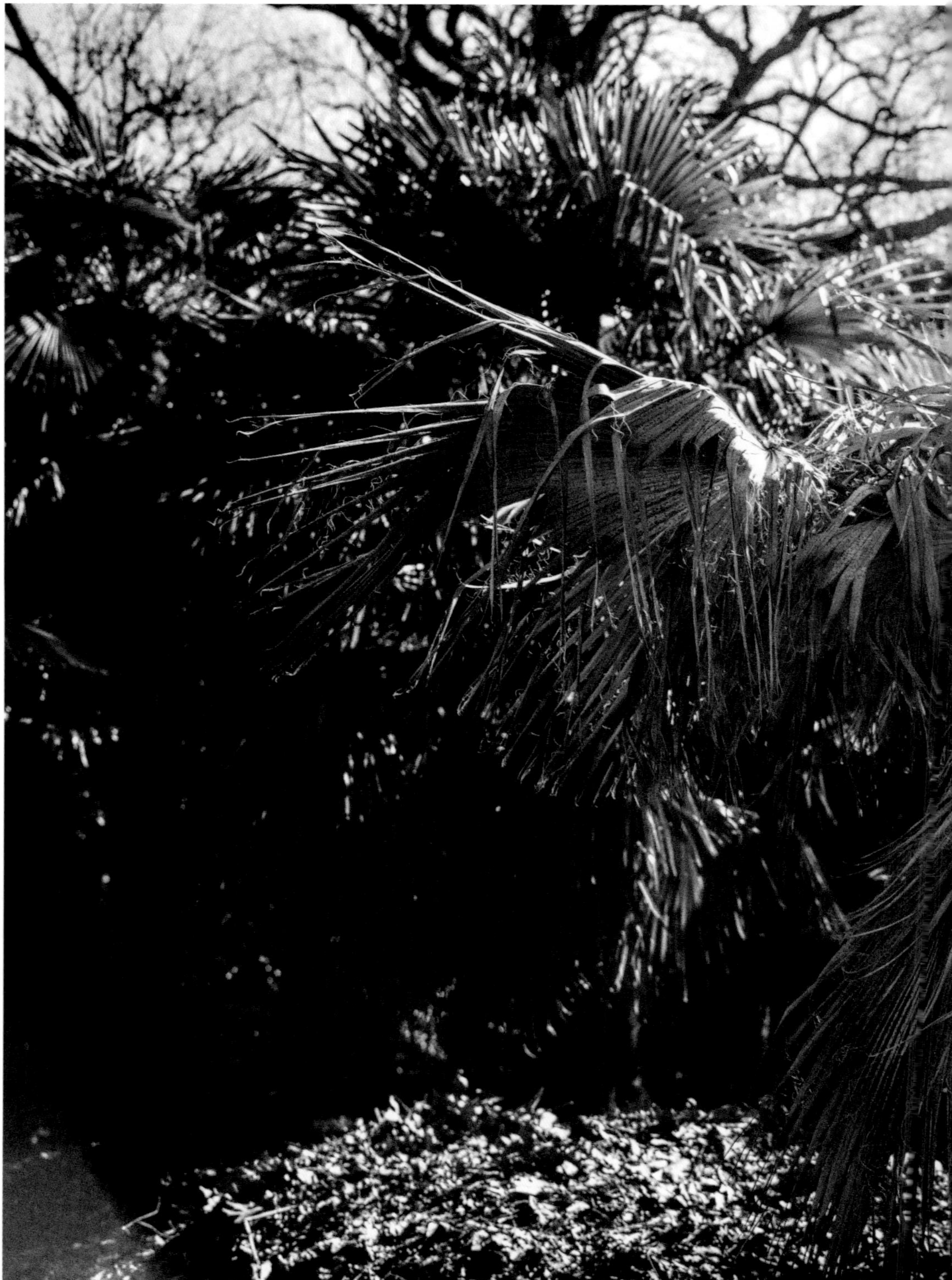

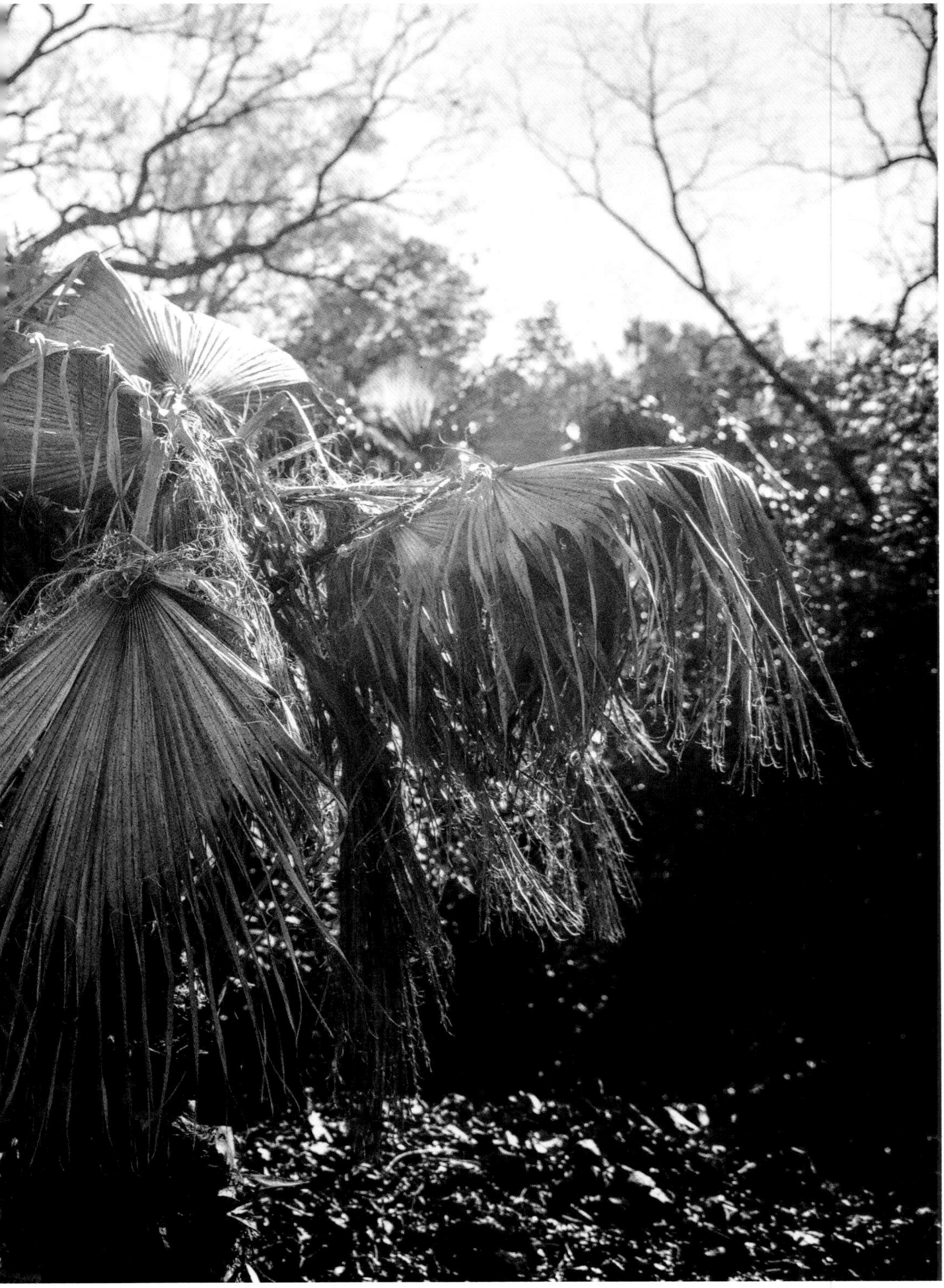

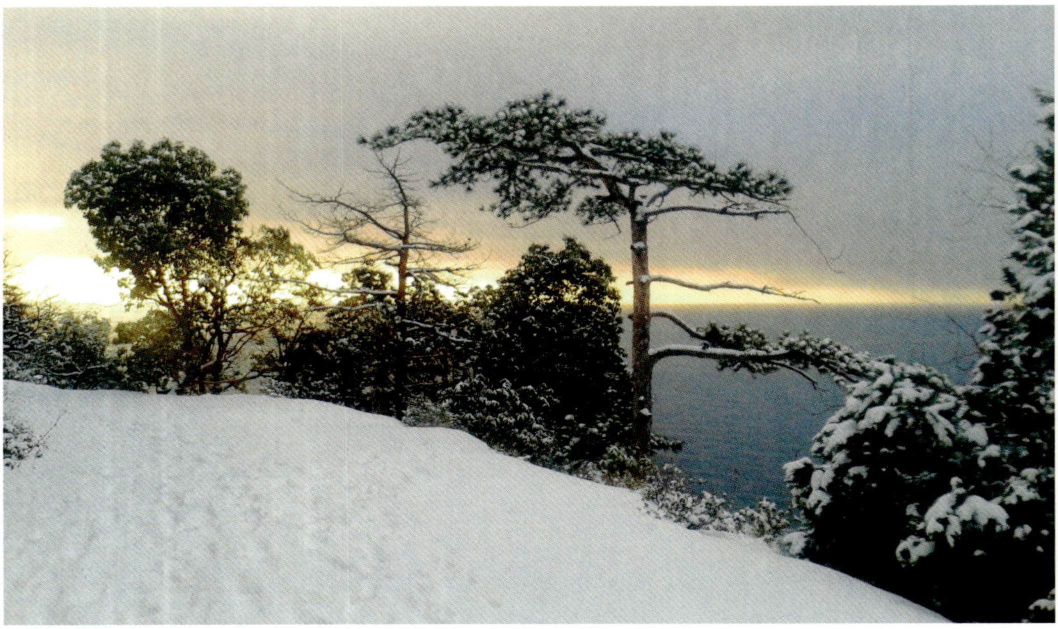

Every time I ask Seva about hopes, ambitions, love, sex or feeling, he gives a giddy little nervous laugh, and does not answer.

One afternoon, on an empty road, I ask him if he finds my questions intrusive, and he assures me not. Yet he rarely responds, except to giggle and blush. Occasionally, though, some things slip out. Beyond a couple of local internet acquaintances, each met only once in person, nobody in his life—friends, family, coworkers—knows that he's gay. He figures that in a couple of years, when he's thirty, he'll have to marry a woman. It seems the only option. More than that, he can't explain.

So what do you know of a new friend, if you have learned too little from words? Any understanding comes through silence, this laughter of non-response, and the gestures of body, face and hands. It's in the tender stroking of moss and bark, the cradling of delicate fronds, the short-sighted squinting into distances.

In our correspondence before my visit, slipped in among talk of weather, work, flowers and family, I asked questions about romance, fear and the future. He answered silently—with photographs:

Have you been in love before?

GOLDEN LIGHT ON A COASTAL CLIFF, TULIPS IN AN ALPINE MEADOW.

What do you dream of?

TWO PALM TREES, A SEASCAPE, WILD CROCUS IN SPRING WOODS.

What kind of man do you like?

SNOW-DROPS, CHRYSANTHEMUMS, MAGNOLIAS, YUCCA.

Have you had lovers, encounters, or crushes?

RUSHING CATARACTS, PINK DIANTHUS, A VALLEY STREWN WITH BOULDERS.

Would you like to live in another country?

AN OLIVE GROVE, STRAY CATS SIPPING MILK, THE MOUTH OF A CAVE.

Are you lonely?

ONE SQUAT PALM DRAPED IN SNOW, A STONE PEAK LOST IN CLOUD.

My laptop has a folder with hundreds of his photographs. I have asked a lot of questions.

We did not tour the palace. Seva had been before, and I was quickly exhausted by the warring neo-Gothic and Mughal motifs on the building's enormous façade, each canceling the other to leave only the impression of ersatz Russian mass. I had no need to enter.

But strolling the terraced oriental gardens along the back of the palace led us to a strange little service door, marked only by an arrow-shaped sign declaring, "Wax Museum." I insisted on entering, despite Seva's protests. Inside, an elderly woman presided over a coffee vending machine in a cupboard-like entranceway. I bought us tickets, and she disappeared behind a black curtain to fiddle with some switches before ushering us through into the display.

The room was large and plain, perhaps a former staff refectory. Along three of the four walls stood a single row of historical Russians, from Ivan the Terrible and Catherine the Great to Tolstoy, Chekhov and a cast of bewigged strangers, all standing in line as if waiting for a taxi at the airport. Along the fourth wall, behind a velvet rope, a display of a different order unfolded: the entire family of the last Tsar, eyes wide open in waxen horror at the collective sight of their own slaughter. Bodies crumpling, fallen or grasping at their red-painted bullet wounds, standing, seated and draped across the floor. It was impossible to ascertain the intention of this terrified spectacle. As a Soviet era celebration of regicide, it betrayed an excess of sympathy. As a post-Soviet recuperation of tsarist humanity, it was squalid, helpless—a panorama of victimhood without the dignity of martyrdom. It seemed an elaborate and ill-reasoned reminder of the entire Russian enterprise: harrowing, messy, indeterminate and unresolved; a certainty of suffering the only inviolable paradigm.

Outside again, the entrance to the landscaped grounds was marked by a tall, pyramidal rock, against which a young couple photographed each other. From there, Seva and I flowed along a darker path, winding among shaded groves and mossy grottoes, bleeding into ever-deeper shadow. I know that we must have walked together there, talking. Instead, I see myself silently following a solitary man into the mysterious lush obscurity of those trails. Recollection makes a Steadicam tracking shot of disembodied metaphorical interest: the passionate disinterest of cinema. In such sequences, the viewer feels simultaneously privileged—so close to the unfolding of life and action—and conveniently cut off, behind the camera, outside the frame, unable to be addressed, beyond the pale of direct engagement or demand. Rebuilt in this way, Seva in the darkling grove is just another chapter to follow the wax museum—a wrought spectacle of specificity and distance, unresolvable despite the representation of rapt immediacy. Remove the moral from a morality play and you are left without indoctrination or narrative. The remainder: signalers without a code, hailing to indicate merely that we are lost.

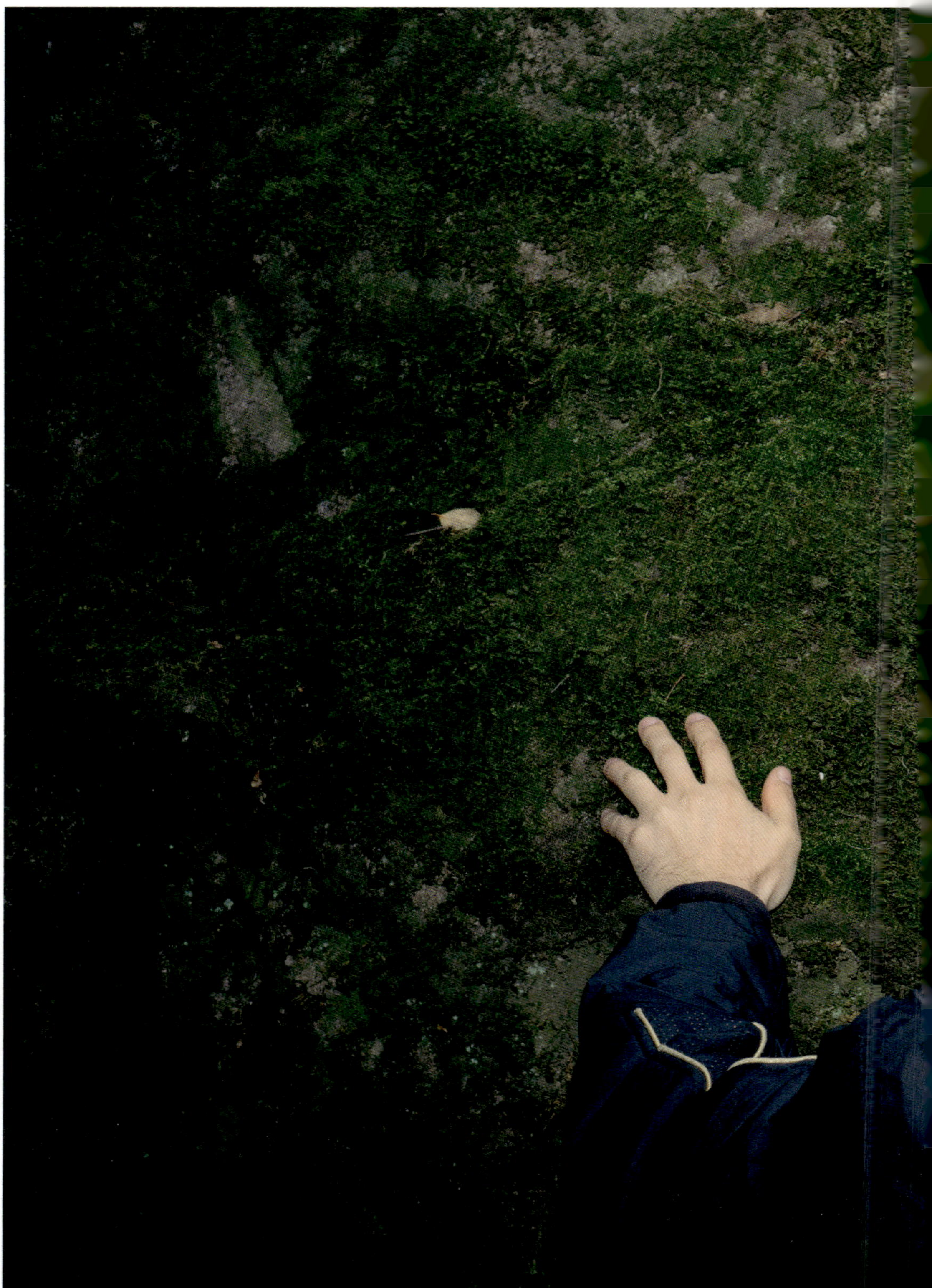

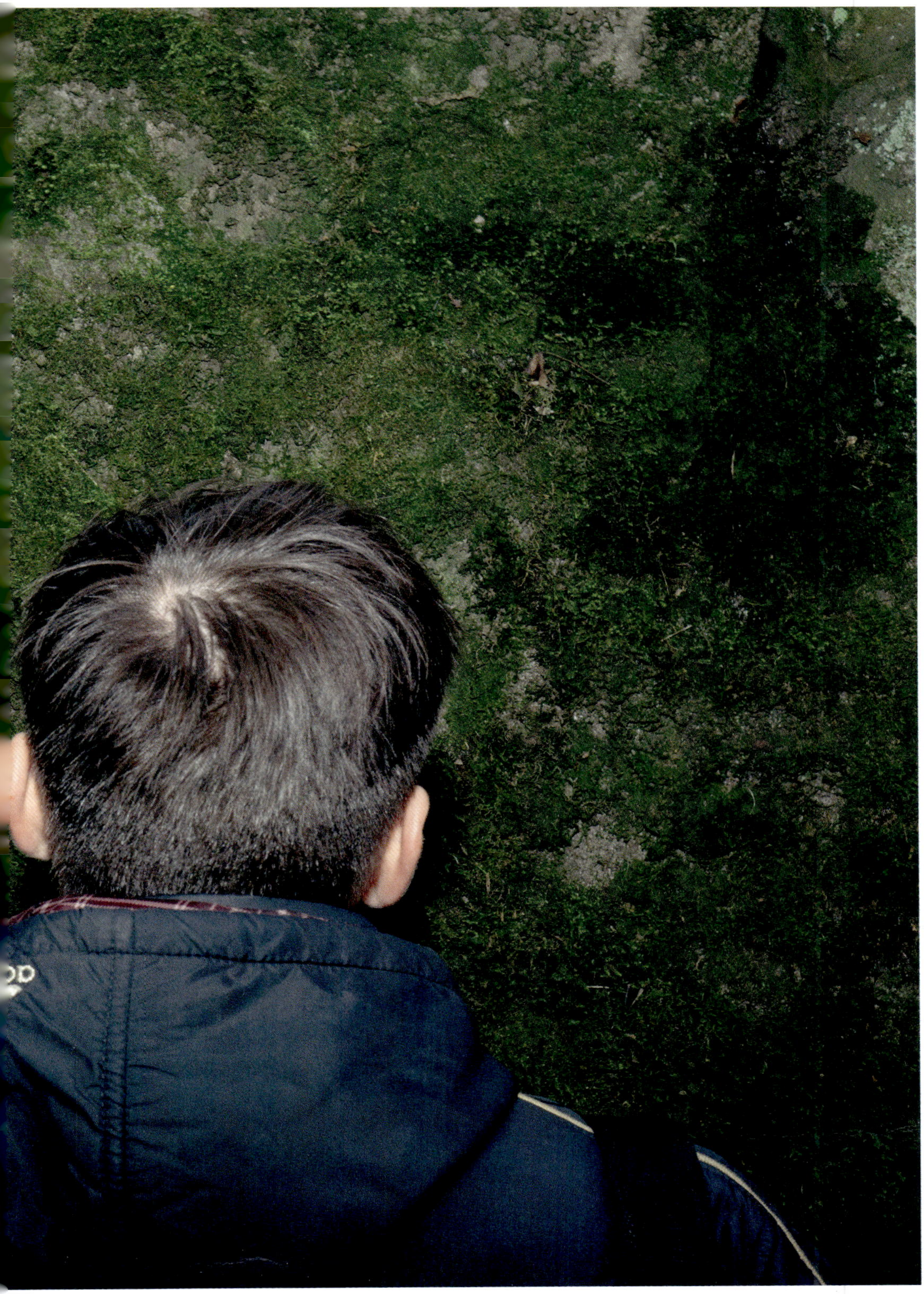

Finally delivered from those crafted sylvan follies to the banality of an open lawn near the bus stop, we came across the item of Seva's most earnest interest: a peculiar small tree, protected by a rickety but decorative wrought iron fence. The specimen, he explained, is over 100 years old, and the only example of its kind on the peninsula.

Is it very rare, I asked?

Not in other places, Seva said. But here, it's not really hardy. Even this one, as you can see, is slowly dying.

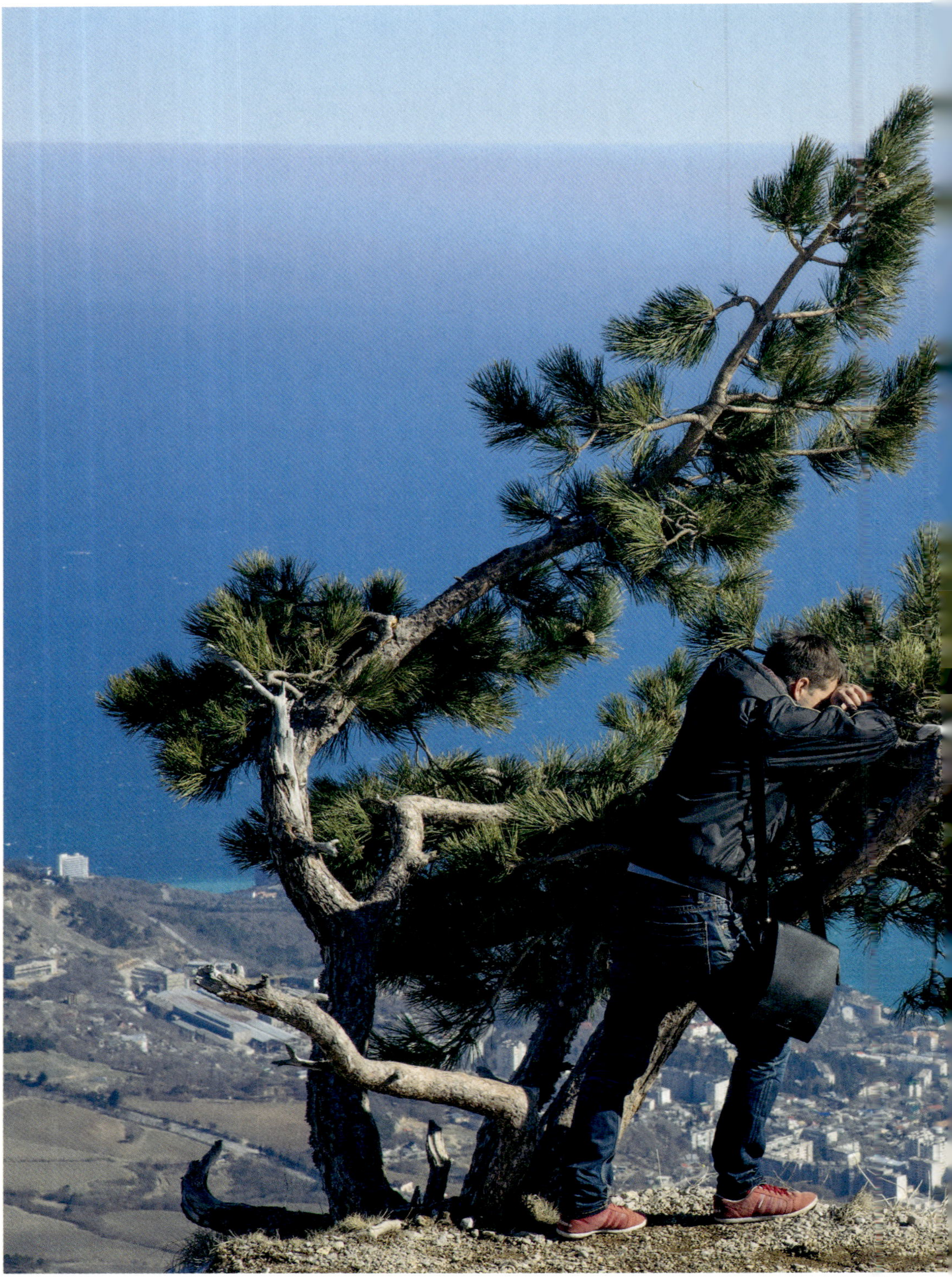

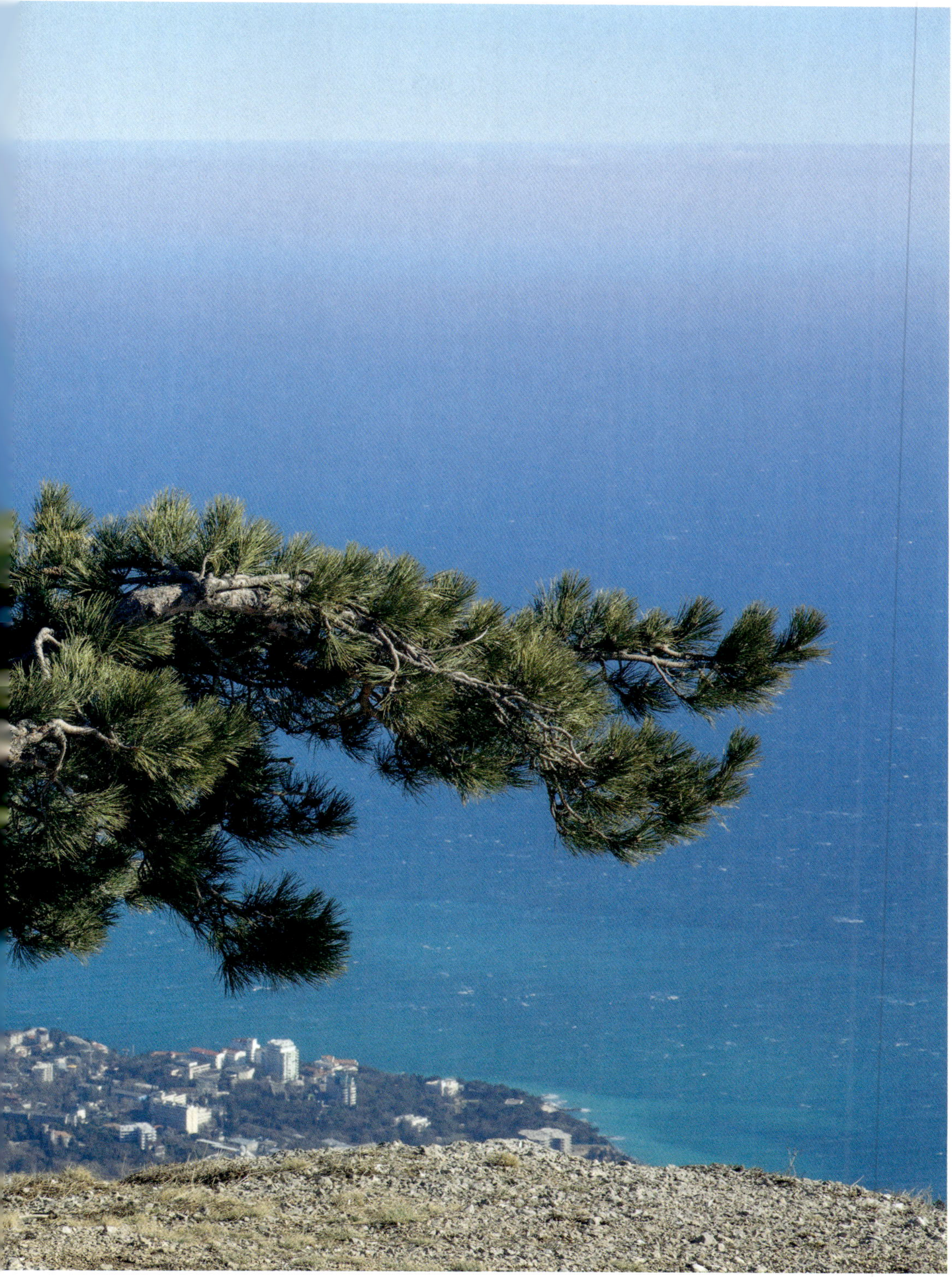

We stood side by side on the icy cement overlook, gazing out towards the inviting warmth of minuscule palm trees and the blue-green of the Black Sea beyond. The sheer cliff-side across a chasm to our left was tilted at the angle of a laptop screen, adjusted to see past the glare. Two figures formed shadows against the far-off rock face, life-sized despite the distance from our bodies to that wall. Is this question of scale an optical fact, or the mind's automatic adjustment of reality to imagination? Do we perceive those shadows in human scale because they are larger than life, but at a great distance? Or do we see them thus because we multiply their diminution by our distance, as if that math brings us all closer?

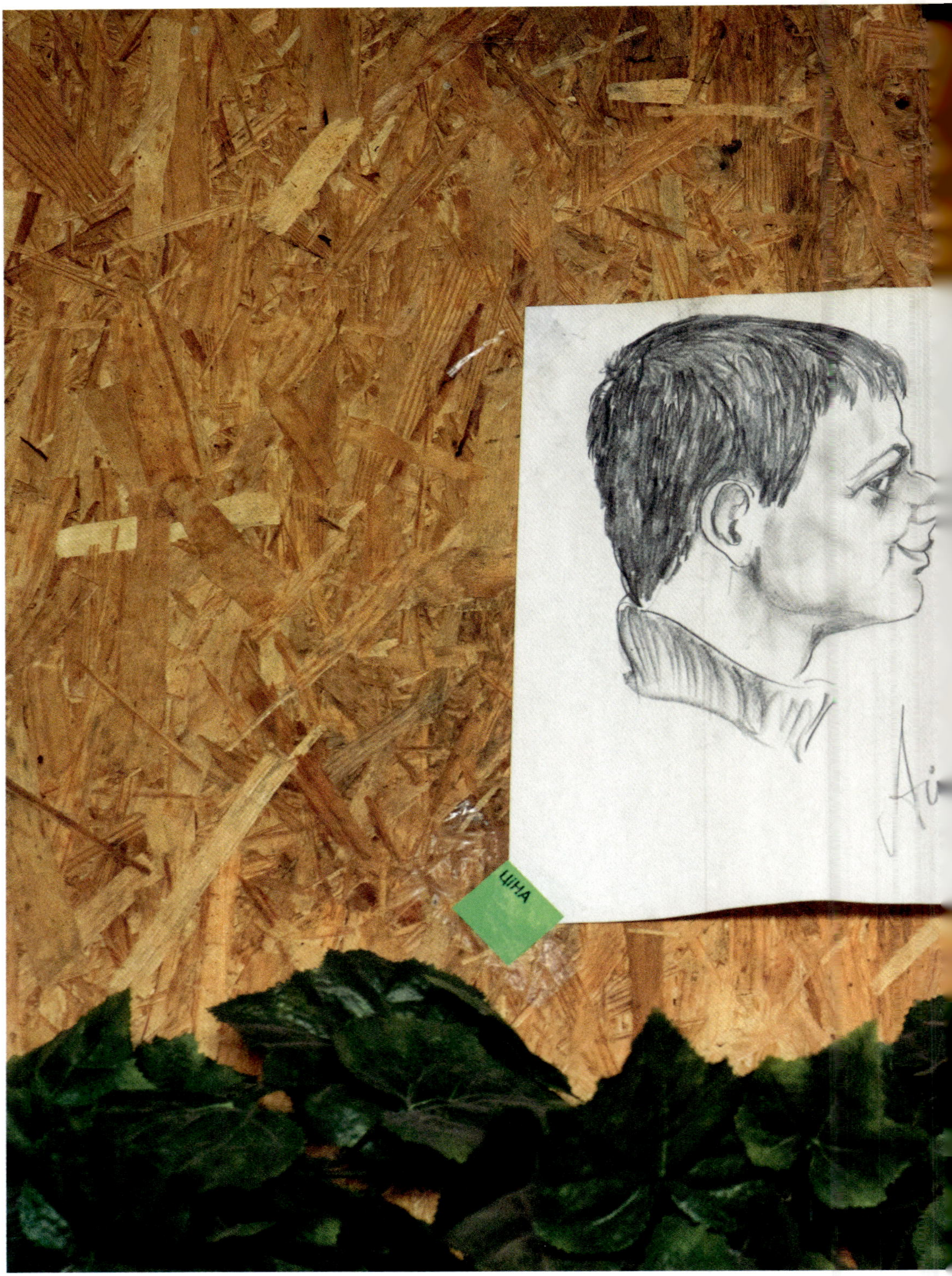
ЦІНА

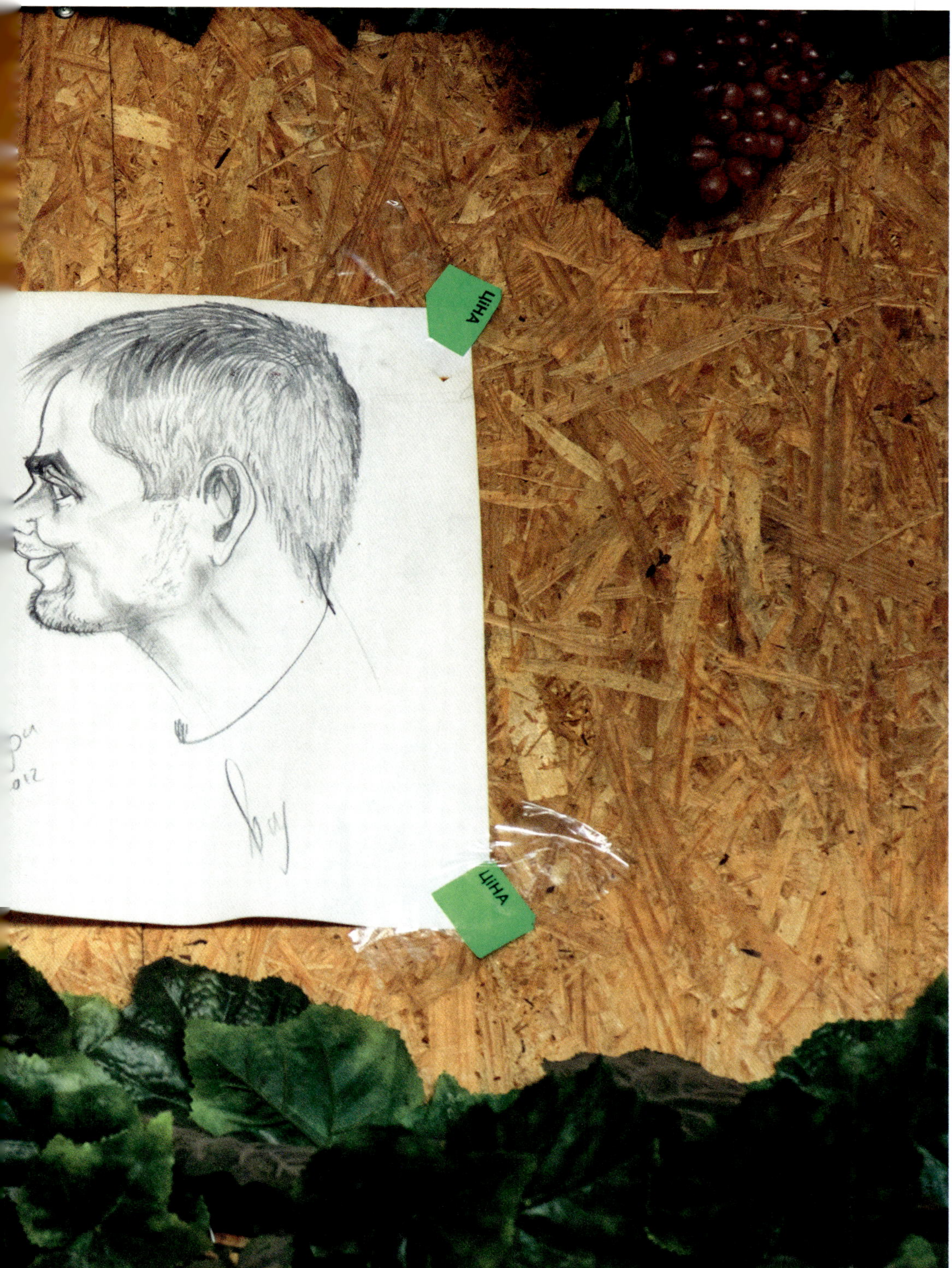
ЦіНА
ЦіНА

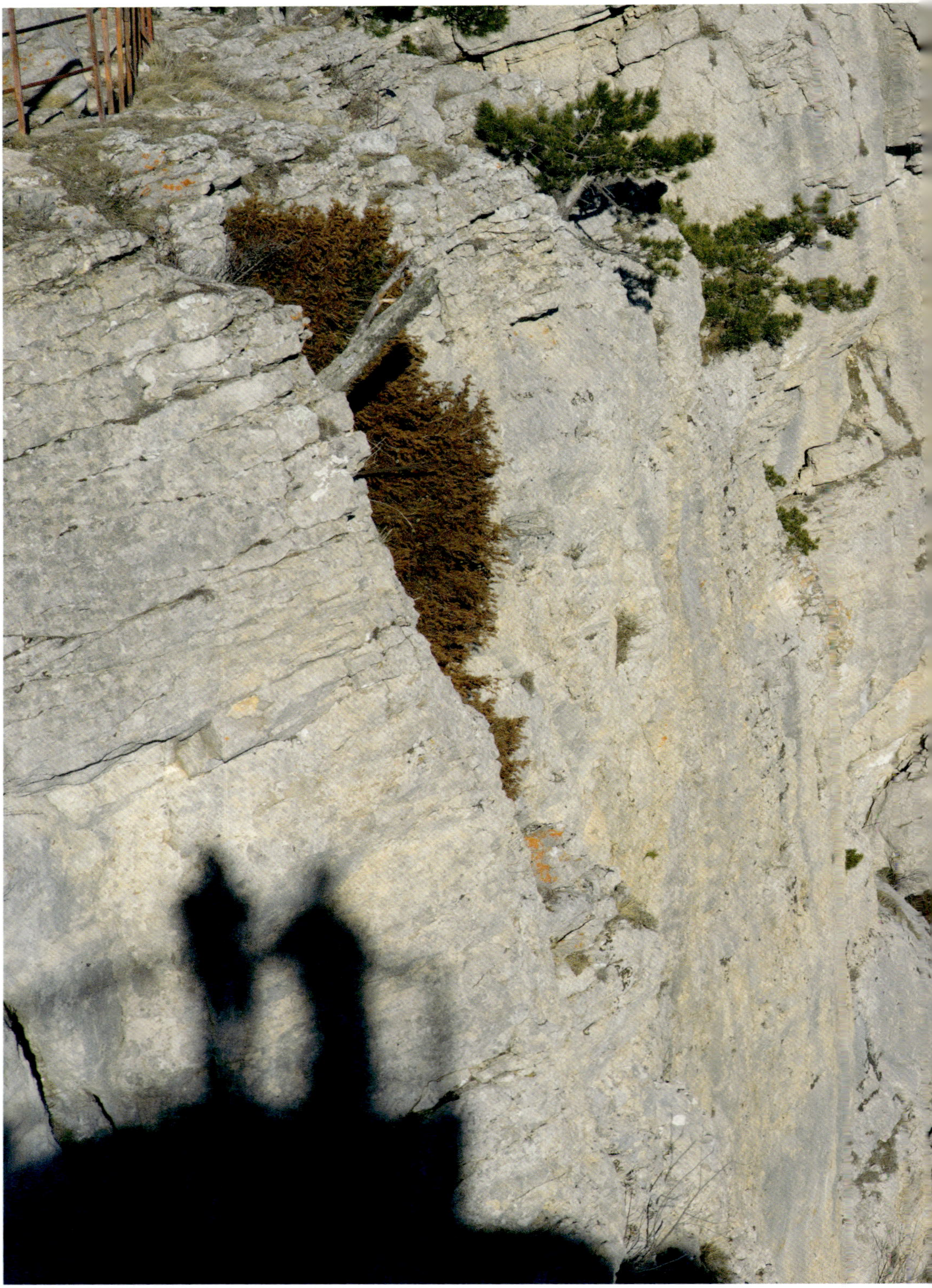

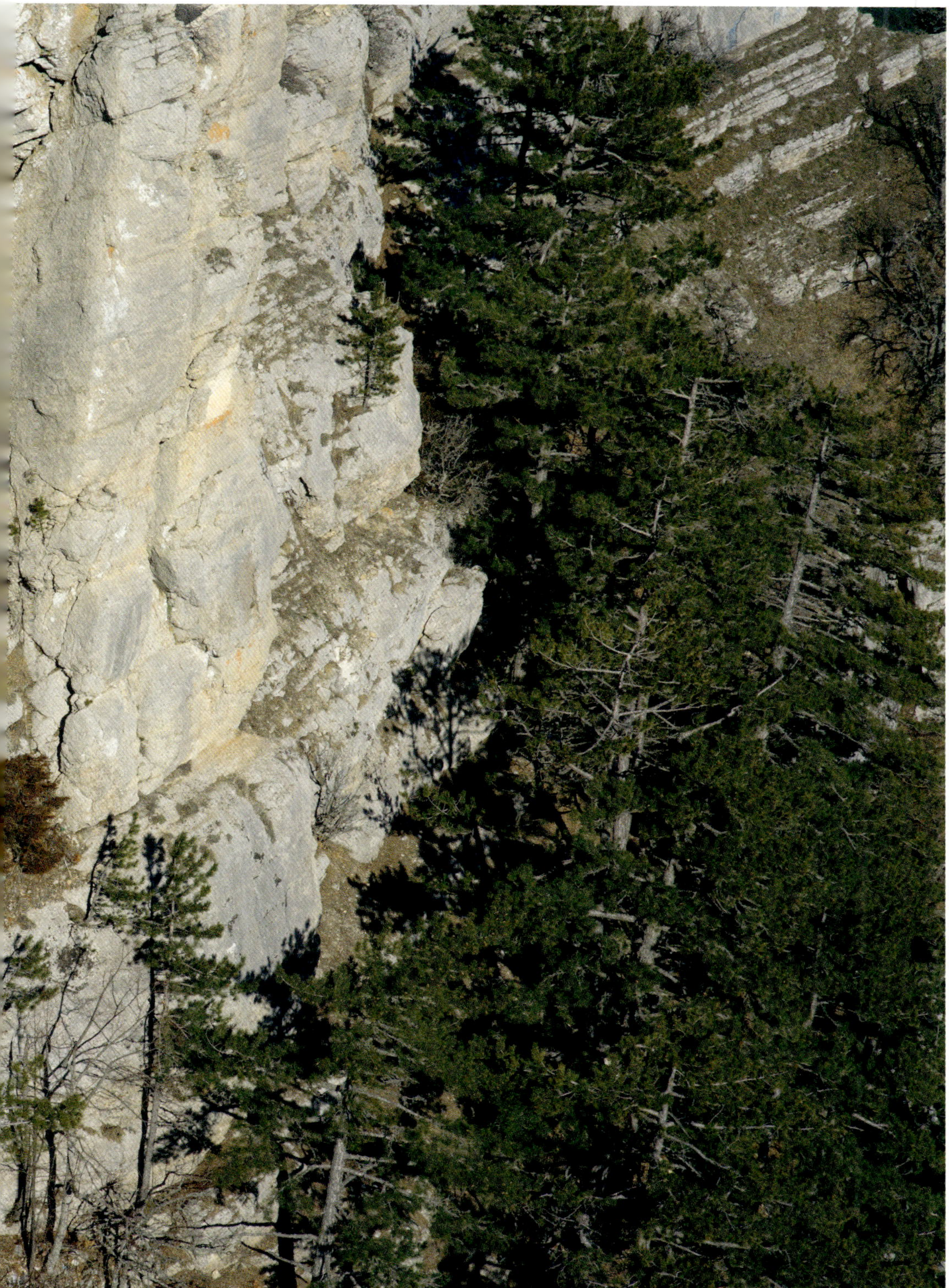

Seva steps out into the purple dusk, burgundy sneakers and blue jeans entering the enhanced shadow of sunset. He dissolves into the gardens through a sympathetic osmosis, taken up through the generous membrane of these subtropics. He does not so much pass by or touch organisms as diffuse through them. In an affront to perspective, superimpositions occur: Seva with the vines, Seva through the hedges, Seva within the olive grove, Seva through his palms. The projector misfires, two slides pressed into the gate at once. Seva and his gardens double exposed, enmeshed. Thus he flows down the slopes and emerges through the palm grove, where it will soon be too dark to see. If Seva were to sleep within the palms, would the morning light reveal him still bound in the thatch of their sturdy little trunks? Seva, though, is sensible. He likely sleeps at home.

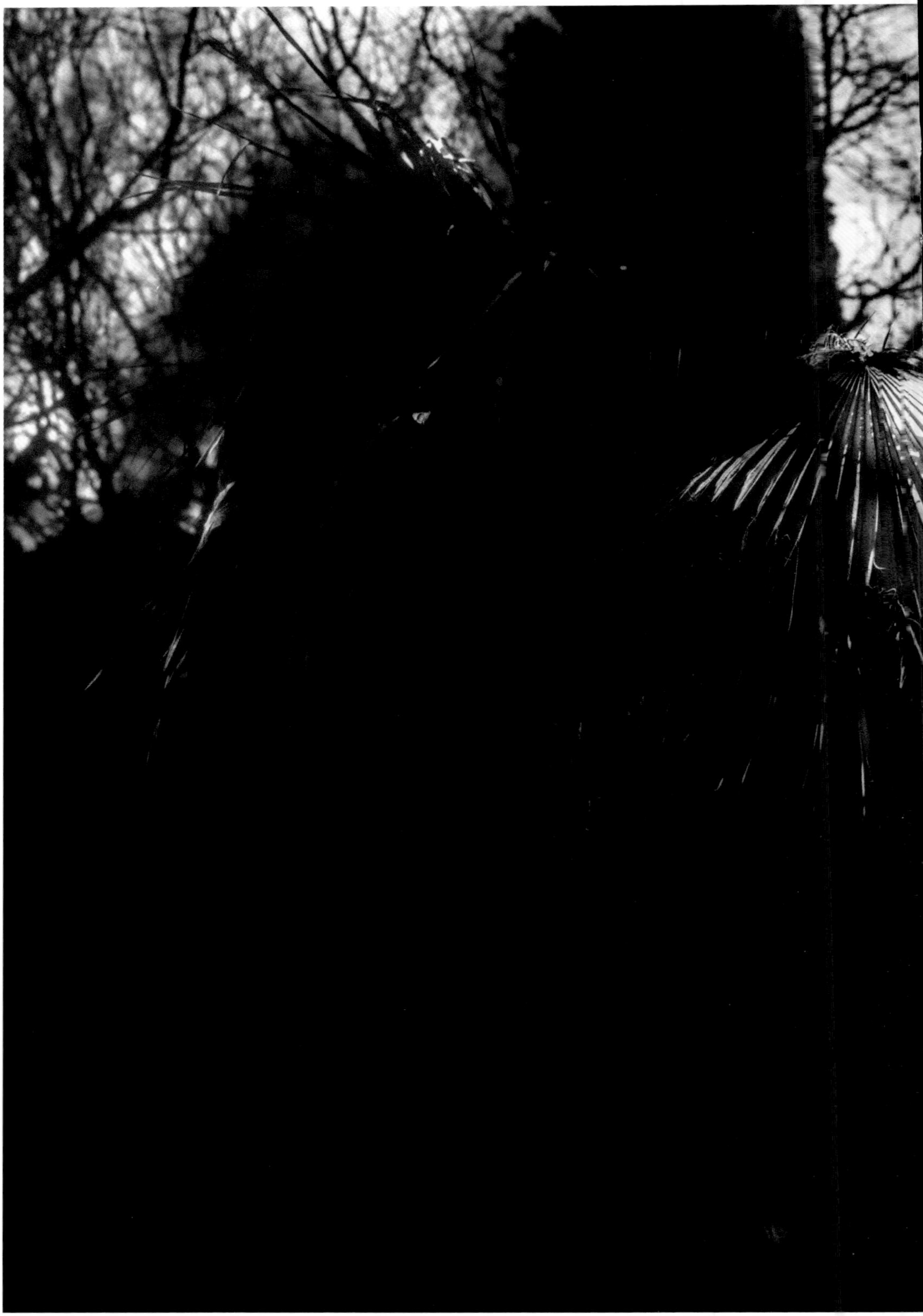

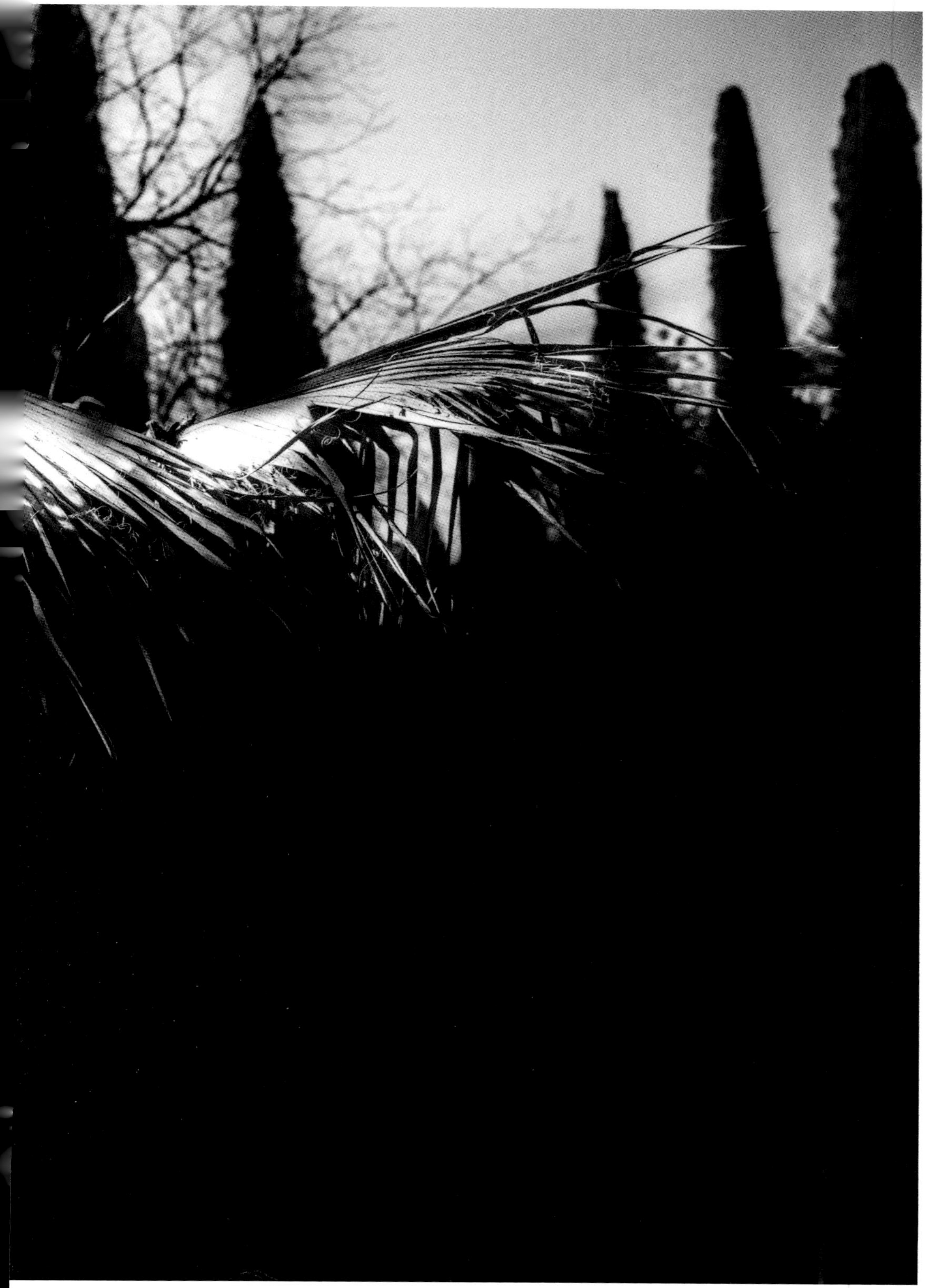

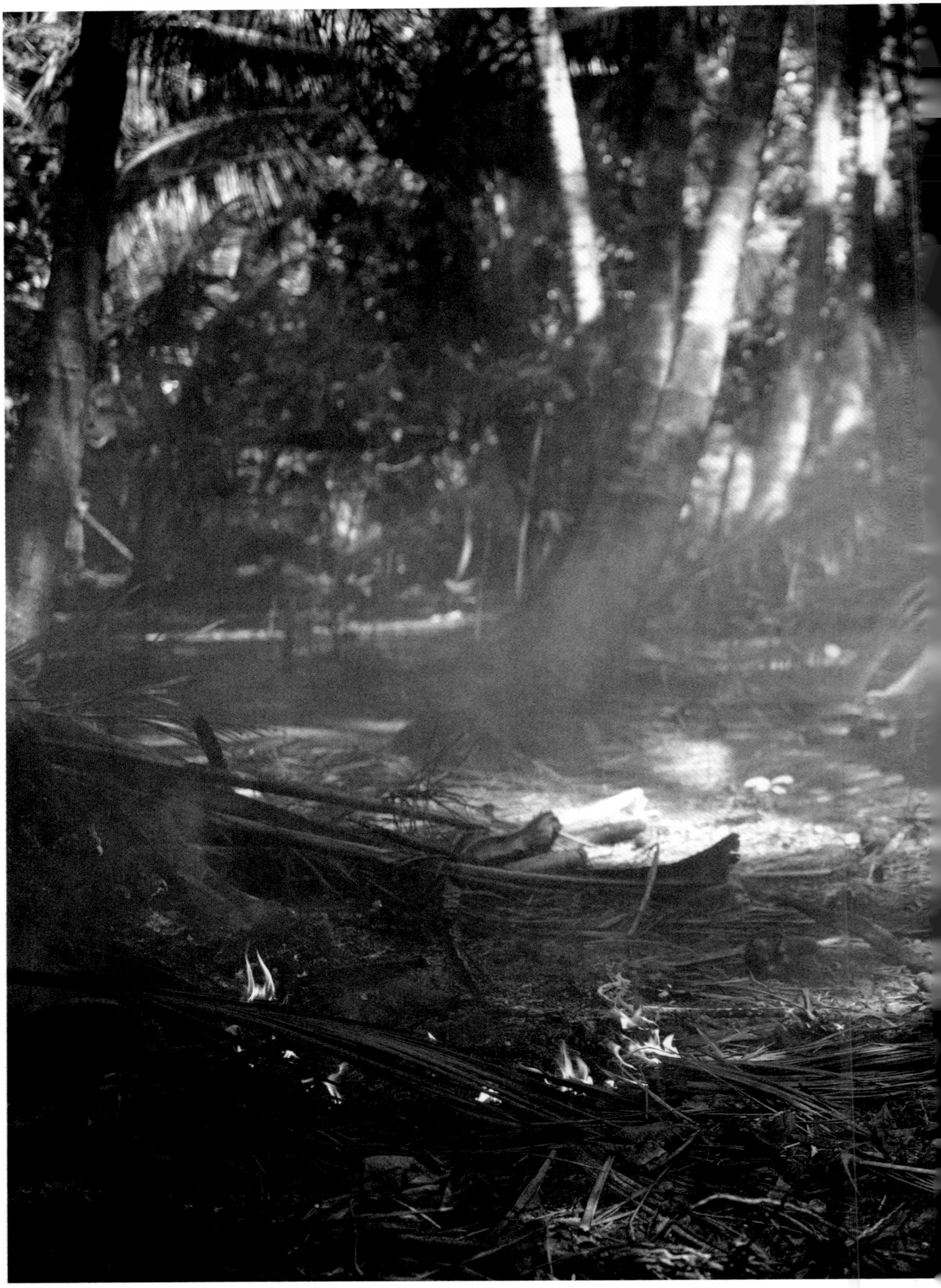

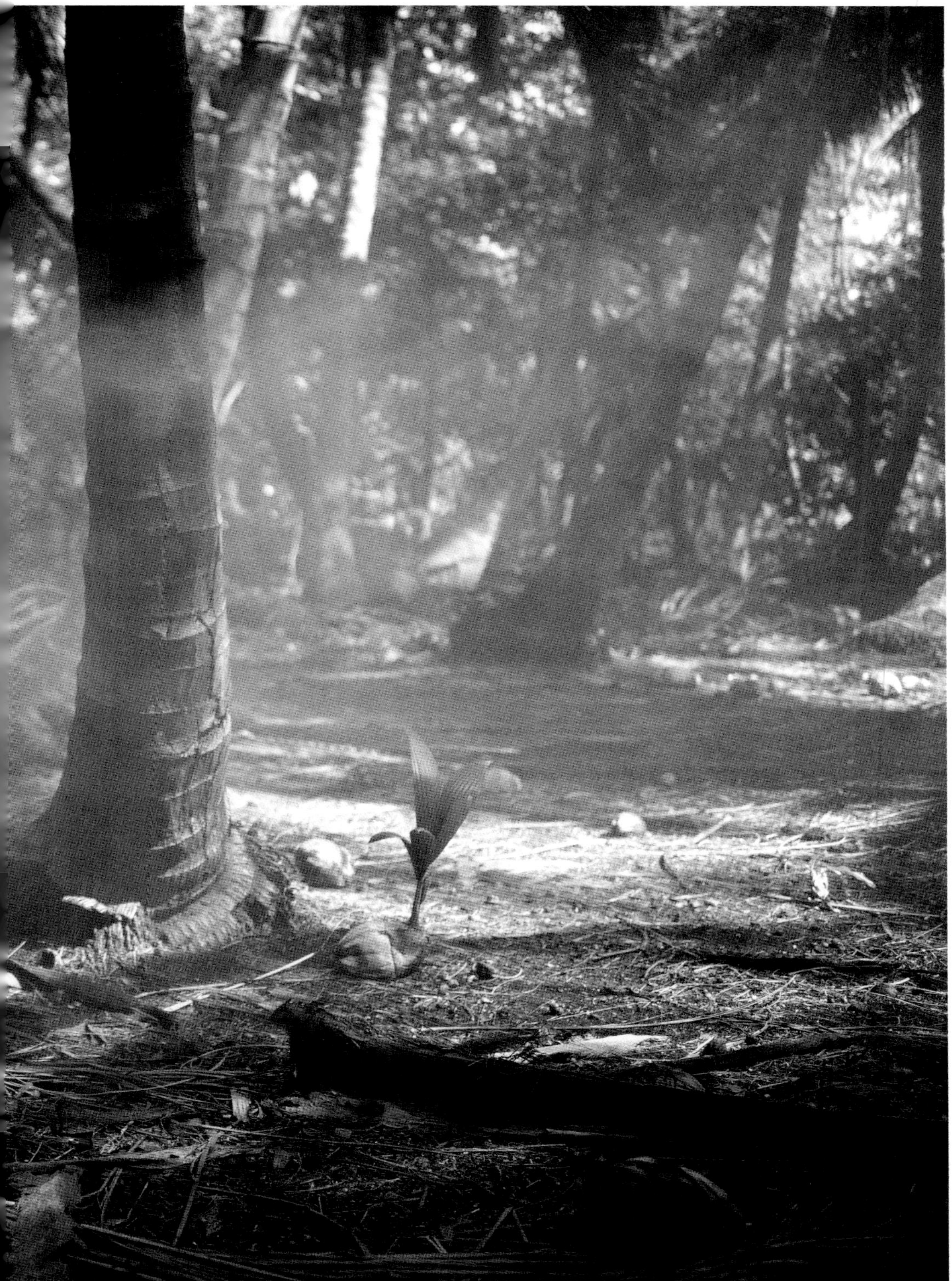

Chapter IX

Another afternoon, arriving at the loose center of the settlement, I rounded a corner of the smallest hut, heading for the washrooms. Momentarily, like a passing artifact in the eye, I perceived Isabel's feet and the lower half of her bright crinkled skirt disappearing into the black of the corridor that runs from the pavilion back towards the kitchens. She was gone as I saw her, but across my path lay the recent wreck of her bicycle, apparently thrown down, or dropped with such haste and inattention that its wheels still spun wildly, parallel to the ground and sky, spokes aglint in the sun. It was an astonishing sight, suggesting panic, flight, abduction and accident all at once.

This dramatic tableau so gripped me that I completely forgot where I had been heading. That is how I found myself wandering away from the huts on a small northward path. And I realized, when I came to myself, that I was looking for Isabel's new house.

But I had no idea, really, where or how far to look. Was it 100 feet or half a mile away? I could not be sure that it even existed. At times, the house seemed purely speculative, at others, all but complete. Tense shifted throughout her descriptions, so as to confound chronology. Once or twice I asked Isabel, where is the house? She would gesture inland, uphill, behind herself (or so it seemed she was always situated, facing the water, waving backwards), accomplishing this movement with a particular loose-limbed vagueness that carried the mystical swagger of a visionary. A conjuring gesture more than a pointing one. I came across a slightly overgrown secondary path and followed it inland, assuming there was an intuition guiding me castleward.

This trail was different from others I had encountered on the island, for though densely overgrown, it seemed dusty and somehow sullied. The foliage here was petty—small, lusterless leaves; spindly branches tangling haphazardly with one another. It felt more like the half-wild land beyond a parking lot than an unspoiled jungle. Yet there was no sign of human presence, to say nothing of Isabel's round fortress. She had spoken of this elusive refuge as a place that could always be anything. Perhaps, it occurred to me, what she meant was the potential to be nothing at all.

Eventually, the ground flattened, the broken jungle receded, and the path disappeared altogether. I was in a clearing, but the space felt darkly gothic. Dusty, flat vines crept sparsely across the ground, beginning to reclaim the field of blackened ash surrounding me. All color had fled the scene.

Ahead lay a ghostly stand of pygmy palms. Their bulbous trunks, covered completely in large hooked thorns, were singed dead from ground to barren tips. This swath of ruined landscape ended in a short promontory that lurched toward the ocean: an outcropping dominated by an immense mango tree. Its giant trunk was charred, but the waxy leaves still throbbed vibrantly under a coat of fine ash. Moving closer, I perceived hundreds of small green fruits, in unbroken clusters across the entire crown of the tree. Covering the vast shaded area beneath it was a jagged, deep carpet of dusty mango pits. I imagined the annual carnage of rotting fruit, aswarm with ants, flies and sugar-drunk crabs, claws waving bright orange flesh above their acid green bodies. And then I remembered Isabel, days before, attesting to the delights of ripe island mangoes. She stood in the grass, eyes closed, elbows bent outward with hands at her mouth, in an ecstatic pantomime of ripping and consuming the perfect fruit. She concluded the demonstration with a grand sweep of the back of her left hand across her mouth, wiping the imagined juices from her dripping chin in sated glory.

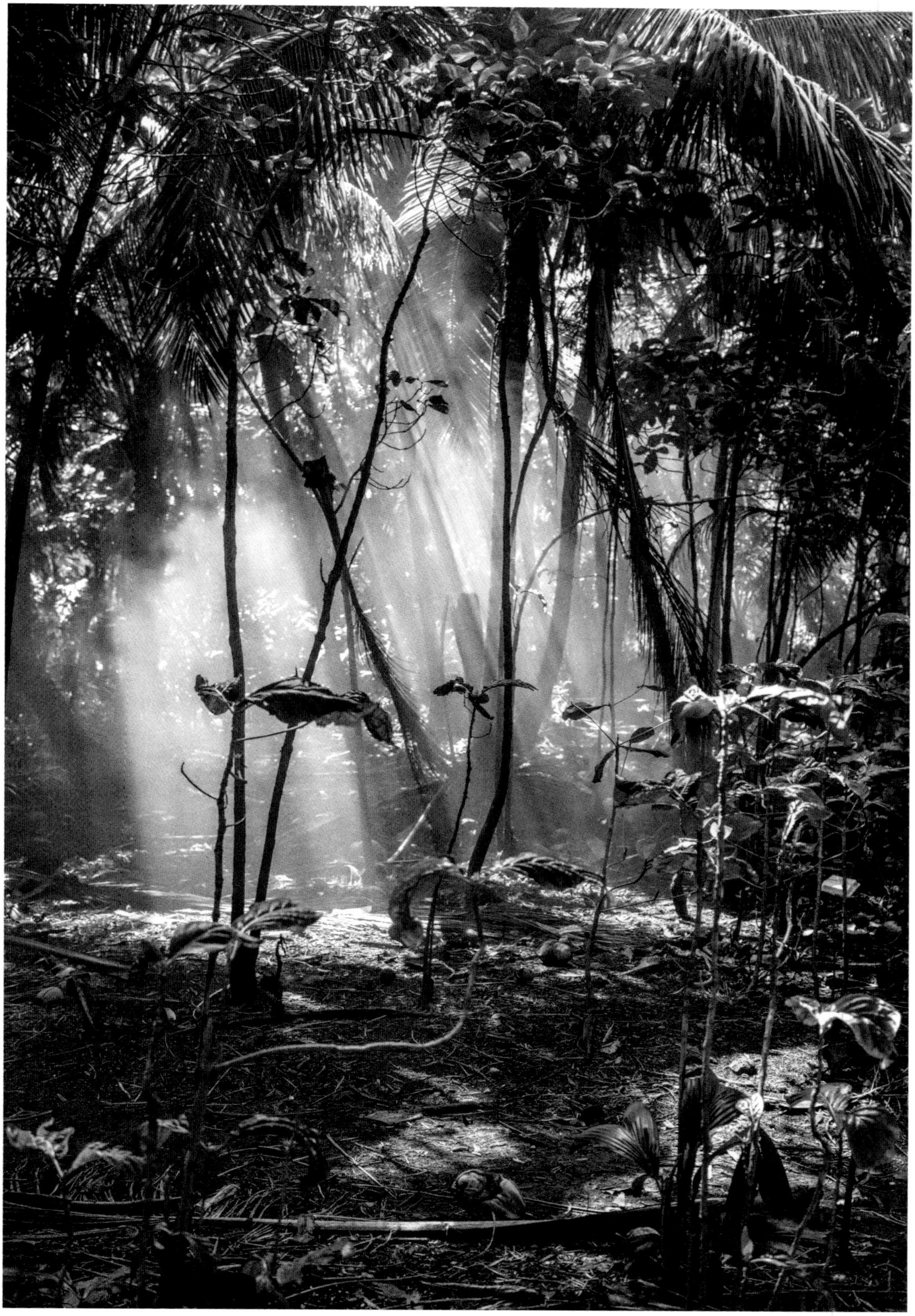

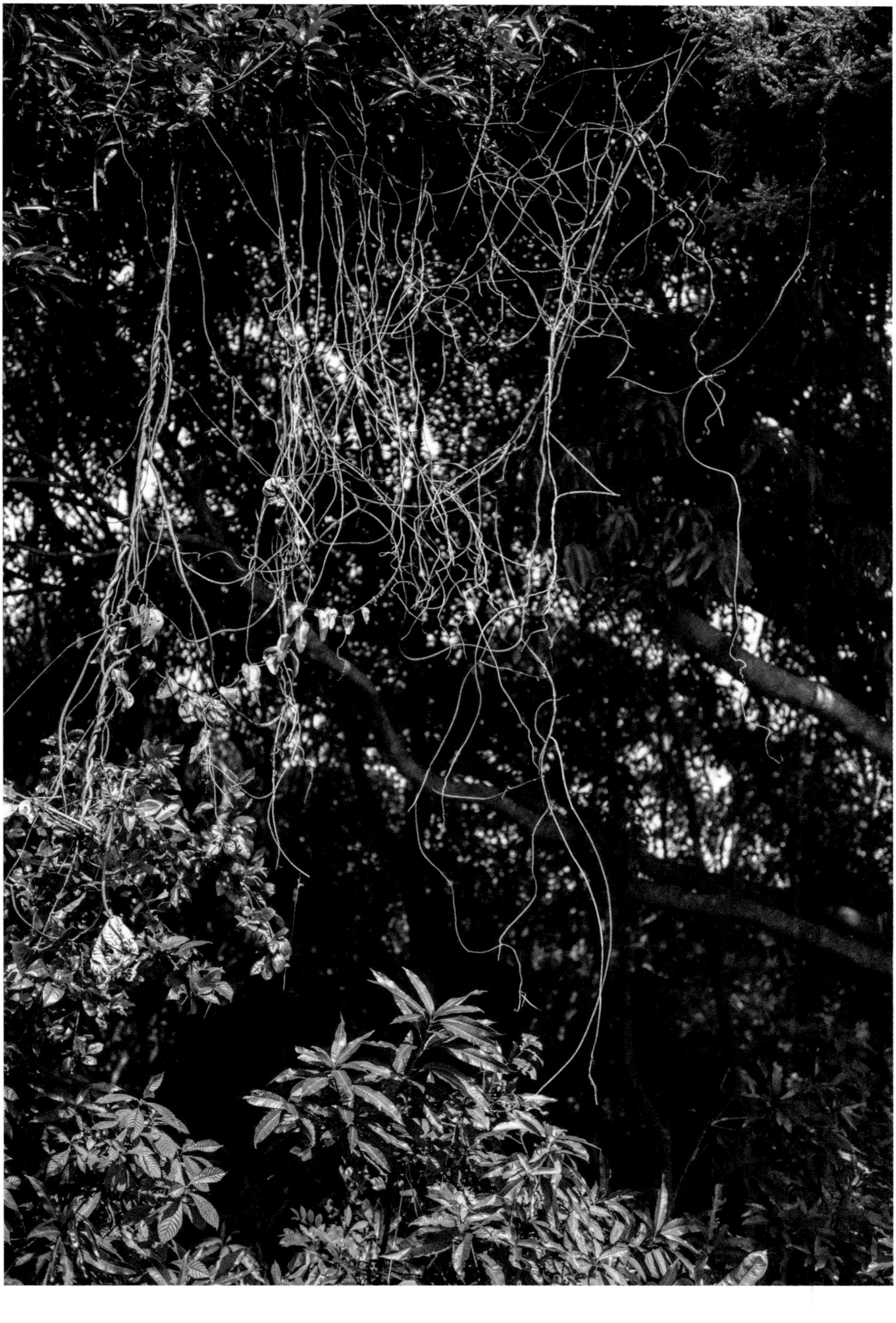

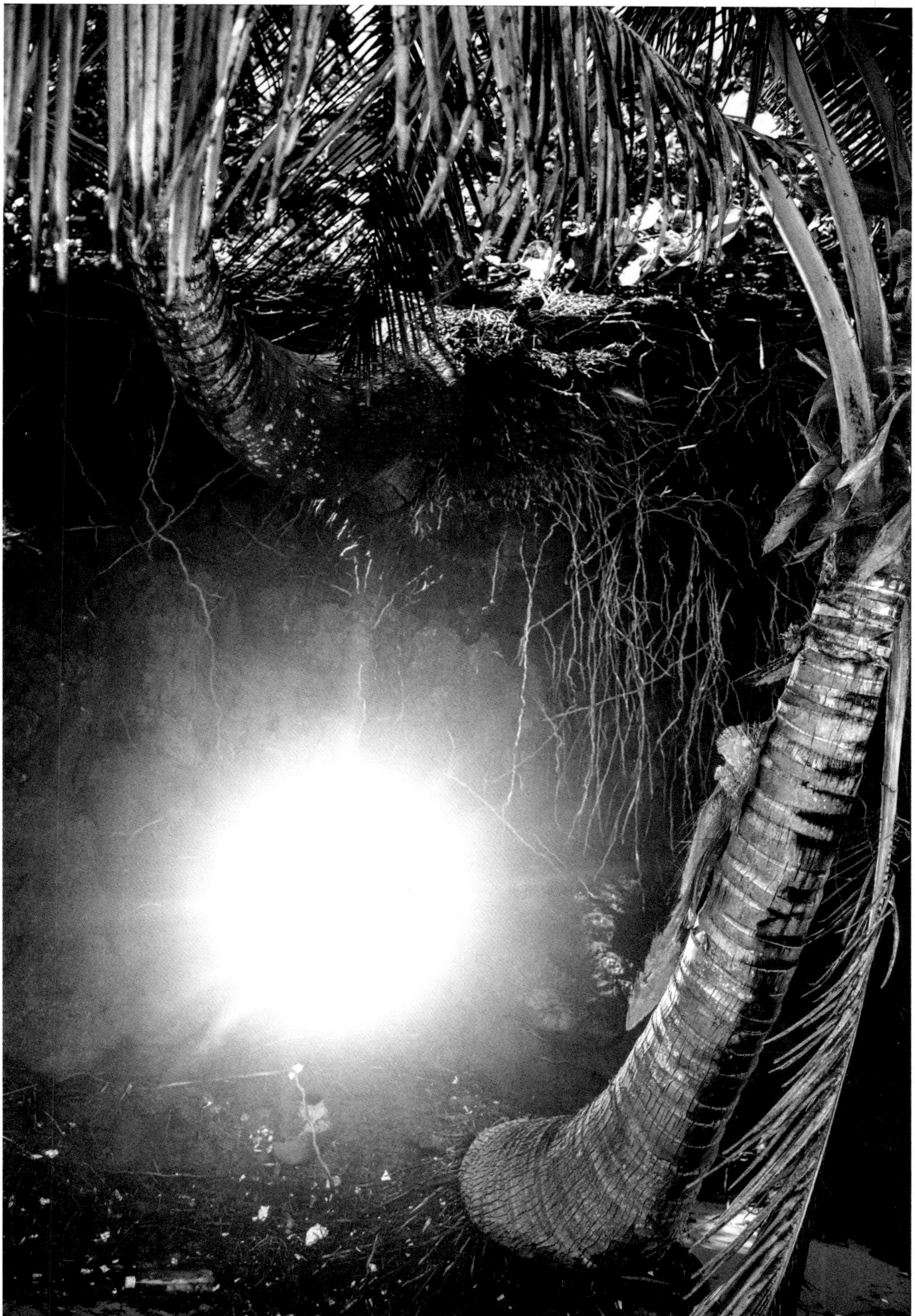

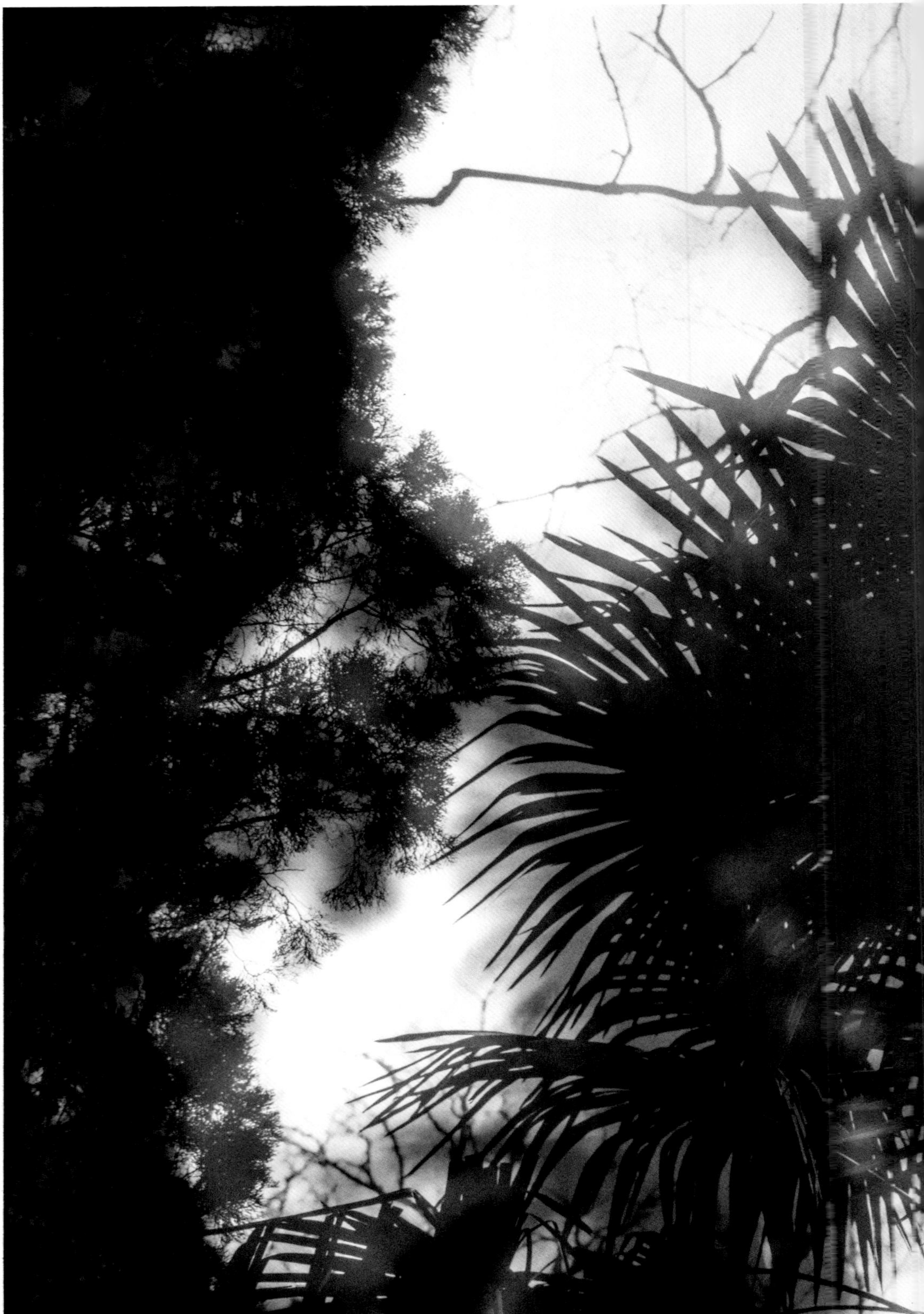

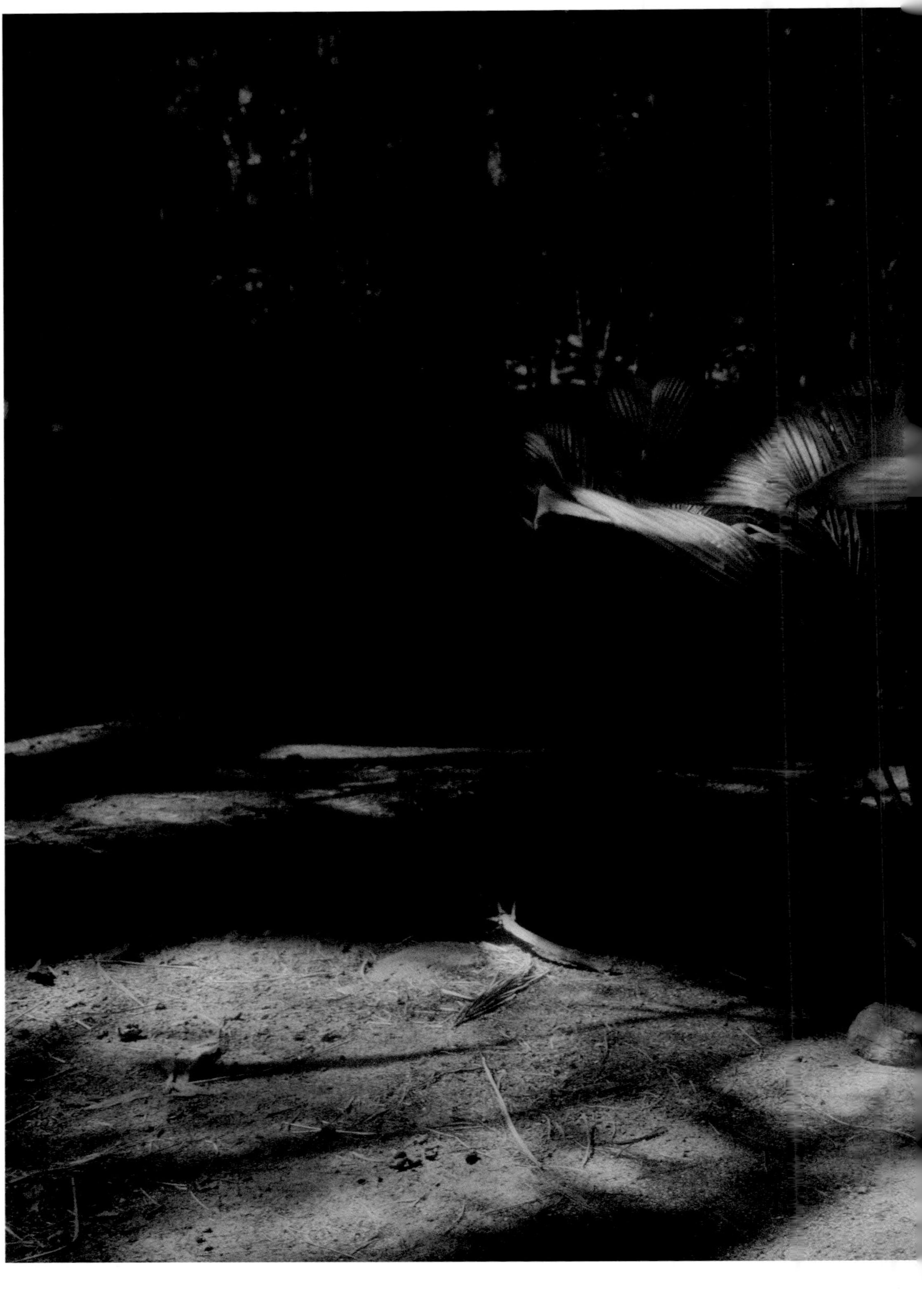

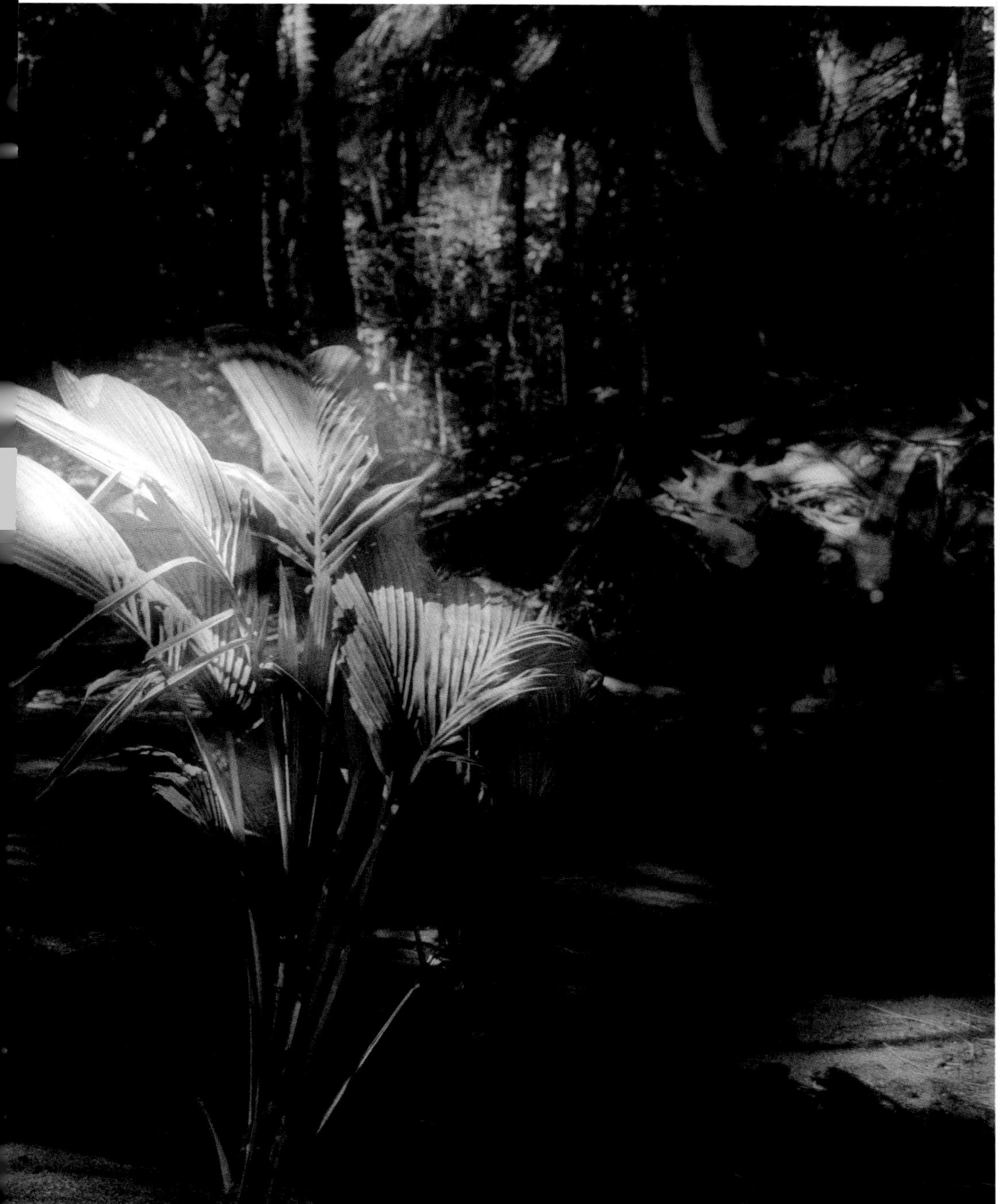

Walking into the hut, I found Isabel standing at a table with her back to me. She turned violently, startled, as if caught at something. Her right arm knocked against the back of a chair, springing a stack of papers from her hands.

The sheets spilled outward, limbs, torsos and chiaroscuro buttocks flashing in the air and settling to the rough-hewn floor. She muttered something hurried and incomprehensible, perhaps in Spanish, and shooed me away. The gesture was of such force that it turned me around. I started to leave, paused in the doorway, looked back. Isabel knelt, arms spread protectively over the strewn pictures, anger and entreaty in her eyes. I said nothing. She whispered in reply: *I am trying to get through.*

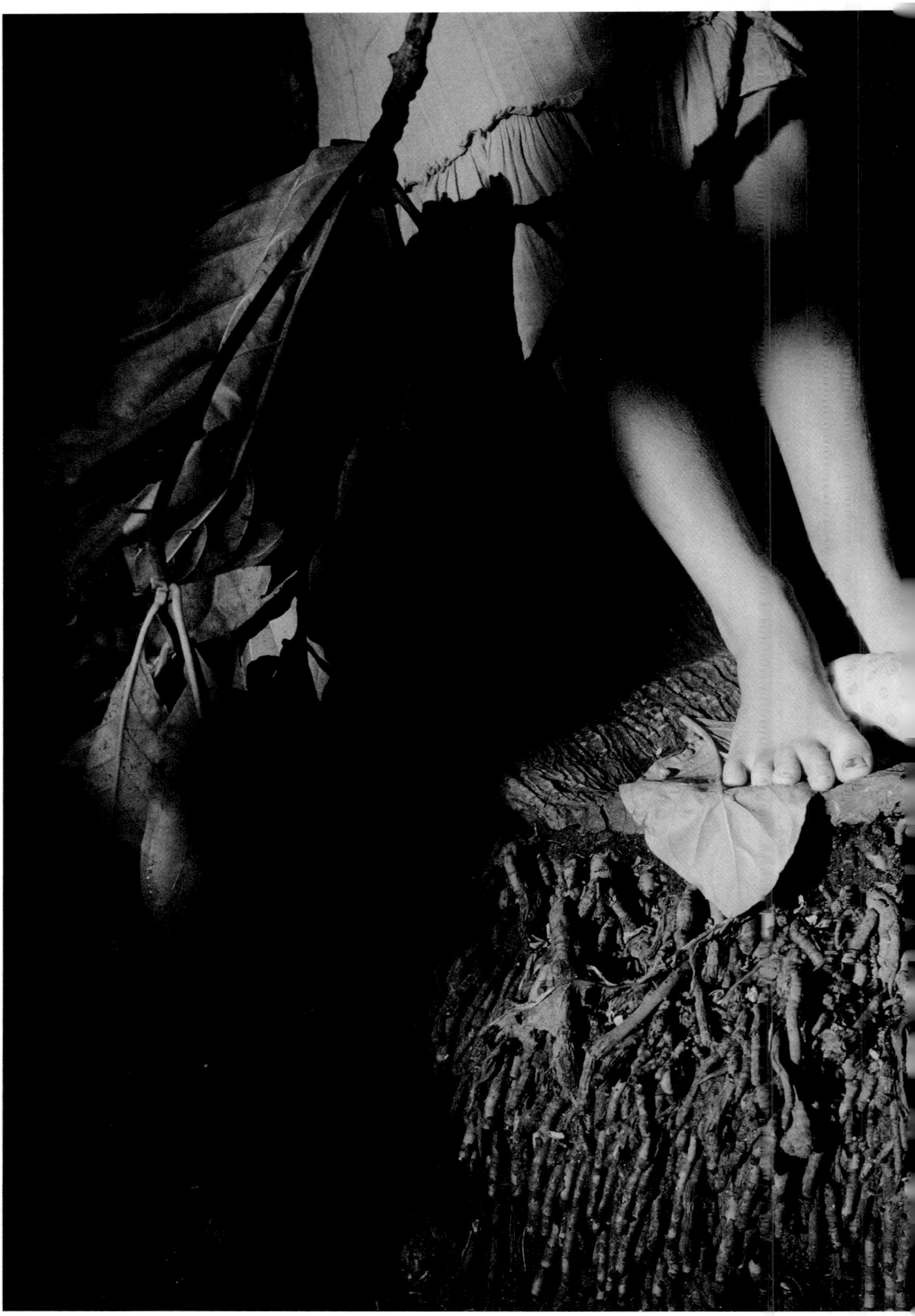

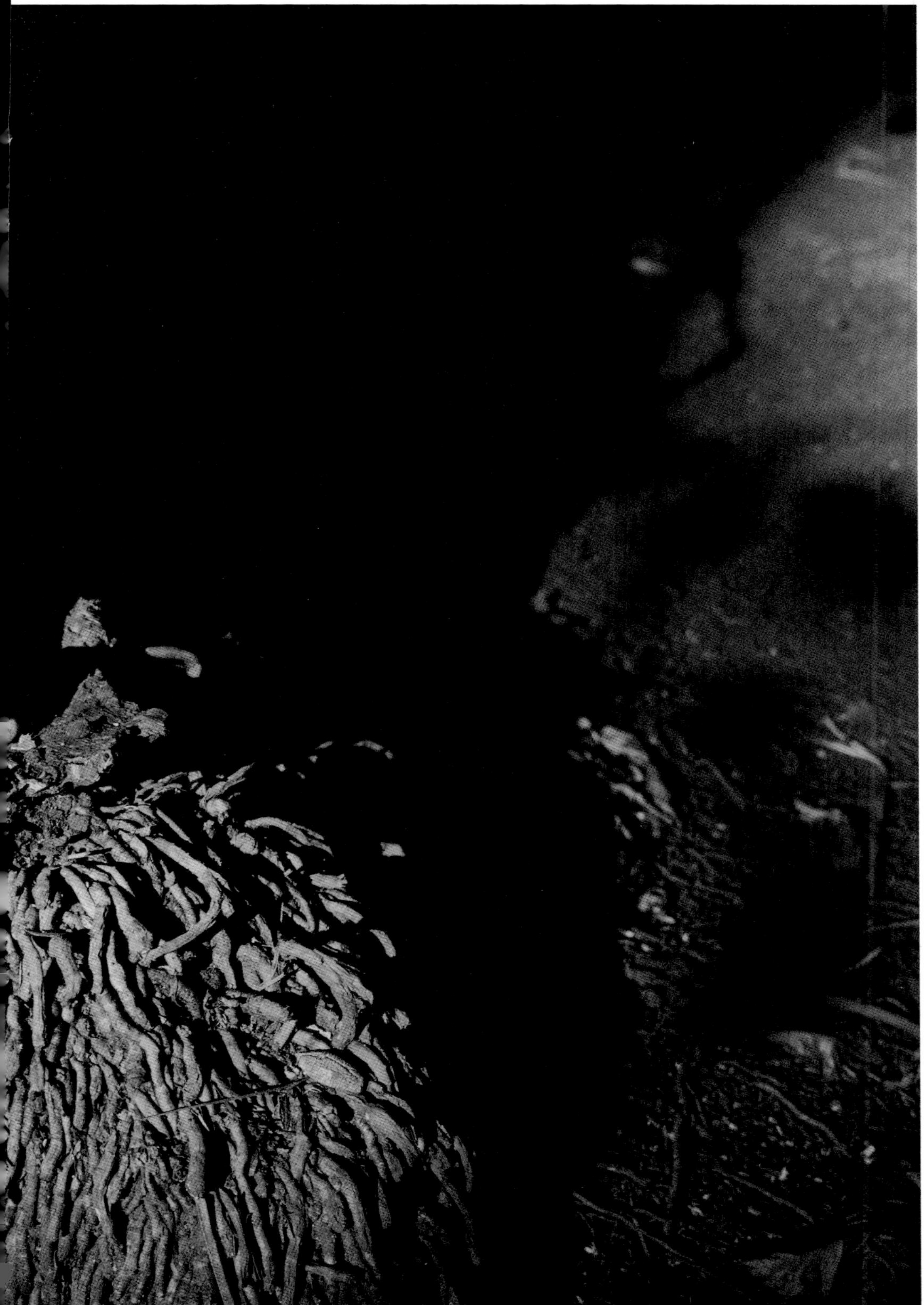

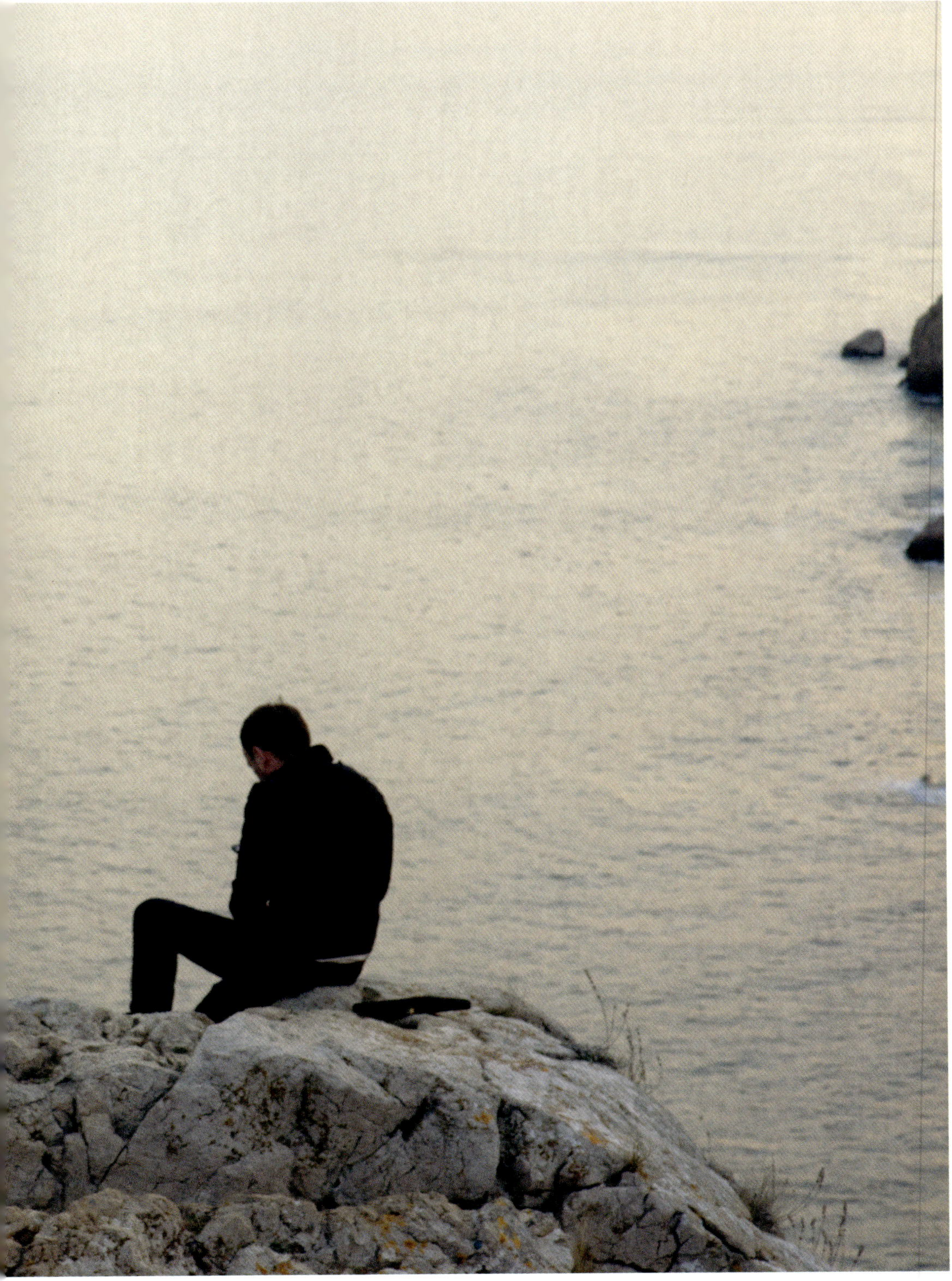

Chapter X

One hour into a thirteen-hour flight, I realize that the entertainment screen in front of me is broken. The flight attendant can't fix it, so he offers to move me to another seat. To his surprise, I decide to stay, opting to pass the hours without movies or TV. I will write, and read a book. I will rest. I decide to consider the absence a gift.

I am happy for hours this way, perhaps even a little smug. But around the seventh hour, when I realize there are still six more, this life of air-bound virtue begins to wear thin. I am tired and bored, and I've finished my book. At some point, without knowing when it started, I discover that I am watching a movie on the screen of the young woman across the aisle, a row ahead of me. It's the kind of film I didn't think I could bring myself to sit through, even on an airplane: a recent melodrama about two terminally ill straight white teens in love. I have never seen two people look so robust and generic while they die. By the time I'm aware of what I'm watching, I'm also aware that, lacking a single word of dialogue, I am following the purest arc of this story. In fact, its silence, scrubbed of the banality and distraction of cheap language, transforms this formula film into something inevitable and epic—a fable or a myth. It comes to life like a mystery play, or the panels of an ancient church fresco. Reductive unlined faces relate the transformative power of martyrdom; allegories signaling in the hyperefficient gestures of cliché.

The young woman unwittingly sharing this film with me, in loose sweats and enormous headphones, is always within my frame, in front-lit quarter profile. As the film reaches its tragic climax, she begins a slow, measured ritual: lifting her left arm slowly from her lap every minute or so to wipe a single tear from each cheek with the back of her hand. The spectacle of her feeling, layered in front of the characters' emotion, creates a kind of baroque frame, more moving than Bernini's sculpted men in balcony seats, observing Saint Teresa in her Ecstasy in the Cornaro Chapel in Rome. Each of my fellow passenger's tears points me towards the tragedy, where stylized gesture and setting produce so many silent tableaux of structured pathos: the optimism and fragility of youth; the transcendence of pure love; the unutterable sadness of loss, and its inevitability. It becomes something other than cinema, and outside of literature. Something I can only call a dumb show.

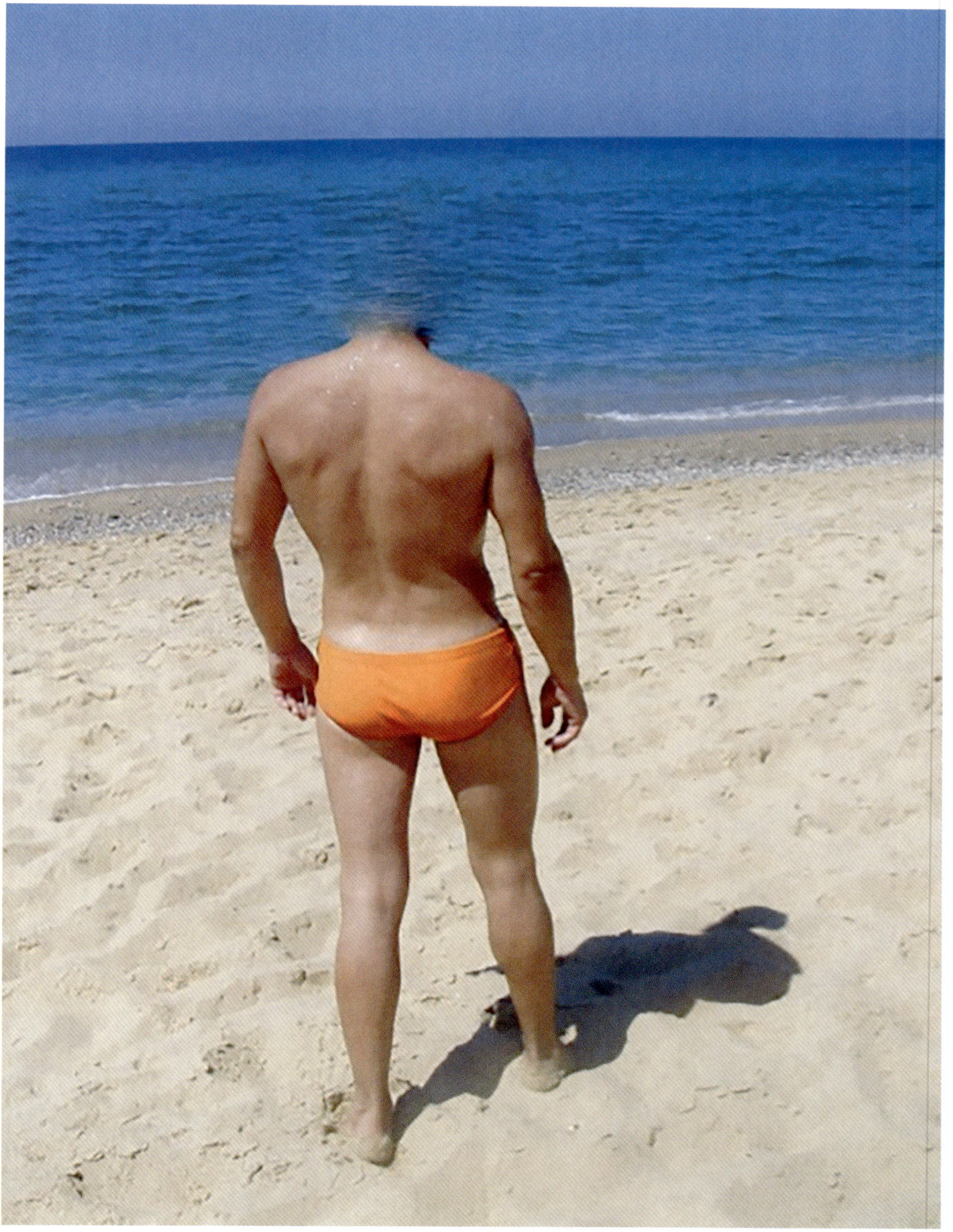

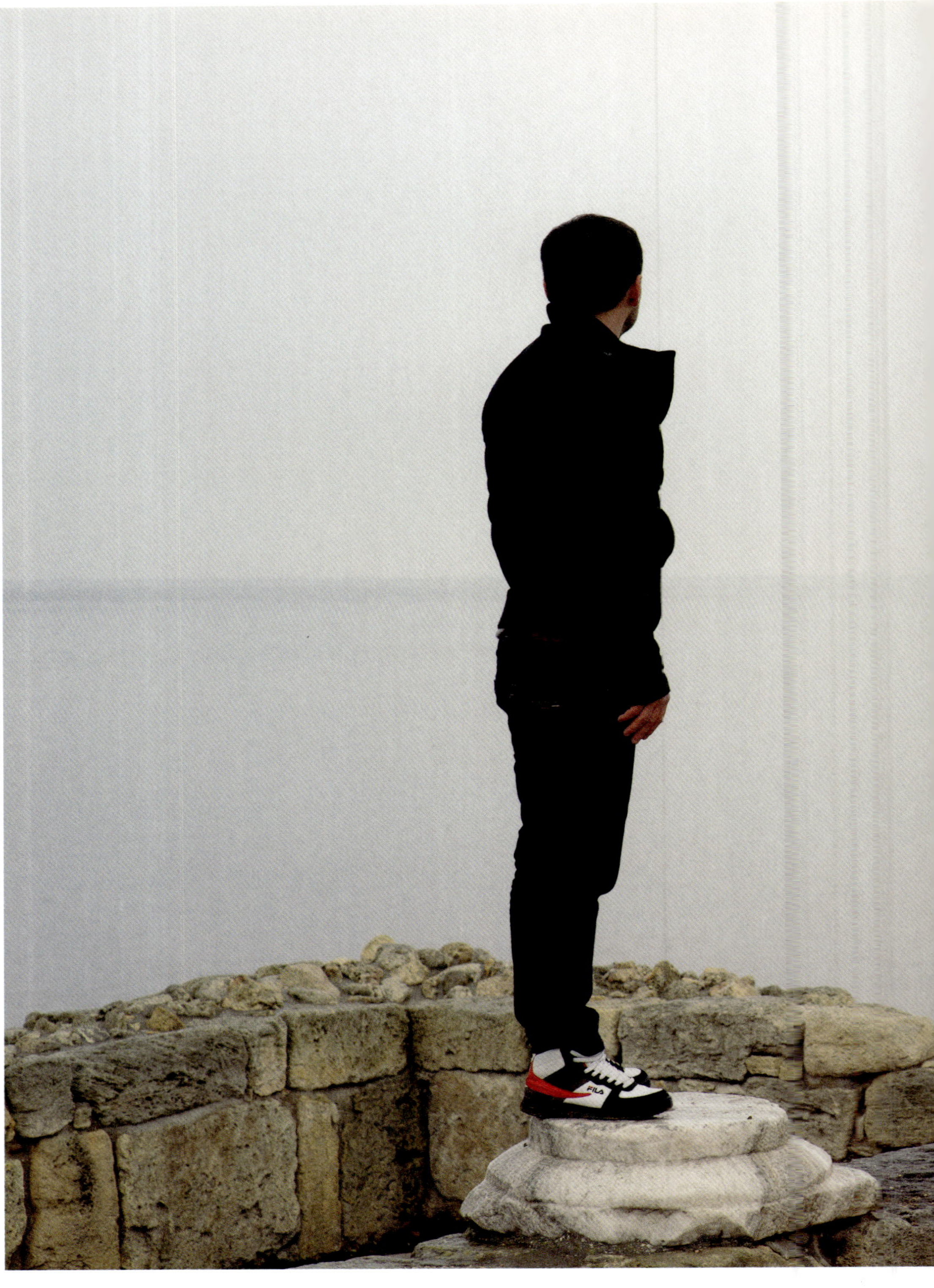
FILA

We walk among the vast ruins, tracing individual paths that intersect, run parallel, draw apart. We drift but remain in orbit. We have only met yesterday, and we speak sparingly. The silence meaning both: *I don't know you at all* and *what is there left to explain?* At these ruins, it seems we are joined by the hulking silence of the past, its distance much greater than that between us, and its presence, its palpability, much sounder than our own.

I can't adequately explain why I wanted to meet Dima, and he doesn't even attempt to explain his own motives. It is, on both sides, a gesture stripped of practical intention, outside of desire, money, or even a conventional idea of friendship. Most likely, we will never meet again. I know only that, like a true romantic, he has renounced the possibility of love.

Dima and I pass row upon row of ancient leveled structures, the traces of a dense Greek outpost that clung to the Crimean coast for over a millennium. One-room houses of varying size shared massive exterior stone walls that can now be jumped over in a hop. Paved roads still cut between them. Along one of these, breaking the flat symmetry of the excavation, appears a deeply sunken structure, plunging 20 feet below ground level. We both stop and look down. The space is long and narrow, cut by small aisles jutting out to each side. In the middle, a low altar rises from the dirt floor, where someone has placed a candle. It is an early Christian chapel, now laid open to the elements, but built with one express purpose: to practice forbidden devotion. An extremely narrow stairway, crudely cut into the site's original stone, leads from street level towards the nave. Halfway down, these stairs dissolve completely. The candle—green wax in a frosted glass cup—is new. How, I wonder, did the supplicant get in there? More mysteriously, how did this worshipper escape?

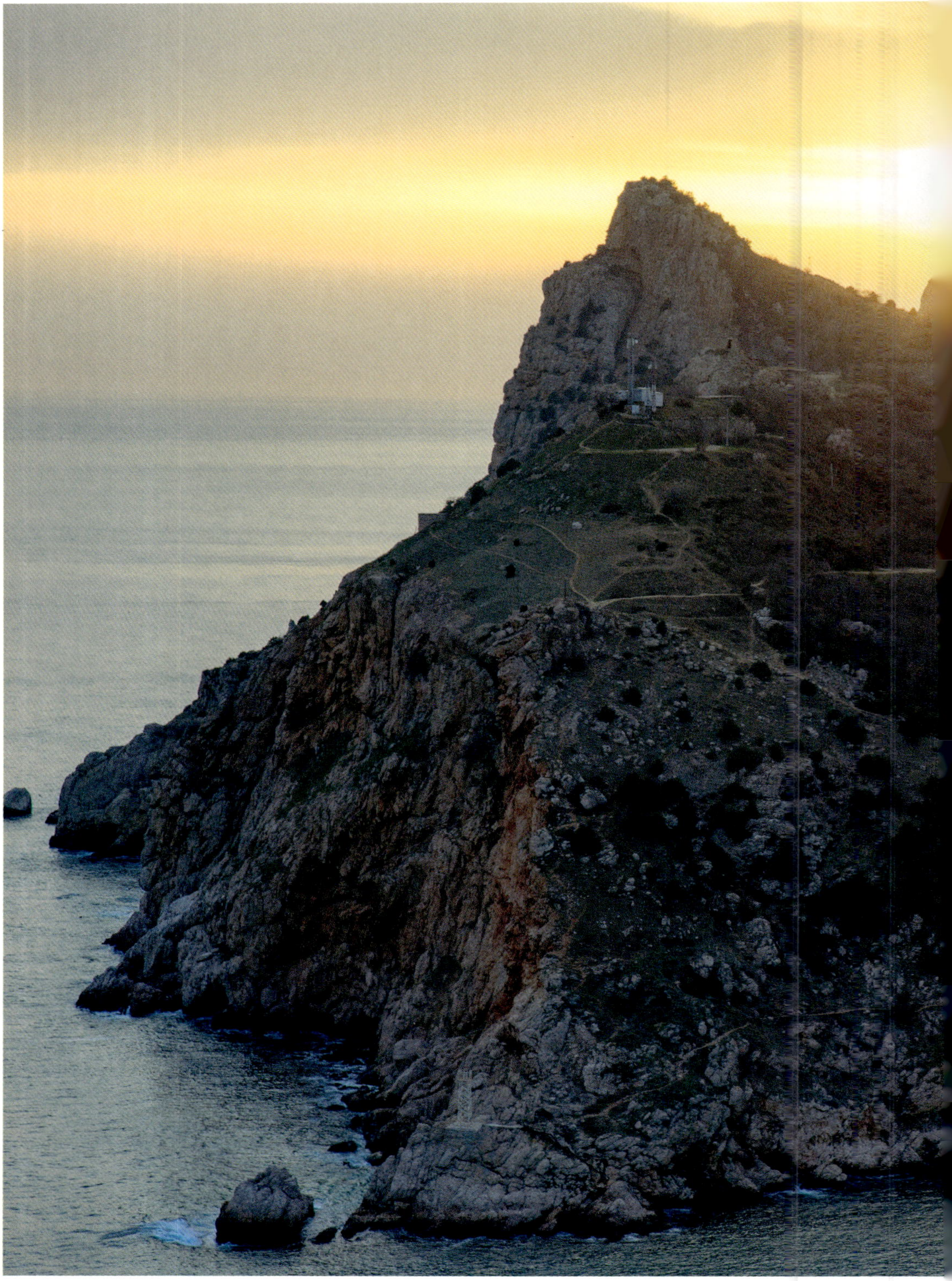

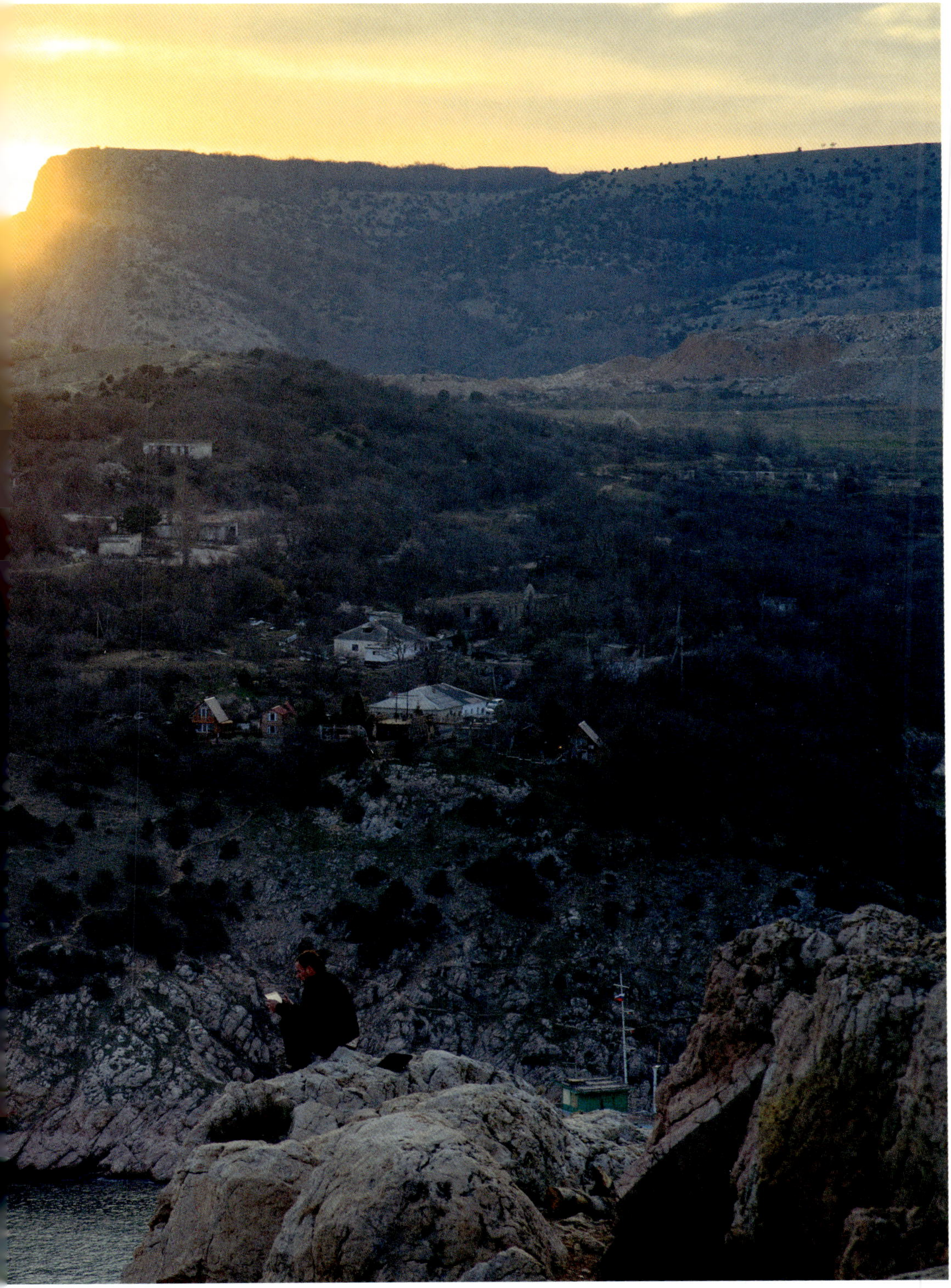

In our correspondence, weeks before, I mentioned that I was a photographer. Dima answered, immediately: "Oh, then! We will have a lot to talk about!" But we didn't so much talk about photography as show photographs in place of talking. Sitting on the cramped bus, side by side, he pulled his tablet out of its leatherette sleeve. He then scrolled through his life, beginning from a recent hike and moving backwards through his last visit home, to the family farm in a Moldovan village, where each house makes its own wine. Dima on a hilltop; Dima by the lake; Dima with his mama; Dima holding grapes. His biography ended where his tablet began: a pastoral idyll signifying origins, family, happiness, nature. It told me nothing of his daily life as a café waiter, or about the one man he moved here to meet, from thousands of miles away, and who eventually threw him over. The tablet bore, instead, a silent mythos of the self he could still be: the vistas and memories that made him, enduringly, silently, only that Dima among so many.

Dima led me up a steep dirt path, past ancient stone turrets supported by haphazard wooden struts, to the natural ramparts that separate the Black Sea from the sheltered harbor below. Behind me, to the north, cast in fresh shadow, the empty tourist town clung to its placid waters, where centuries of alien armies have fortified themselves against local forces. In the west, the sun hung bracketed between two peaks. Dima and I, in silent agreement, moved away from each other, to contemplate the day's last dramas alone. My companion clambered onto the outermost cliff-top promontory, kneeling in silhouette against the desaturating blue of the ocean. He looked south over the water, eyes squinting hard as if sufficient effort would discover Turkey. After a while, his head dropped down to contemplate the infinitesimal waves hitting sheer rocks hundreds of feet below. Eventually, he rocked back gently on his ass, tucking in his chin to take in his phone, glowing blue in the space between his knees.

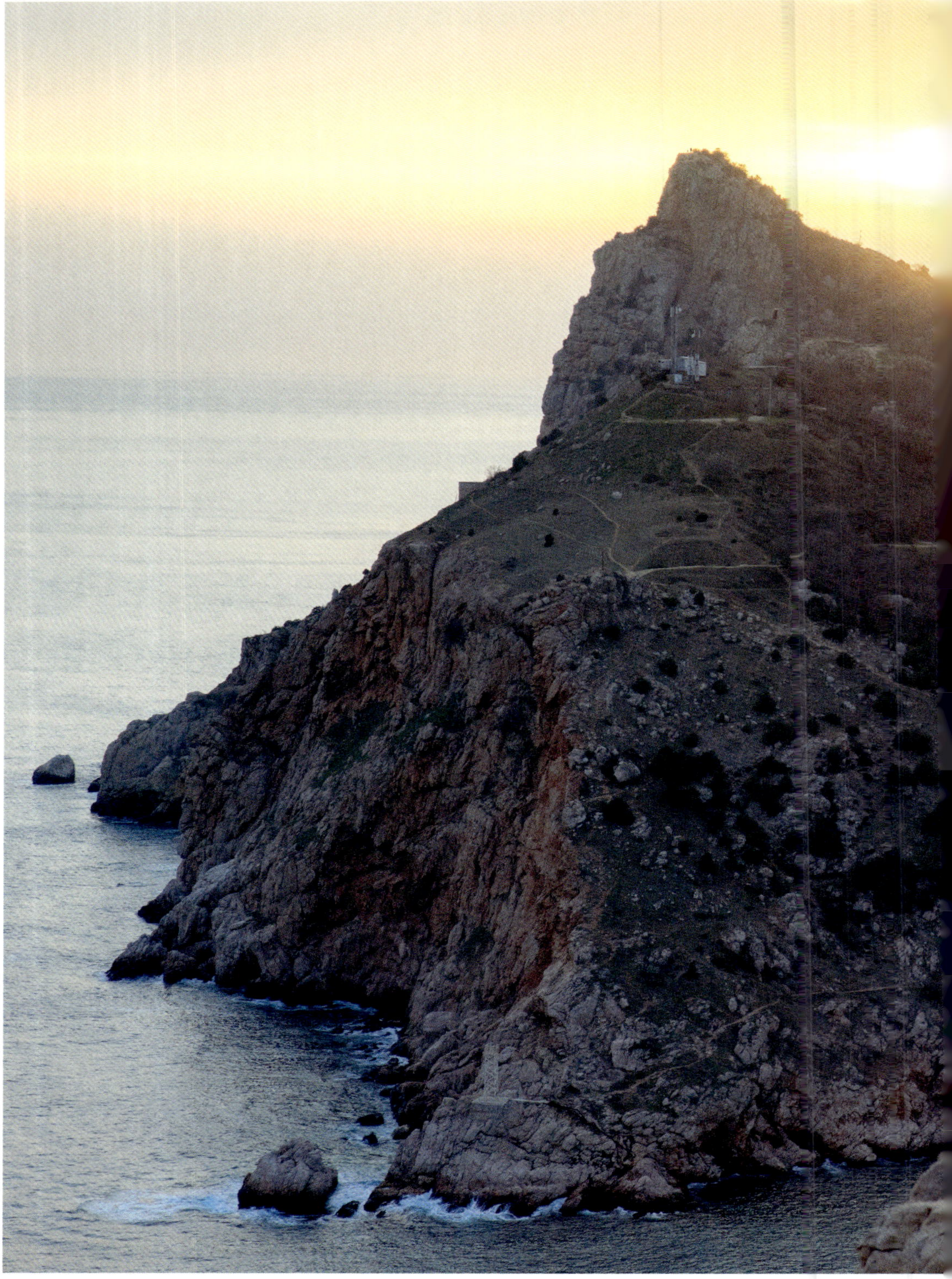

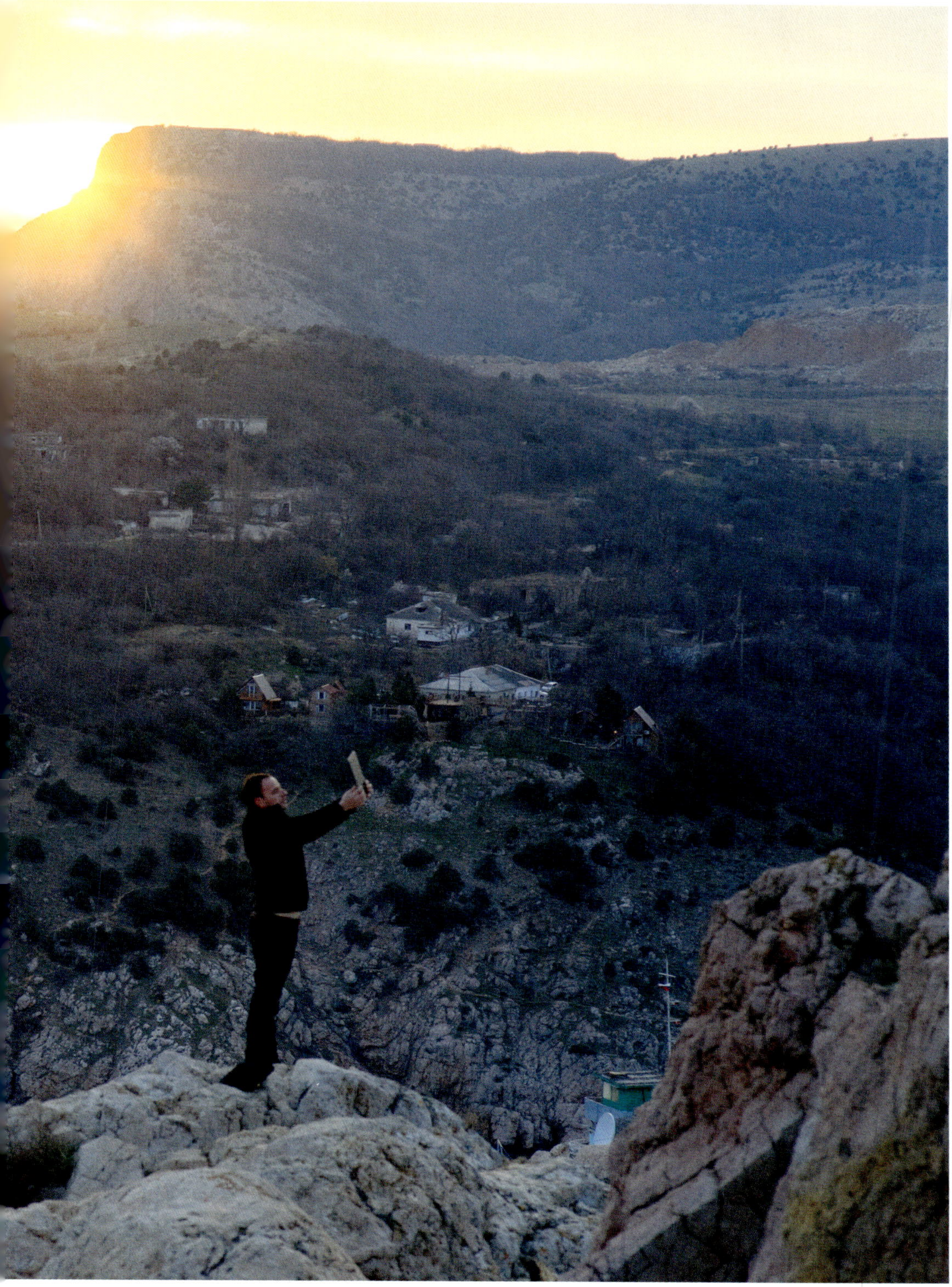

I stood upon my own small promontory, contemplating Dima and the West as everything around us grew dark. In my overladen state, each gesture seemed to issue its own communiqué. Squinting at the horizon: a question of the future. Contemplating the surf below: the offering of death. Looking at his phone: am I utterly alone? They probably said none of those things. But I have seen too many pictures and I no longer know better. The image of the gaze is often a question of destiny; an imaginary refraction of fate beyond words. I turned away to stare eastward, where the high cliffs folded in on themselves densely, foreclosing contemplation. My body shielding the late sun from my camera's screen, I reviewed the photographs of Dima and the infinite I had just made—light, shadow, form and gesture extrapolated from the tableaux of moments ago. Already, his meaning was inscrutable. Despite the scale of the site, the light's tenderness eroded distance. The scene was intimate, internal, and not for me. My screen revealed merely what I saw: rock and sea and sun and Dima, together on a flat surface, conspiring to signal something to an interpreter unknown.

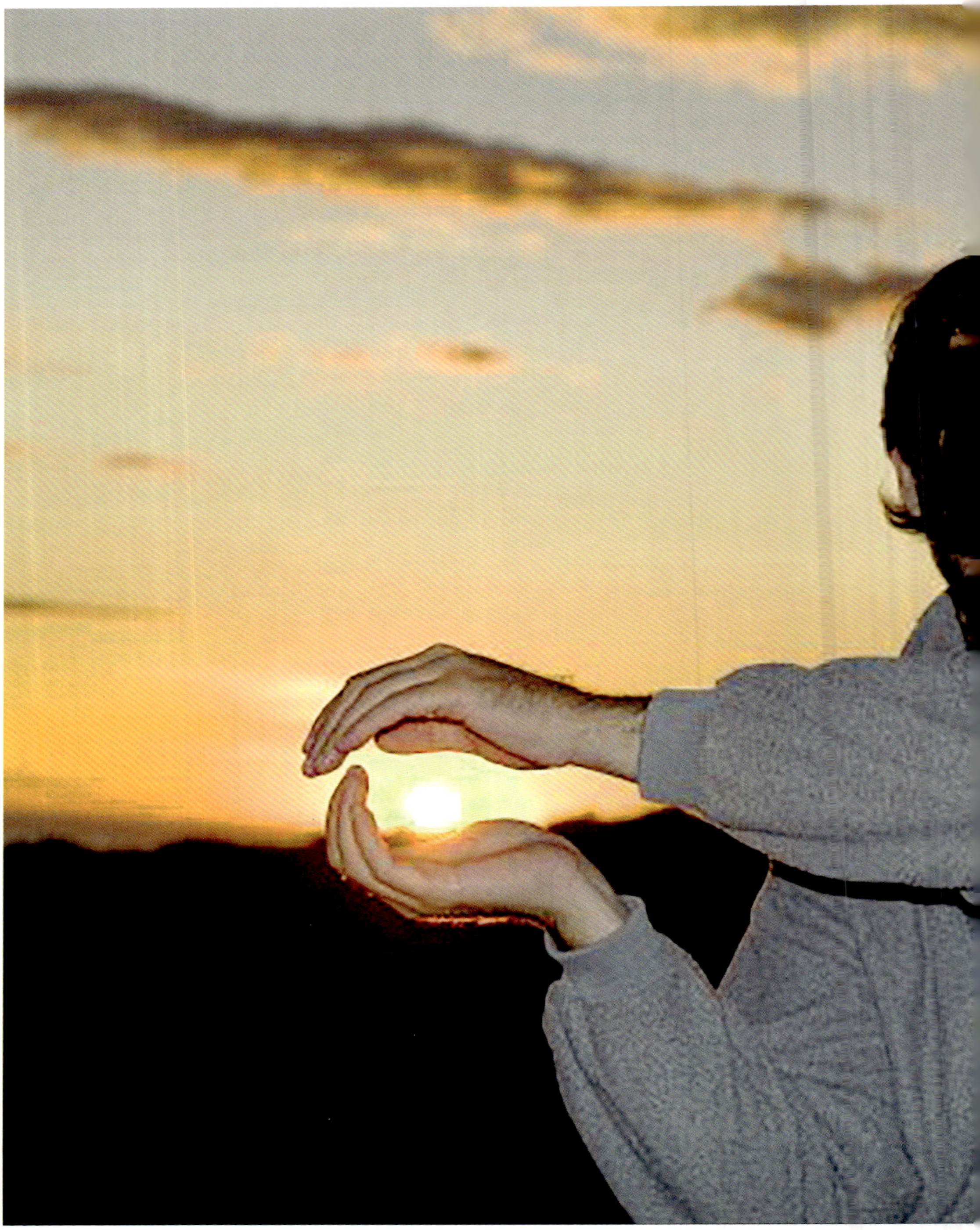

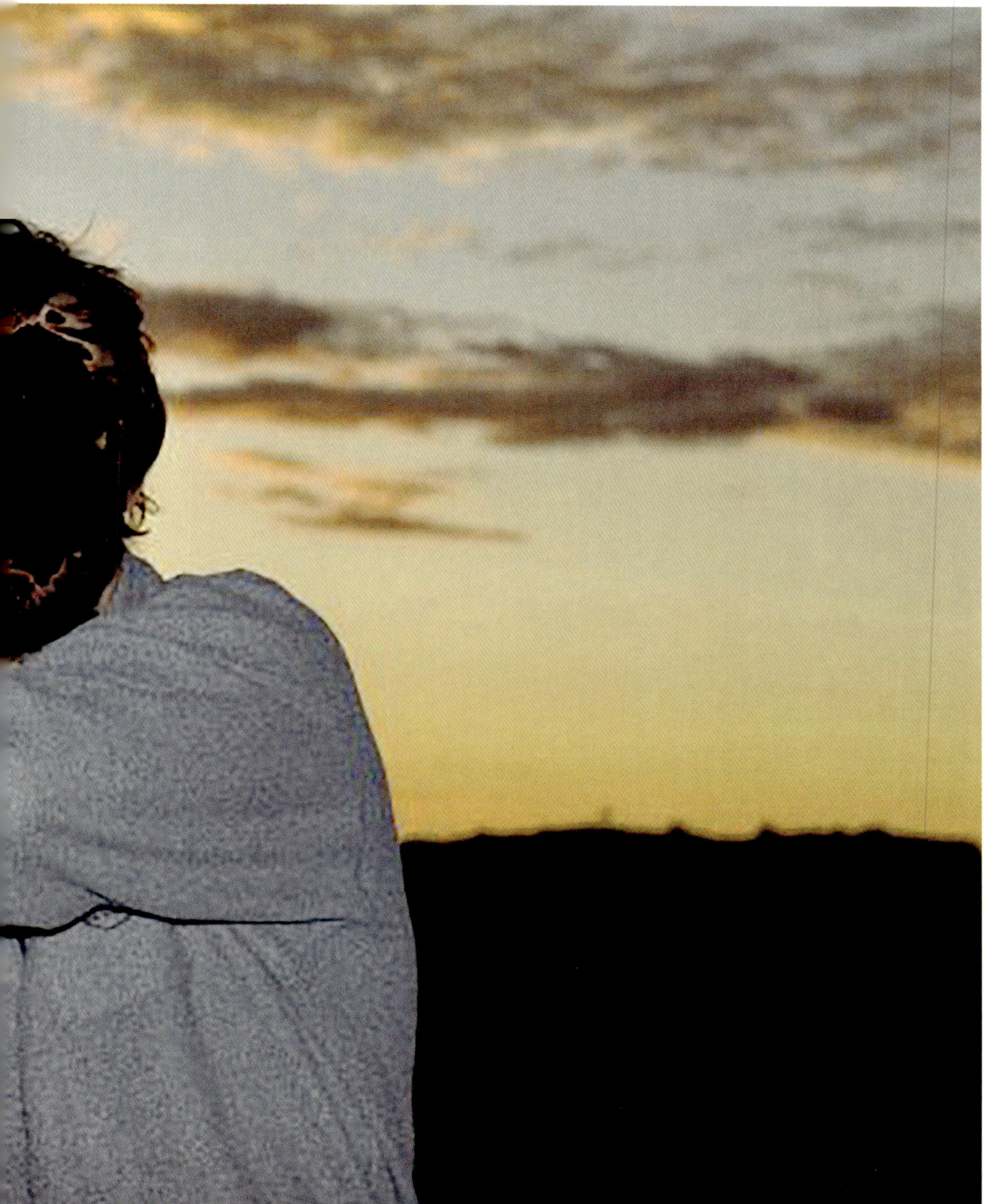

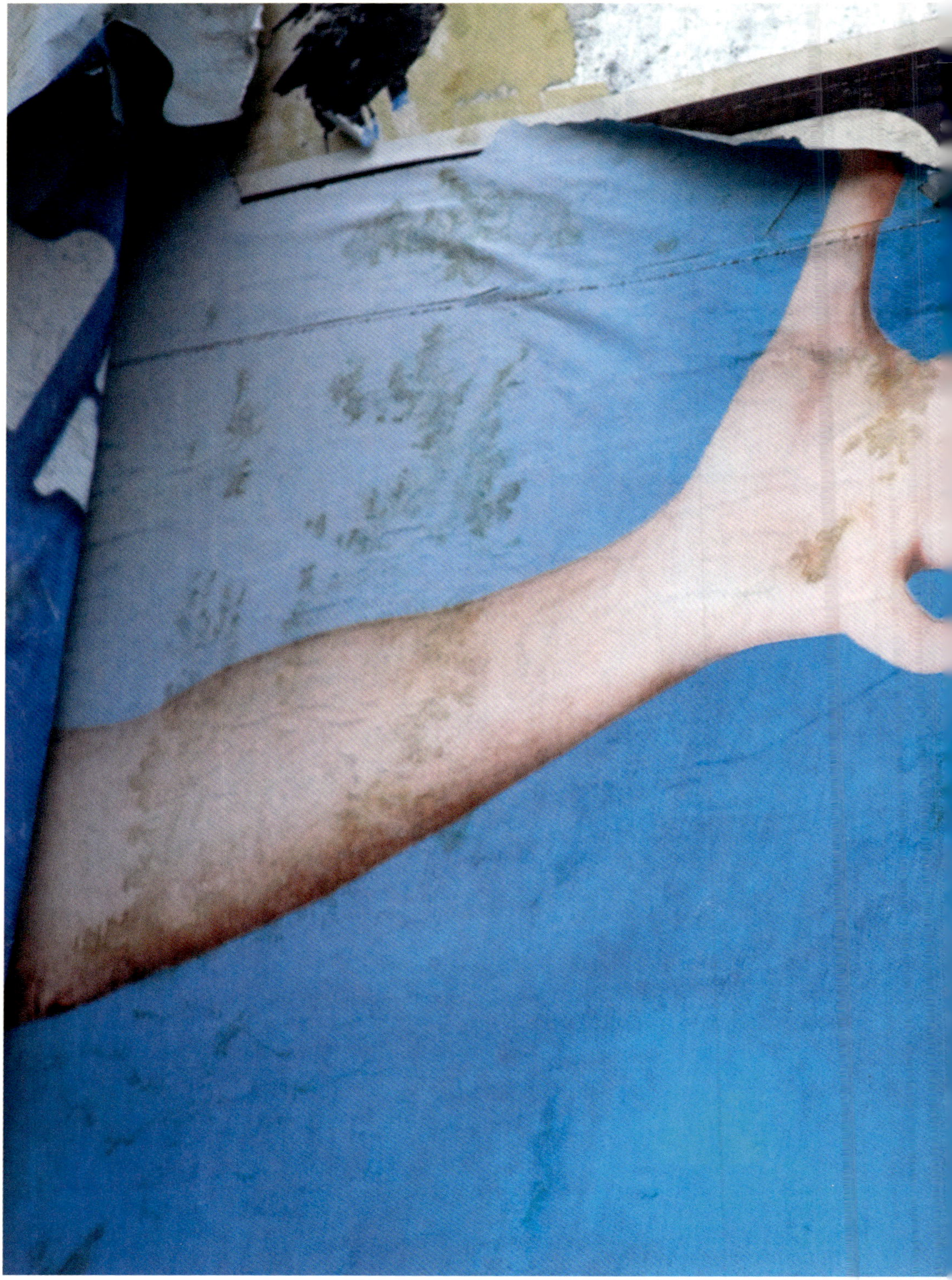

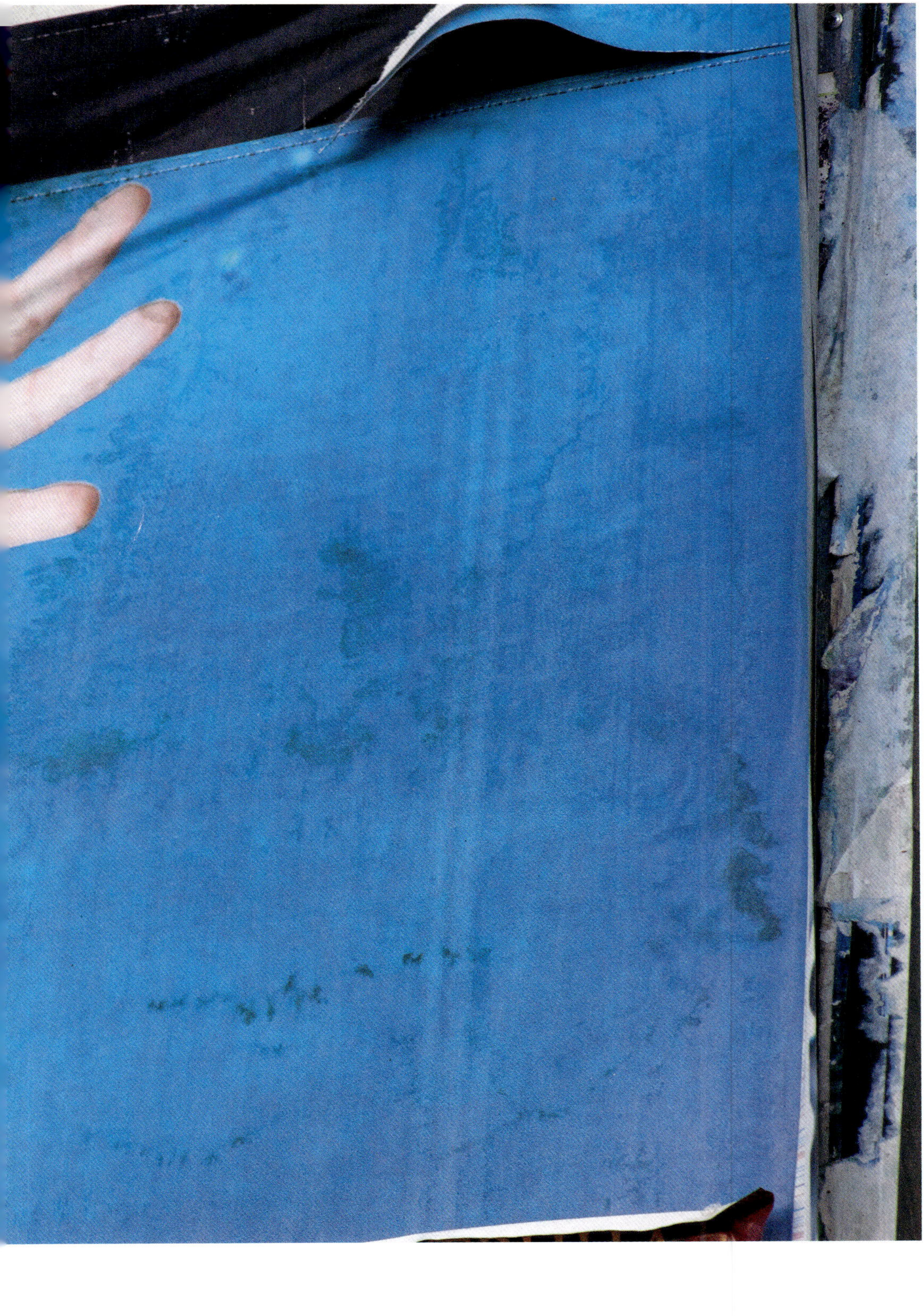

In the late 19th Century, Étienne-Jules Marey developed machines to register heart rate and detect lies, devices to animate and simulate the flight of birds, and cameras to record high-speed images of movement on plates of film. These latter included runners, leapers, pole-vaulters and cyclists, as well as a single sequence of nine frames depicting a handsome French gentleman of 1891 in a handlebar moustache, eyes shut tight against the bright sun, enunciating three words: *je vous aime*. These words, etched in blackened silver, are silent. But I hear them in my head. What was to be learned from this act of science, and to whom is the statement addressed? To Marey, who never heard them, or to me who thinks he hears them every time he views these frames? This sequence should break your heart. The proof of love materialized, documented, absent, unheard. The speaking lover, blinded by direct light, cannot see his object; the other lover, muffled under a dark cloth, will never hear the words. Perpetual intimacy, suspended in the logic of solitude: a photograph. *I love you as if under ice.*

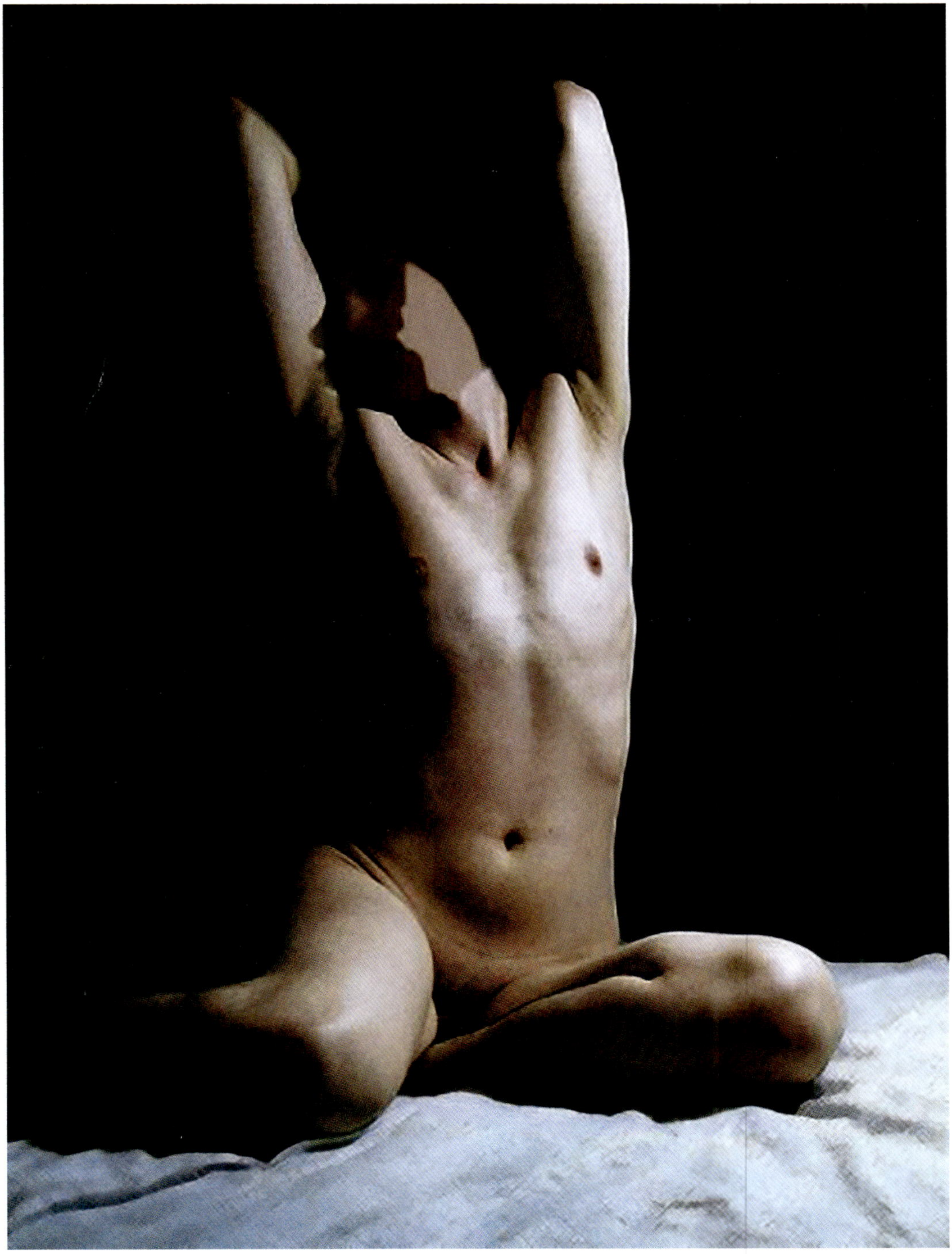

Every evening, alone after work, Vadim boots up and logs in, opening windows within windows, tunneling through flatness. Often, at first, there are only the clipped epistles of text, delivering stripped essentials: age, location, weather, hobbies, sexual priorities, family, job. But most evenings, despite the previous night's disavowals, Vadim eventually finds himself seeing himself again: a tiny, grainy, shirtless torso bathed in the formaldehyde-yellow light of his webcam's struggling sensor. In an adjoining window, a correspondent body, but not always another slight one like his own, startled little nipples clinging to the pale pink field of his hairless chest. Sometimes, a bolder companion peers out—or is it in?—with worried eyes. Sometimes, Vadim does the same, reframing his ethereal self upwards. This exchange of gazes is inevitably restless. If he stares into the other's eyes, he is not looking into the lens. Locking gazes is thus impossible, and the stranger's body competes with an estranged erotic vision of himself.

The first intimacy is solitude, but we are taught to forget it, to seek out others in place of ourselves. Vadim's correspondents have off-brand computers with broken microphones, or have turned the sound off because their mothers, like his, are asleep in the next room. They meet in silence, each offering himself to himself, as much as to each other. They share a language of curt waves, fugitive smiles, some pointing at objects or parts; a transcontinental shrugging of hands, the raising of eyebrows, some shaking of heads. Questions, compliments and requests are typed back and forth, like intertitles for an existential drama. Occasionally, things escalate: flashes of erection, a tongue traveling lushly over dry lips, a hand tracing slowly across a chest. At such moments, Vadim likes to expose one armpit, hand braced behind his head to give the bicep a small bulge. This pose, he believes, is masculine and sensuous.

Very rarely, there is a kind of consummation: watching each other get off. Not so much encounters as paired offerings; two solitary transmissions launched in homage to the touch. A spectacle of the flowering of solitude. Neither lover ever leaves his chair. One thing both know: each of them is alone.

The ghost image, burned into the screen before purchase, reads: *you are always alone* or *you are never alone*. The ghost image is difficult to read. The surface of this ice has two sides.

One afternoon in L., I noticed that a street-level window near the sidewalk ahead of me was open. My eye caught the slight billowing of a gauze curtain and perhaps a touch of movement. Looking to my side, not a foot from me as I passed, a girl—barely thirteen, chubby-cheeked, bespectacled—stood in the window, looking outward, framed from the waist up. Just below her heart she held an illuminated tablet, screen facing towards the street. There I glimpsed the upper half of an older woman holding a toddler forward in the frame, as if to present her, at the level of her belly. The baby, the woman and the girl all saw me through the window as I passed.

Returning to the capital, in an open stretch of road between mountain passes, the bus stopped abruptly by the side of the road. Two young women with blond ponytails descended onto the verge, quickly separating to left and right as they charged urgently into the low scrub and sparse trees of the adjoining landscape. The entire population of the packed bus watched in blank fascination as the women—one in a snow-white parka, the other in a safety-orange windbreaker—unsuccessfully sought camouflage on the semi-barren slope. Finally, about forty feet from the road and fully exposed to the bored, rapt eyes of every passenger, the two women squatted in opposite directions to urinate. Nobody laughed or moved or spoke during this desperate episode. The women returned on board; the journey continued.

Across from me on this bus sat two other women in brightly-colored hijabs, knees huddled together to support a single tablet. They were deaf, and their four sets of fingers hovered antically over the device, alternately tapping and swiping *on* and signing *above* the screen. They flipped fervently through photographs I couldn't quite see, swooping in, and pulling back, looking and speaking all at once, without uttering a single sound. Touch, gesture and language melting the ice of the screen. A spell to swipe away the desolation of the surface.

The surface is a river and the world courses through it. The surface is a stream of light, or it is dark, inert, a blank. Everything and immaterial, or nothing at all in your hands. In fact, each tablet is graven, illegibly, before we even begin. The message can be unlocked with a private passcode, late at night, face bathed in the ritual blue screen-light of immanent enlightenment. The words rise to the surface, a fish from the deep, knocking against the icy glass with a glint of silvery sheen and disappearing back to darkness without capture. By morning, the words are a faulty memory: *we are alone together*. Or was it: *together, we are alone?*

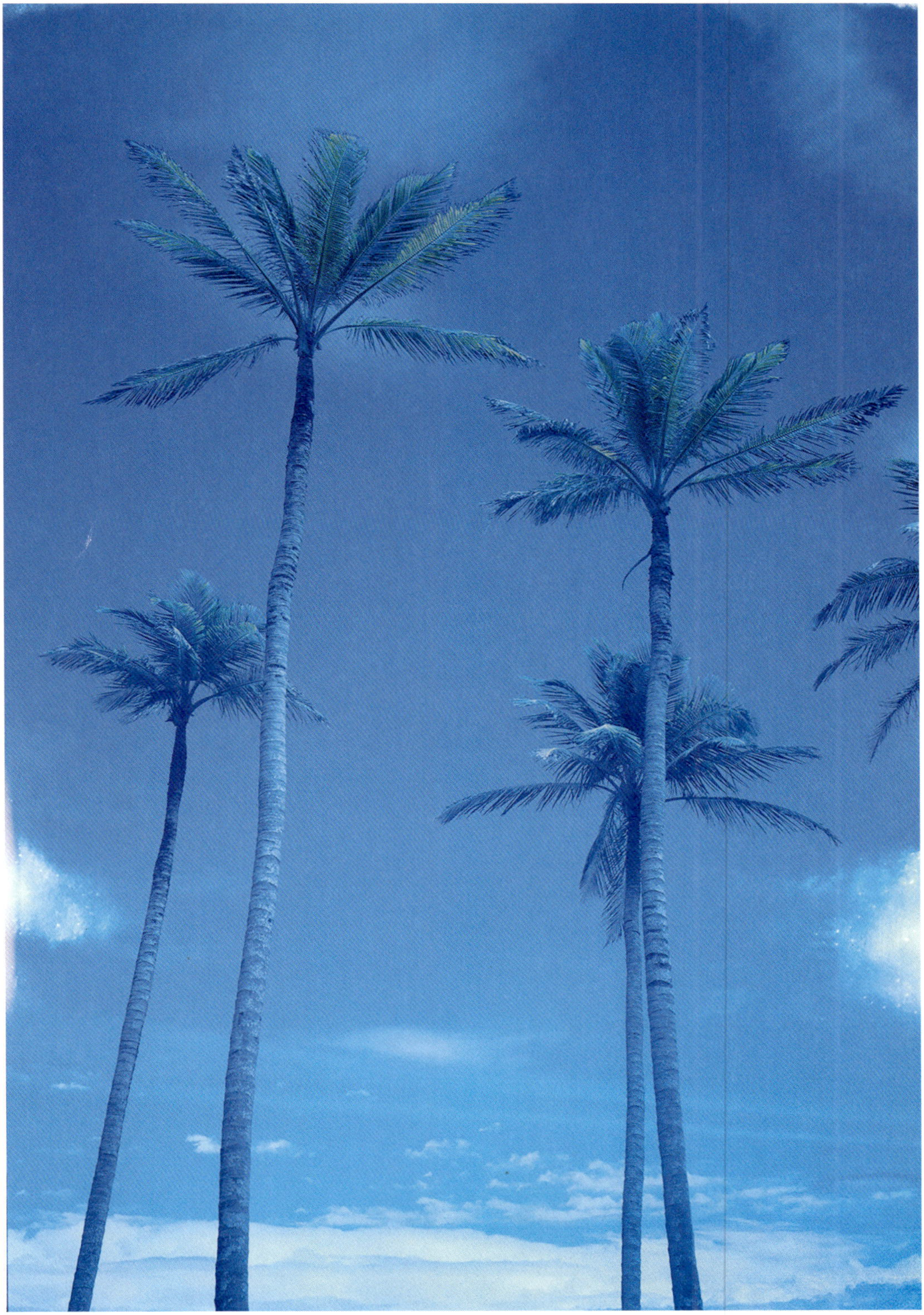

Chapter XI

Late in the afternoon, in search of a beer, I came across Isabel in the pavilion. She smiled crookedly, perhaps happy to see me, or perhaps out of misaligned politeness. I asked about her day; but she had been thinking about something else—her dream from the night before. Unbidden, she told me. More than usual, her English was a little troubled—broken and emphatic at the same time, with a dramatic surfeit of extra pronouns. She was speaking the dream, not telling it. I let it flow past in silence. That night, asleep, I saw it again. My dream became the cinema of Isabel's dream, and here is what I saw:

In Managua, when Isabel leaves for the airport, it's as if she is being stripped of things, ceremonially, one at a time. Walking out the door, stepping on to the bus, entering the terminal, checking in, handing over documents, emptying pockets: common acts amounting to a dumb show of sacrifice. Each action becomes a gesture; each gesture a rite: renunciation.

Alone after all that, in a diamond-paned atrium, Isabel pauses for the barest moment, and for that kaleidoscopic instant I see her from all sides. Then, she begins to walk towards the plane: from the bright vestibule to a broad hallway, narrowing to ever-duller passageways. Isabel moves with tremendous lightness, as if sprung from an Egyptian frieze into the freedom of modern space. She whips across mechanical walkways, swinging effortlessly through automatic doorways. Finally, before another sanitary and inestimably long jetway, there is a last checkpoint: an unmanned stanchion, a gateway left to the chemical winds of these endless interiors, flanked by stainless steel bins, above which hang immense laminated signs, commanding her to leave all liquid things behind. Looking down, Isabel realizes she isn't holding anything at all. She freezes; backs up a step, unsteadily. A stranger, in a dark raincoat and trailing a bag, sweeps through the doors behind and passes, undeterred. Now she is absolutely alone for what I understand to be the first time. A sudden panic courses through her. What were those things that might have been called hers? It did not matter. She is without them. It is impossible to undo those rites of surrender, to pass back through all of those thresholds. There is no ritual for reburdening. All she can do is carry on, weightlessly, down the carpet that sweeps towards a capsule that takes her away, with a faint hush, on fixed wings. Equatorial sun strikes the fuselage as it climbs. Isabel evaporating in arid incandescence.

She finished the dream, and a moment of rigid stillness passed between us. Finally, Isabel shifted, ran her flattened palms in opposite arcs along the table, and held my eyes, not quite casually, as if to say: don't you understand? Thus challenged, I clung to facts. When had she last visited Managua, I asked. The banality of the question seemed to disgust her, as if I had proved too stupid, even, for her simple dream.

"Two weeks ago," she replied, acidly.

But a moment later, her face abruptly brightened: "I bought the most beautiful piñata there, for the birthday party!" She was grinning now, almost indecently. "A donkey, of course. Completely covered in sequins!" Here, she illustrated its splendor with her own body, in a sweeping gesture from neck to knee. "But do you know," she continued, "when you buy a piñata, it comes completely empty! I had no idea. And there were the children, battering that shiny donkey apart, and nothing inside! *Desastre!*" Here she laughed and rose in one fluid movement.

Pushing herself up with her palms from the table, Isabel turned and carried herself away.

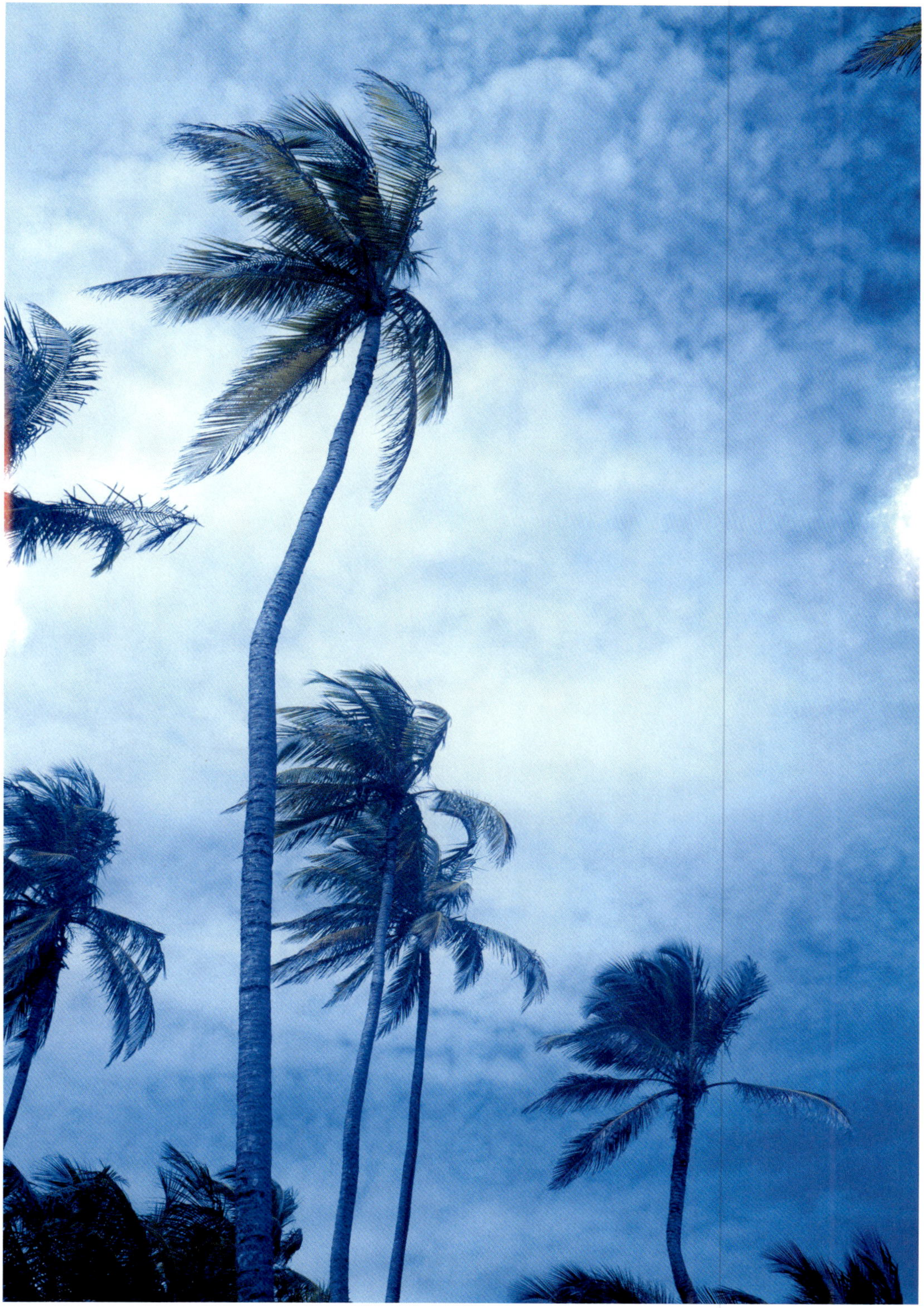

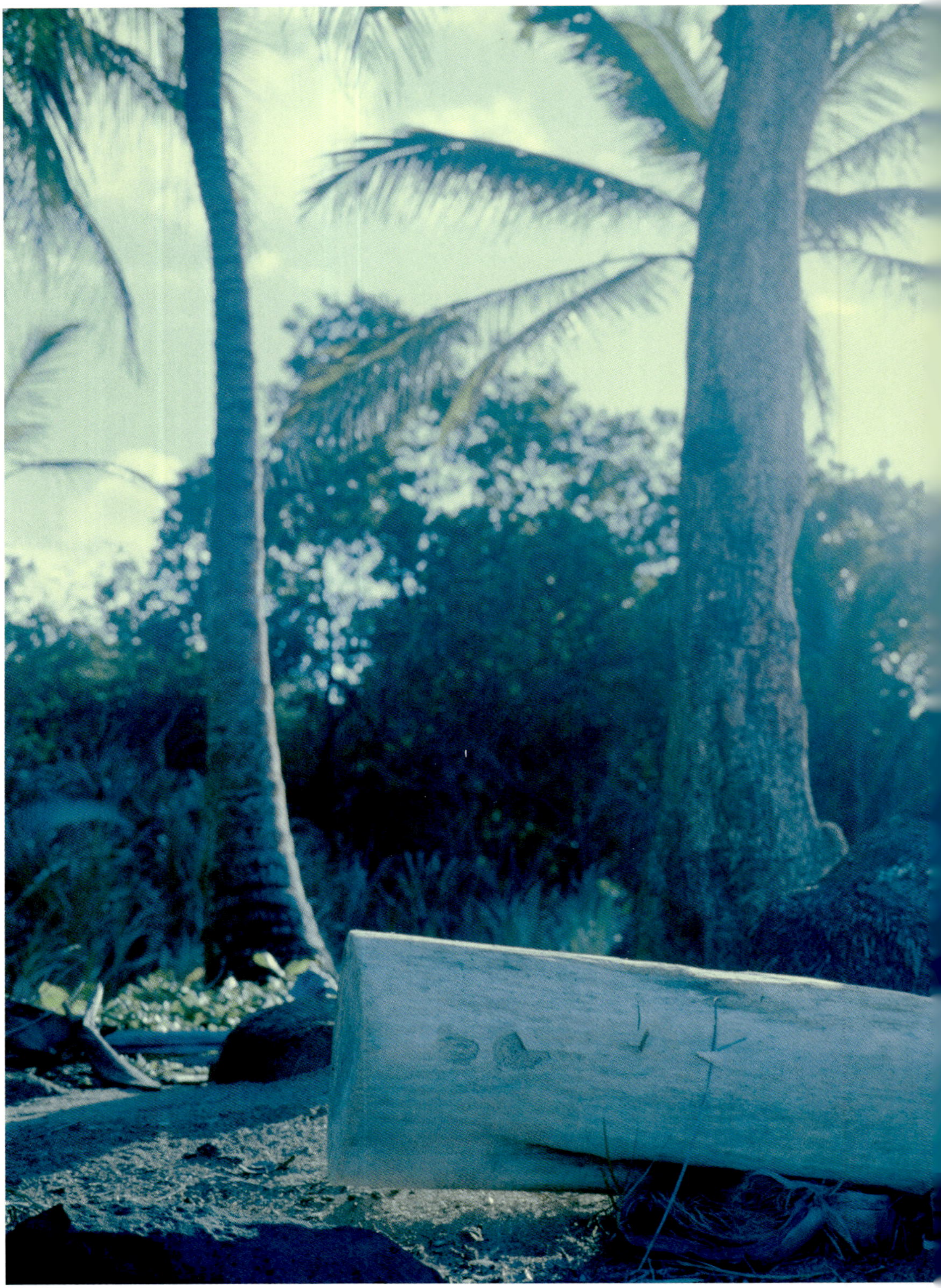

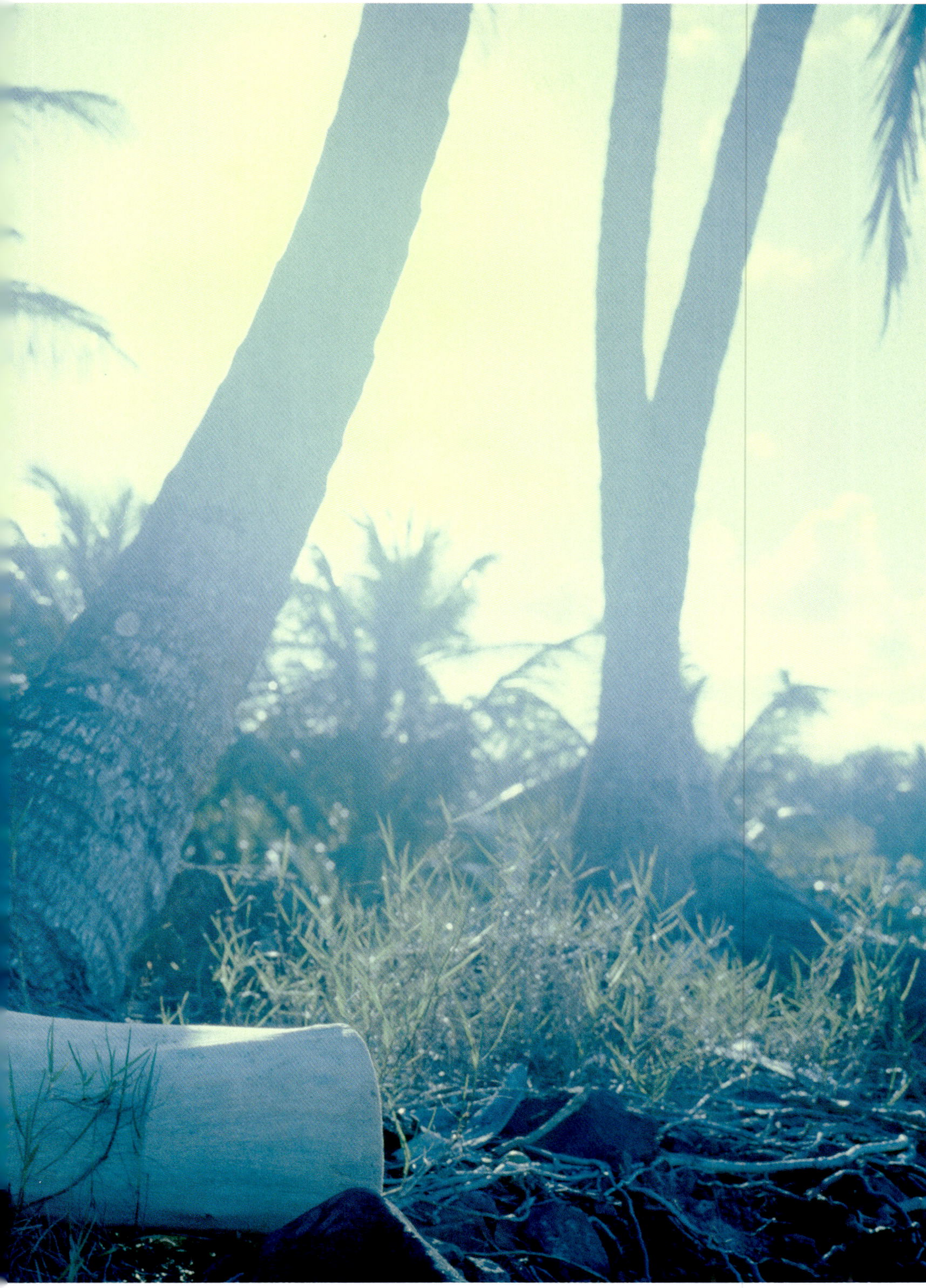

I depart early morning, threading the channel eastward, towards the clouded sunrise, through the shallow reef that leads to open water. Accelerating southward in safer depths, I look back at the beach, and there is Isabel, erect at the water's edge. I see that she is standing, arms stretched above her head. I can make out the rioting colors of her skirt. But her face is beyond my vision. Is she following my receding white boat, or taking in the bright, tender, blue-grey horizon beyond? Caught in the prism of her perception, do I signify or simply disappear? I will never know.

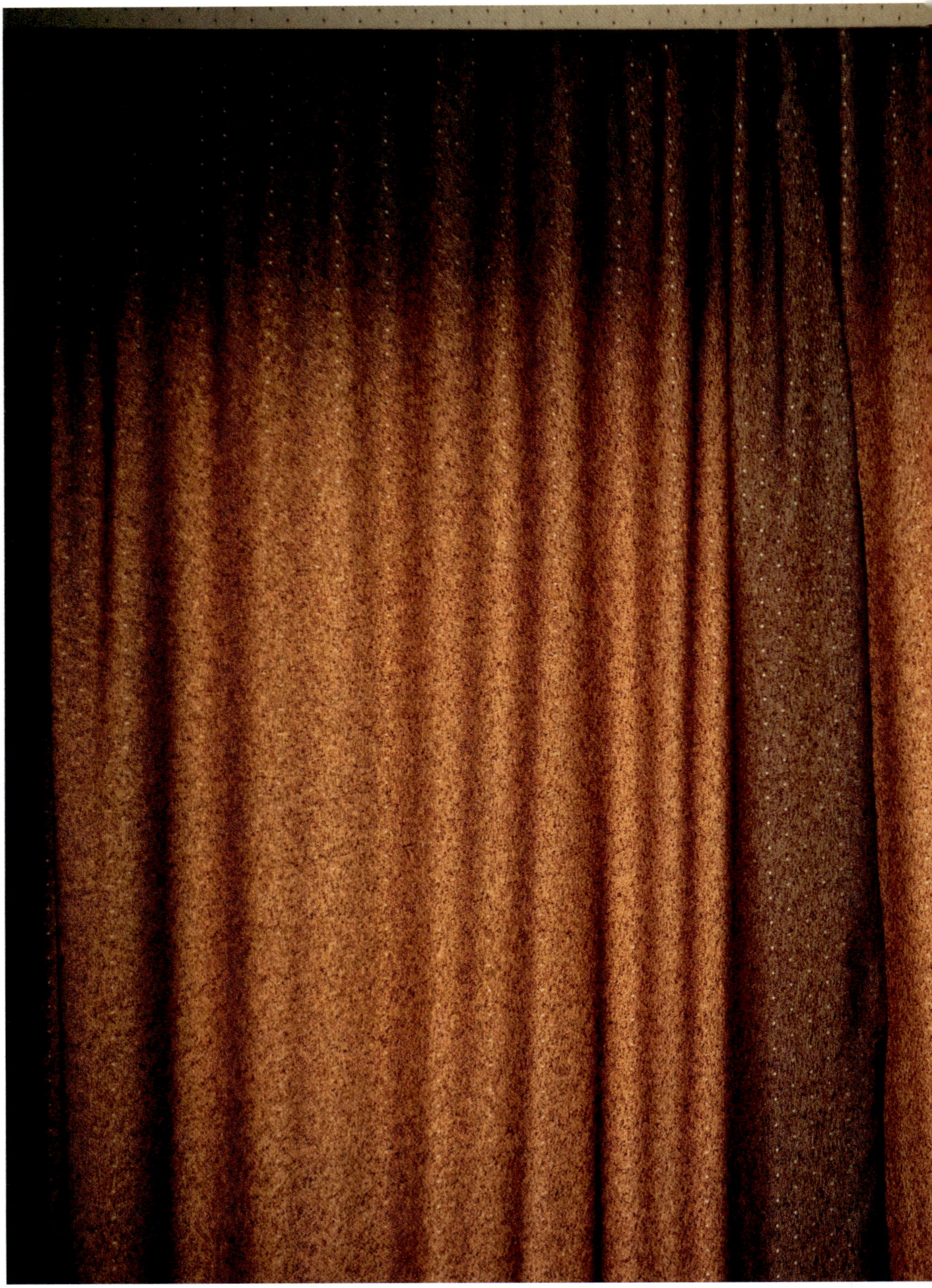

Chapter XII

I would like to assert the following:
In 1841, a few months before his death in a duel, the great Russian Romantic poet Mikhail Lermontov wrote a verse about solitude, love and the internet.

This strange missive from another age begins with the poet imagining himself as a young officer, shot in the heart and dying alone in a dusty valley in Dagestan, the borderlands of southwestern Russia. He falls asleep, his life bleeding out in the pitiless sun, and dreams of a glittering Saint Petersburg soirée. A group of carefree young women banters lightly about the absent soldier. But one among them drifts off into a melancholy daydream. There, she sees the officer, lying dead in a dale in Dagestan, the wound still smoking and his blood beginning to run cold.

Let me make it clear. In this poem, the poet imagines that he is a dying soldier, who dreams about a distant woman, dreaming she sees him dead. A remarkable telescoping of projections, in which the further the imagination moves out into the world, the deeper it retreats inside.

And thus it repeats, the ritual of running away from oneself and returning, of dashing into oblivion, which is death but also the world, and retreating to oneself, a solitude; trapped against a mirrored wall.

Sometimes, people lead lives of solipsism by force, and other times, by hazard. Sometimes, in our world, even the idea of the other is interior. Still, we extend ourselves in the ways we can; running out from ourselves, and vanishing inward; venturing out, and disappearing.

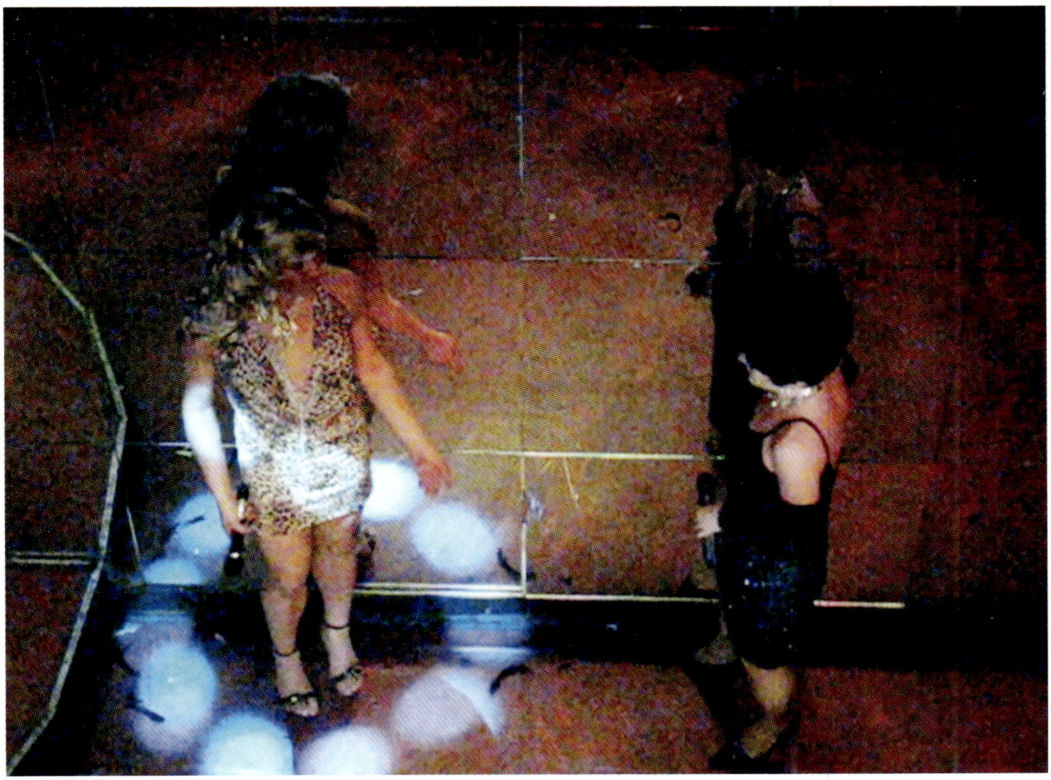

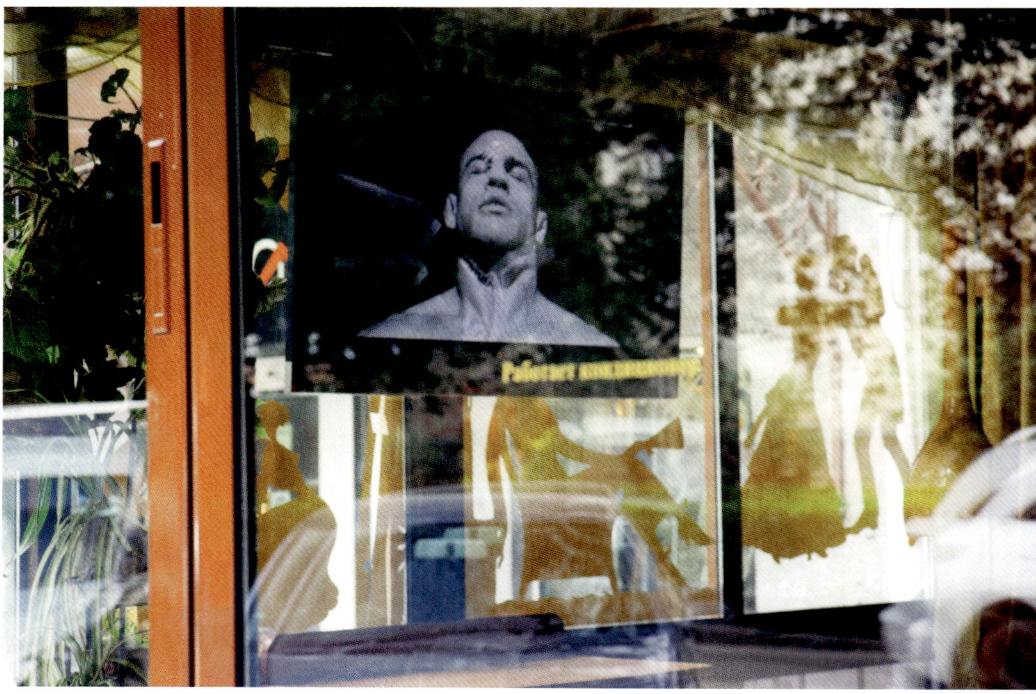

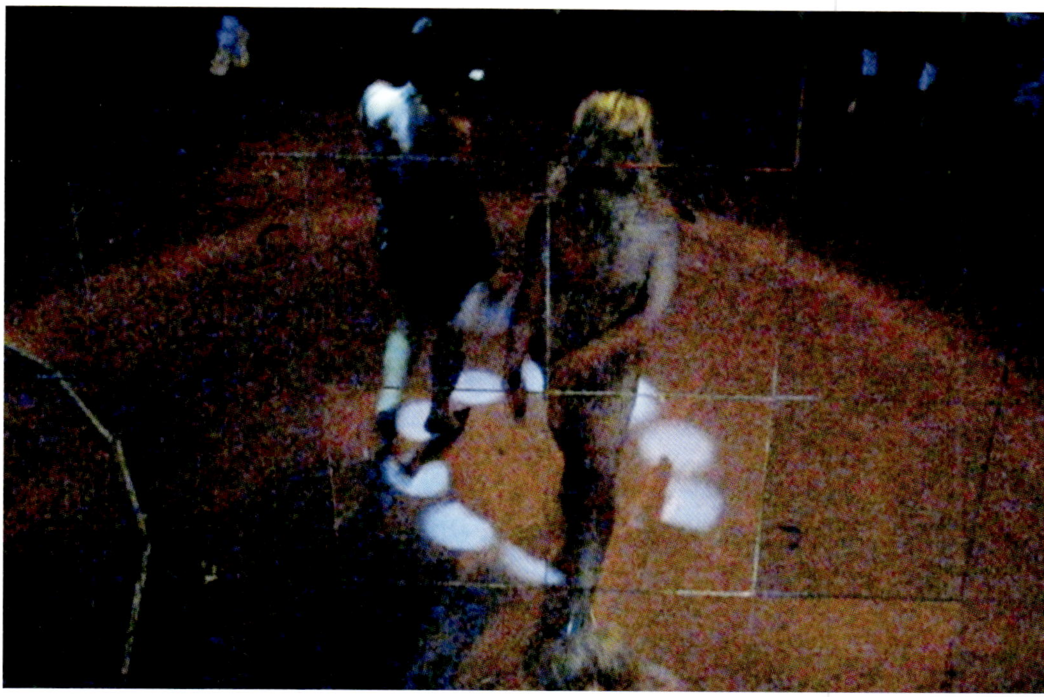

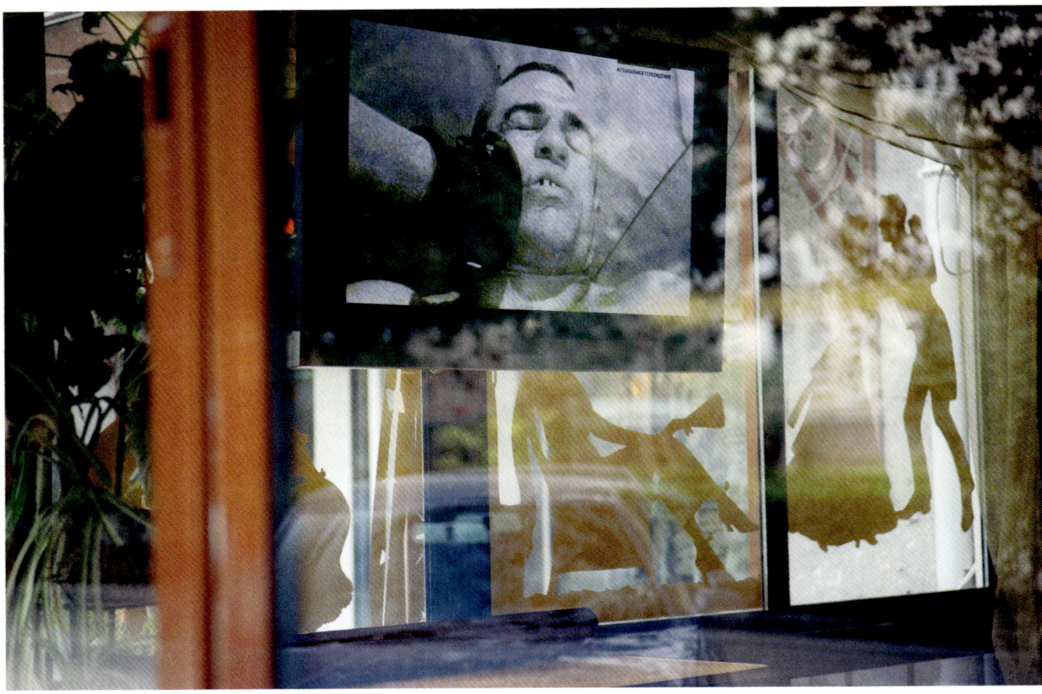

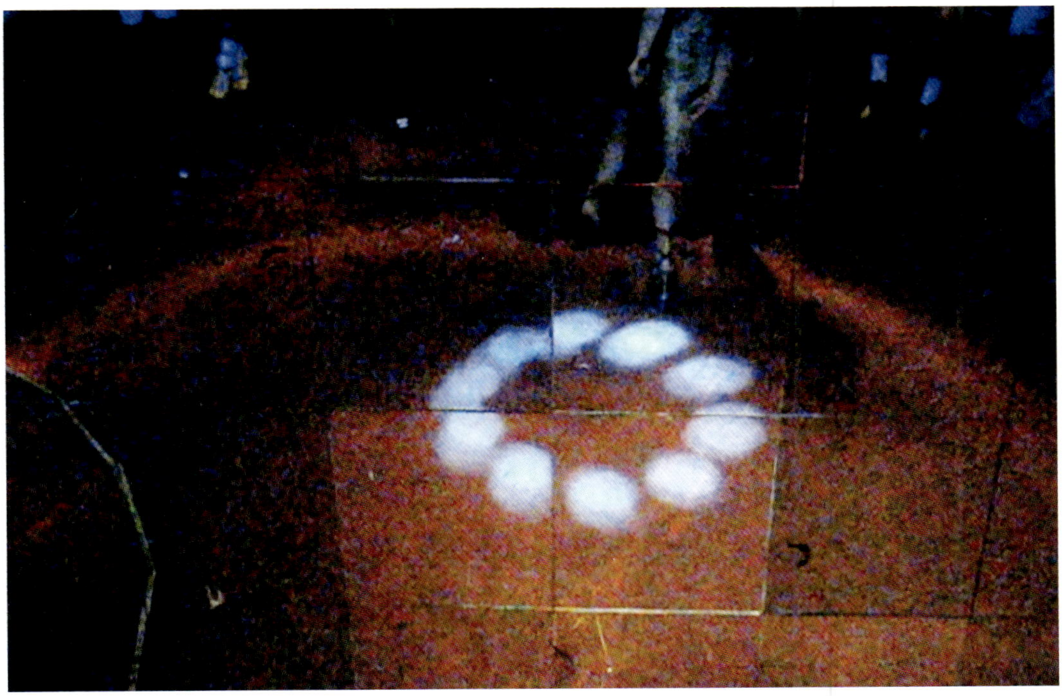

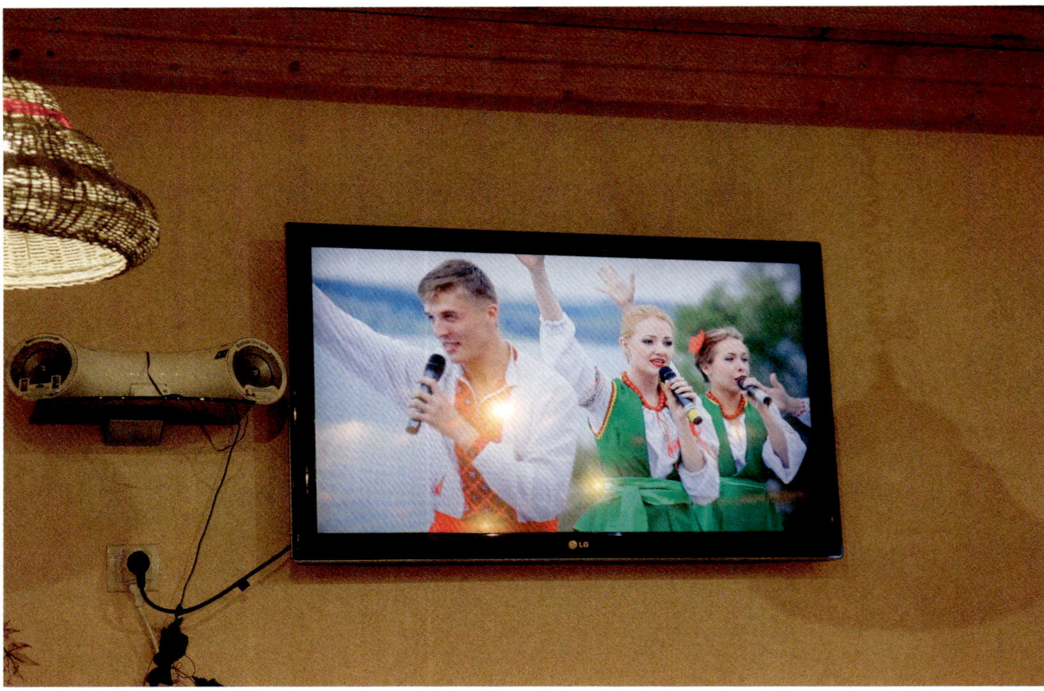

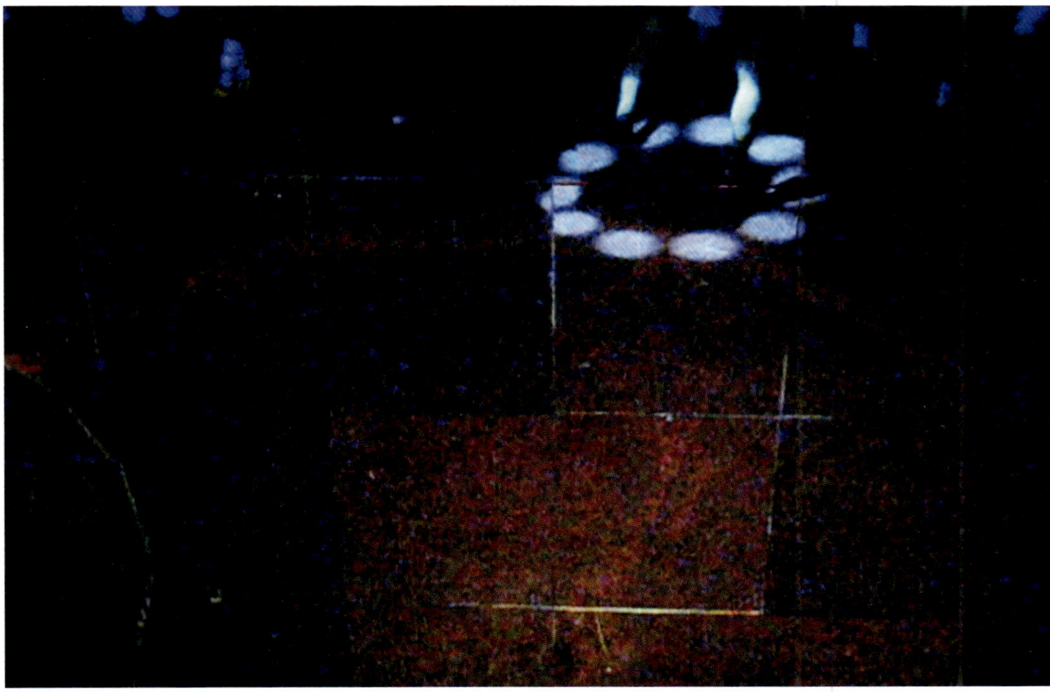

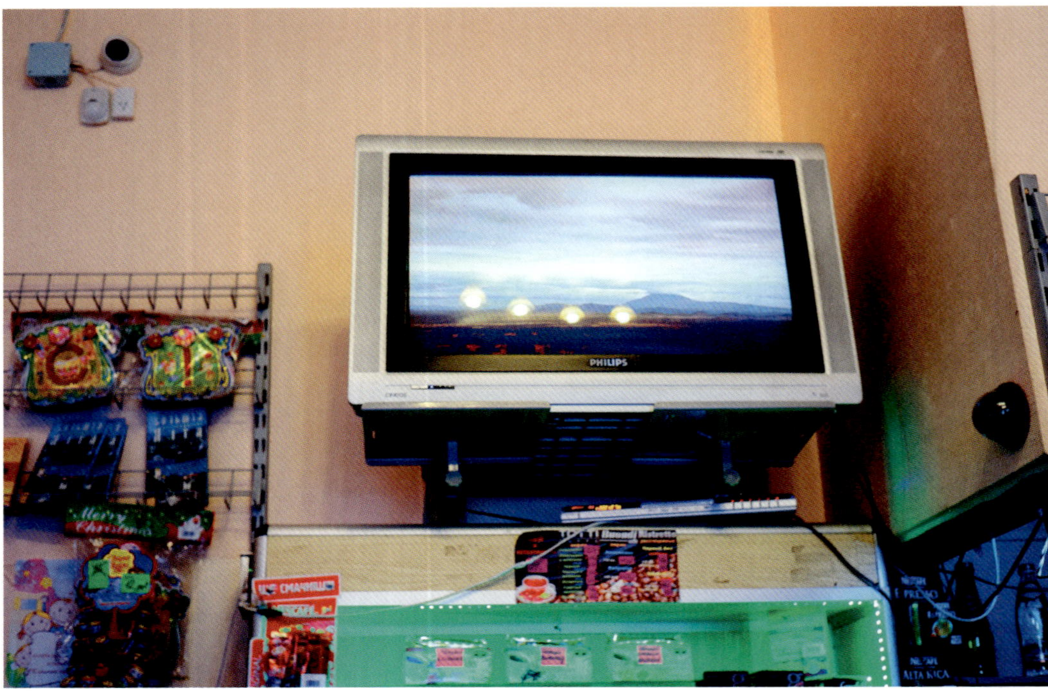
PHILIPS
Merry Christmas

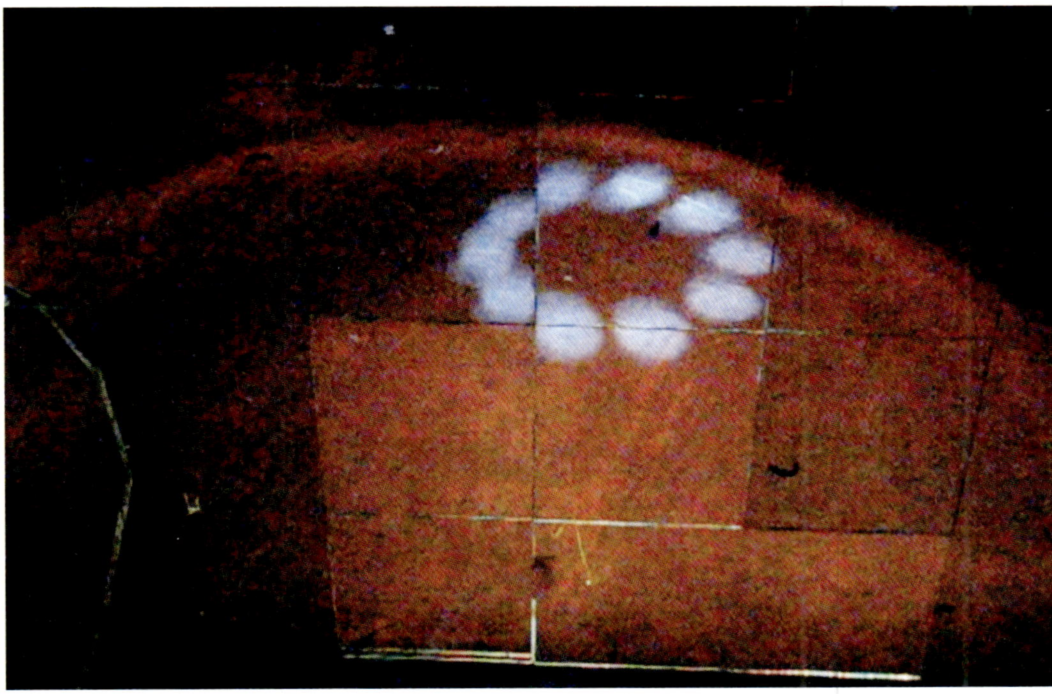

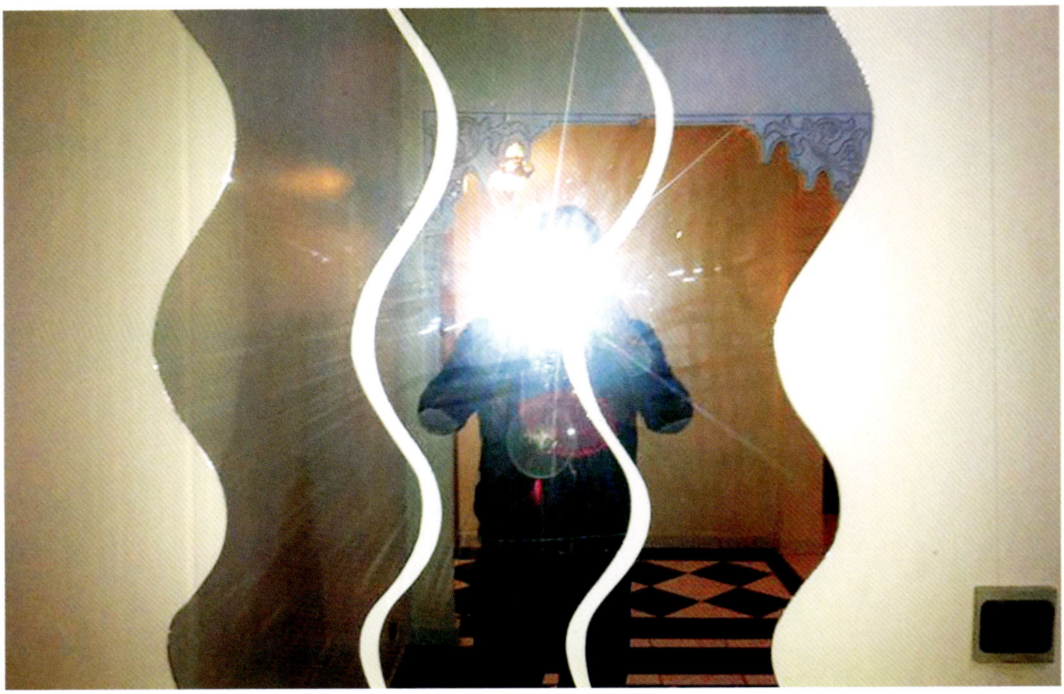

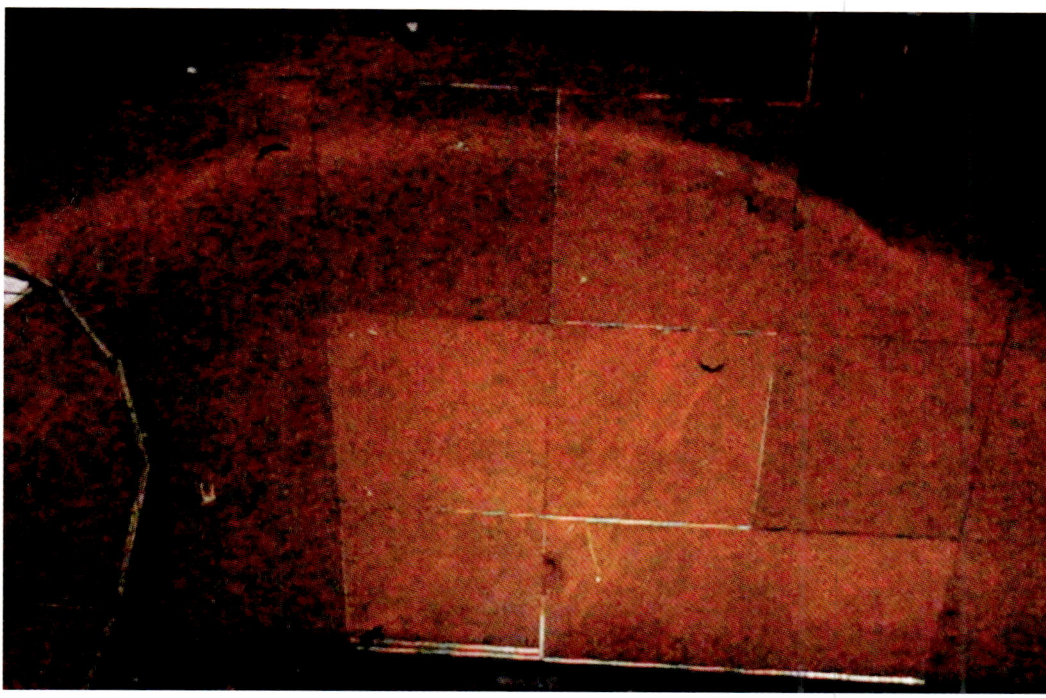

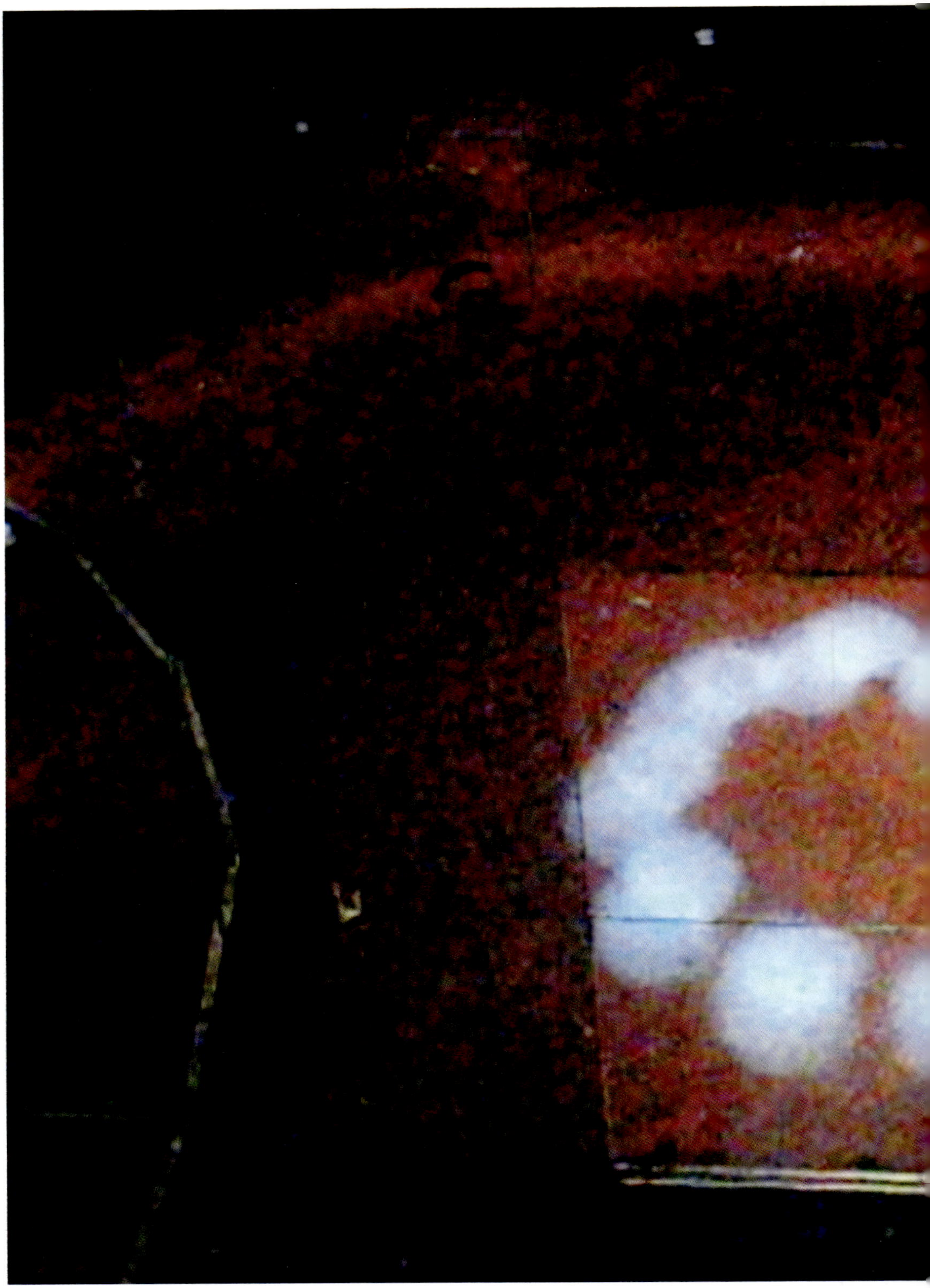

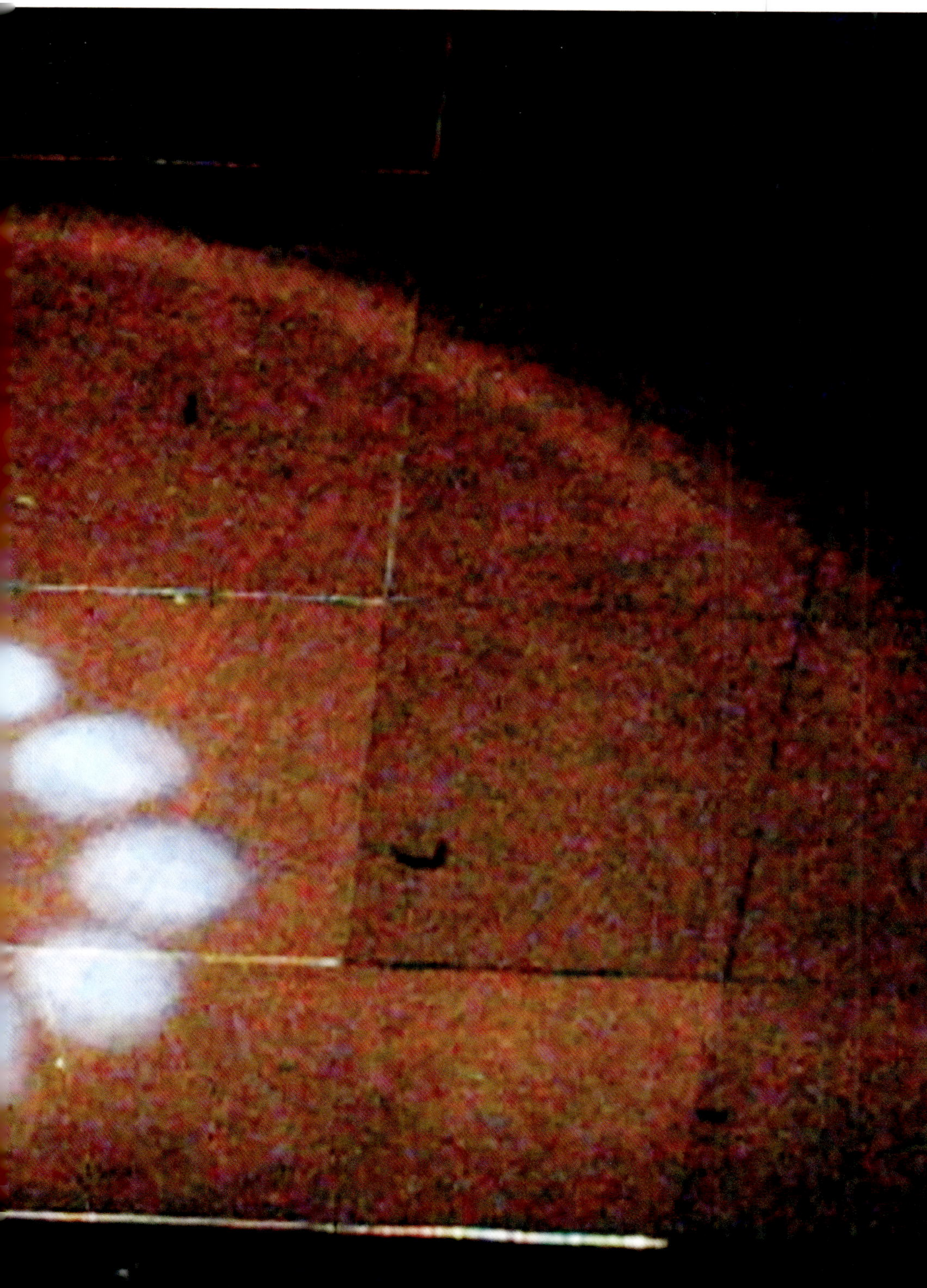

In the end, we never met. Stepa and I had corresponded, on and off, for over a year. But when I passed briefly through his town, the timing was off. We spoke on the phone, we texted, and then I was gone. Once, months later, he messaged me in English:

it is a pity
but in Crimea really a lot of very good and beautiful places, we have a very mild climate, and we have almost nothing to fear, we have very few poisonous insects, predators and other dangers. We can say that there is nothing to be afraid of us) except some people are not very adequate, but such is everywhere) And of course, everything is much more beautiful in the spring when everything is green and flowers. Summer is also good - warm but not too hot, the mountains and the sea)In the sea is not something that the ocean-there is very safe, no not poisonous jellyfish or predatory fish, and of course the wave-very small and tender. And at night there comes a pleasant freshness, and if you know the place is so beautiful stars that can sit up all night looking at the sky and breathe clean and fresh air. Highly recommend to visit the Crimea from spring to autumn - winter because we only ugly to go, because we have quite warm and the snow is long and melts quickly turning into mud and dirt that lasts from December until the spring thaw when it begins to dry out
I have a singular anything interesting) Slightly changed the mood, to collect the bike who wanted ride and realized that they no longer want it)))
In general, sold, what is collected through the autumn, is now saving up money again, but on a motorcycle)
other than that, all is quiet - work, home.
met an interesting guy, and now we meet. He recently returned from Kiev

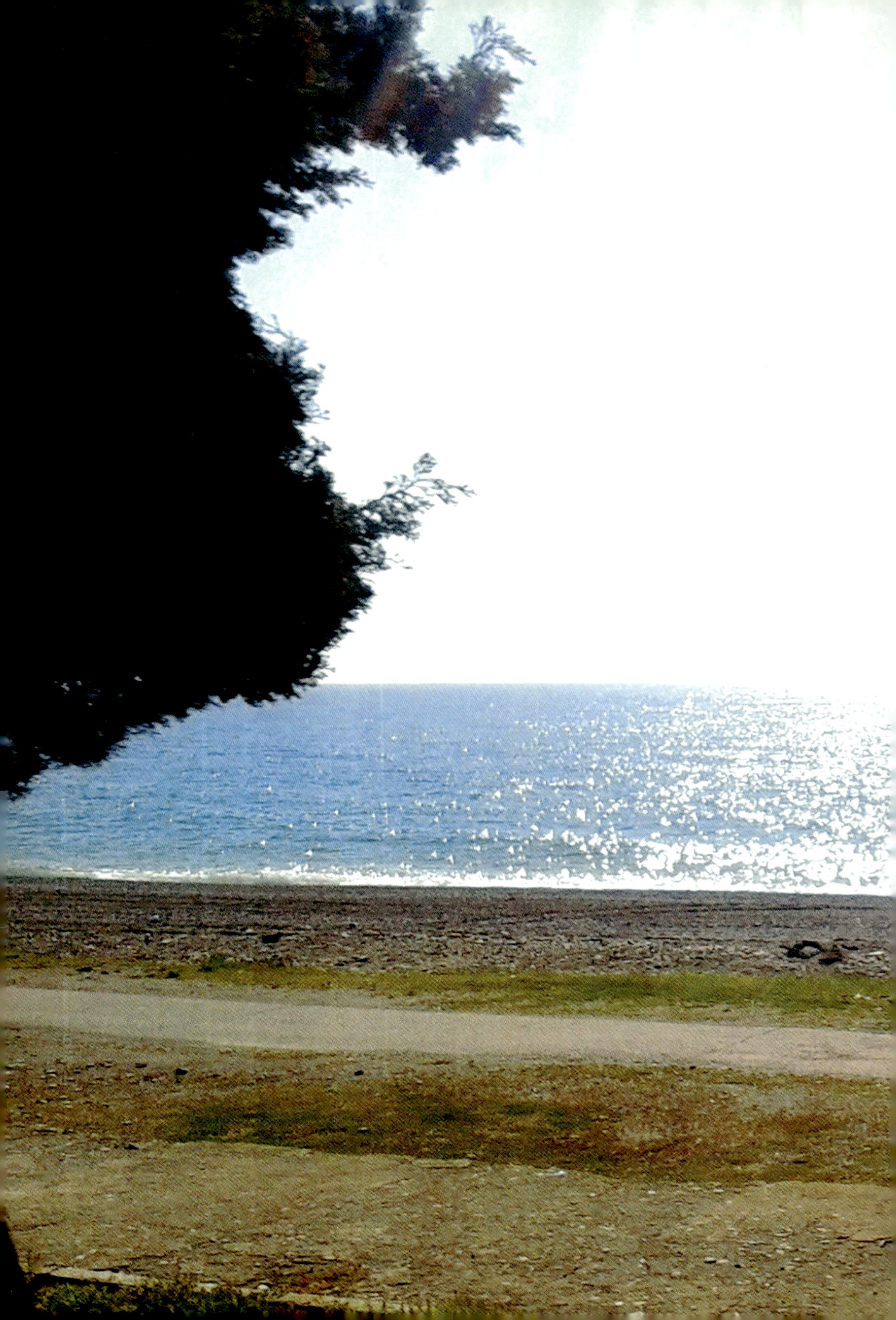

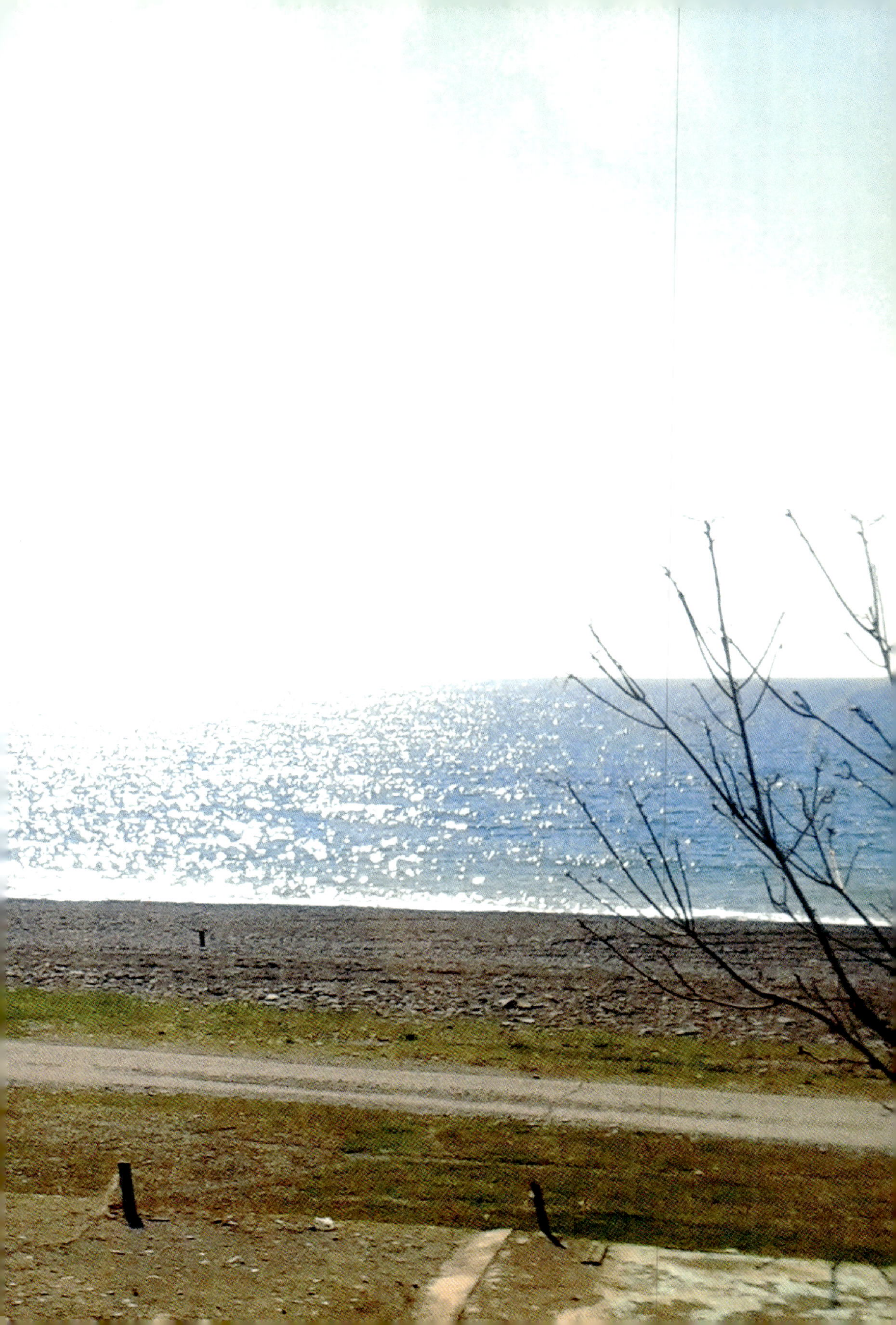

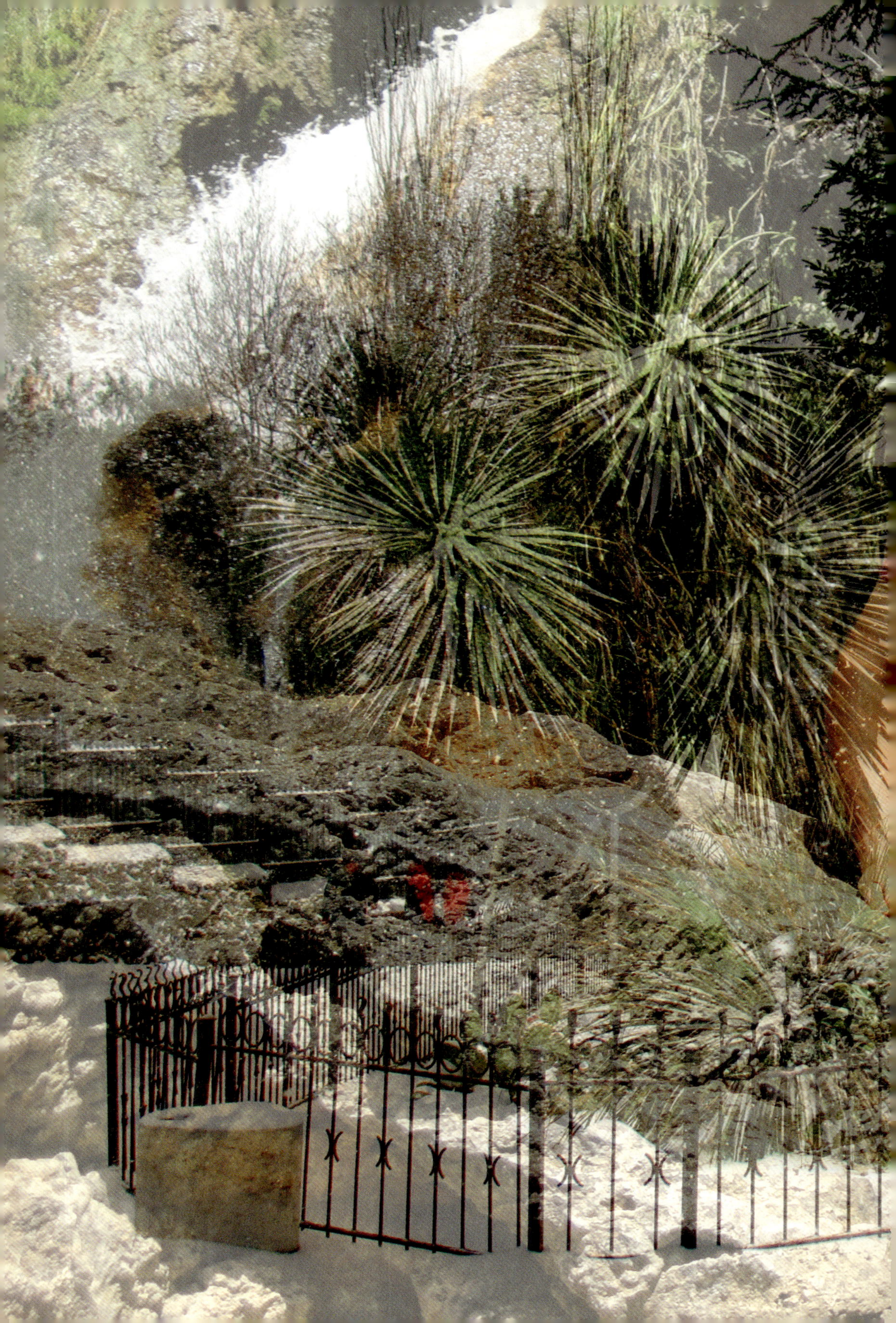

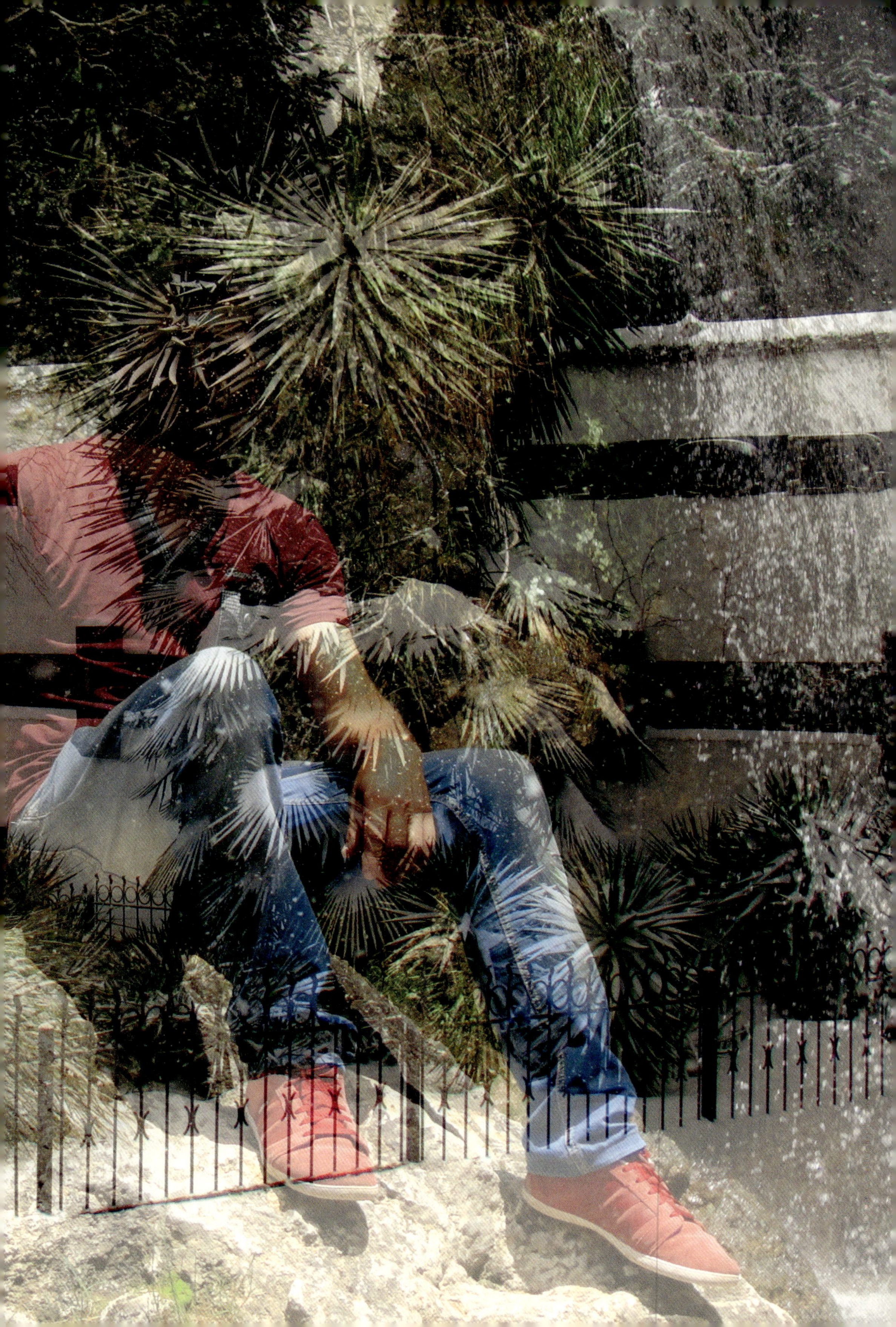

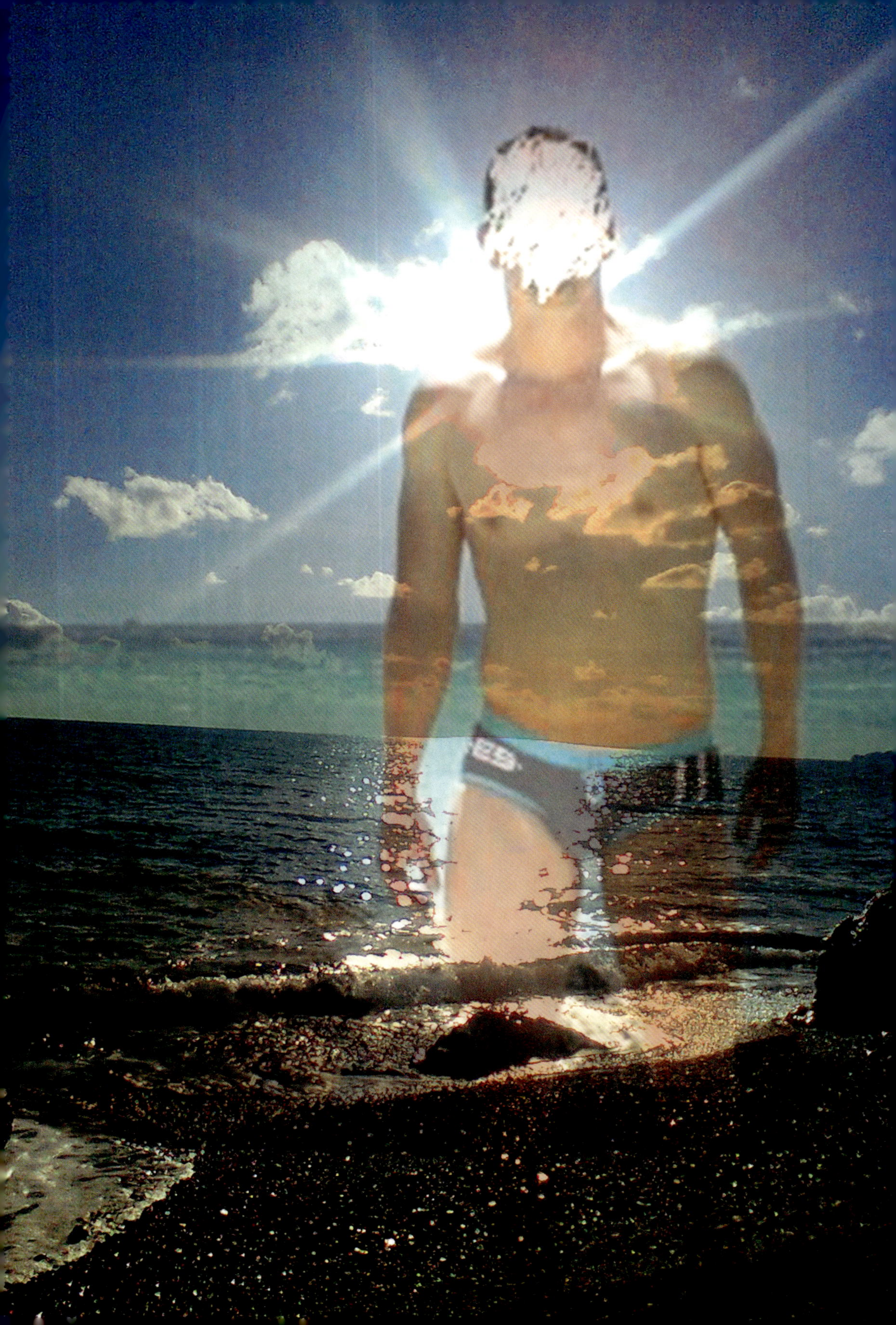

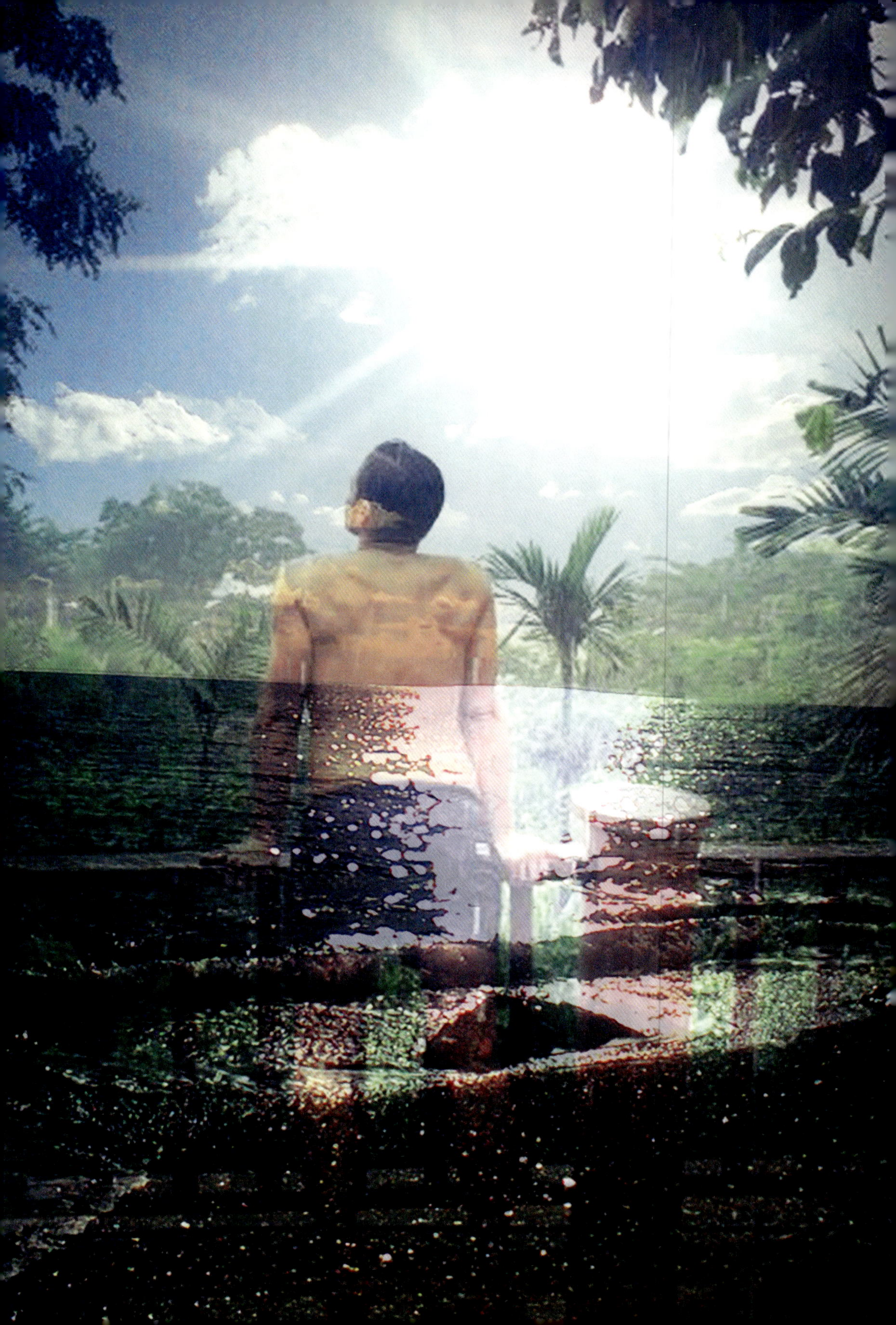

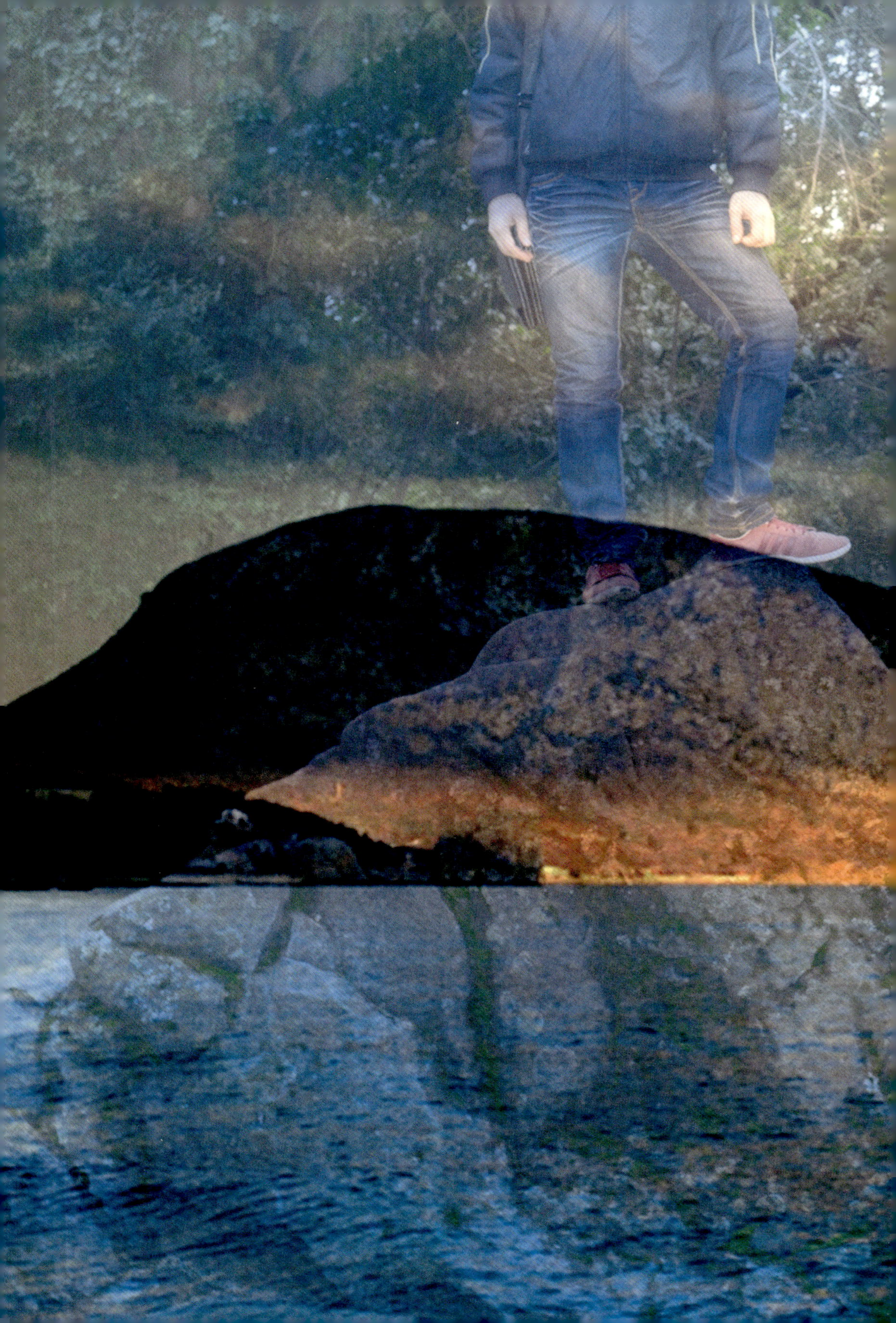

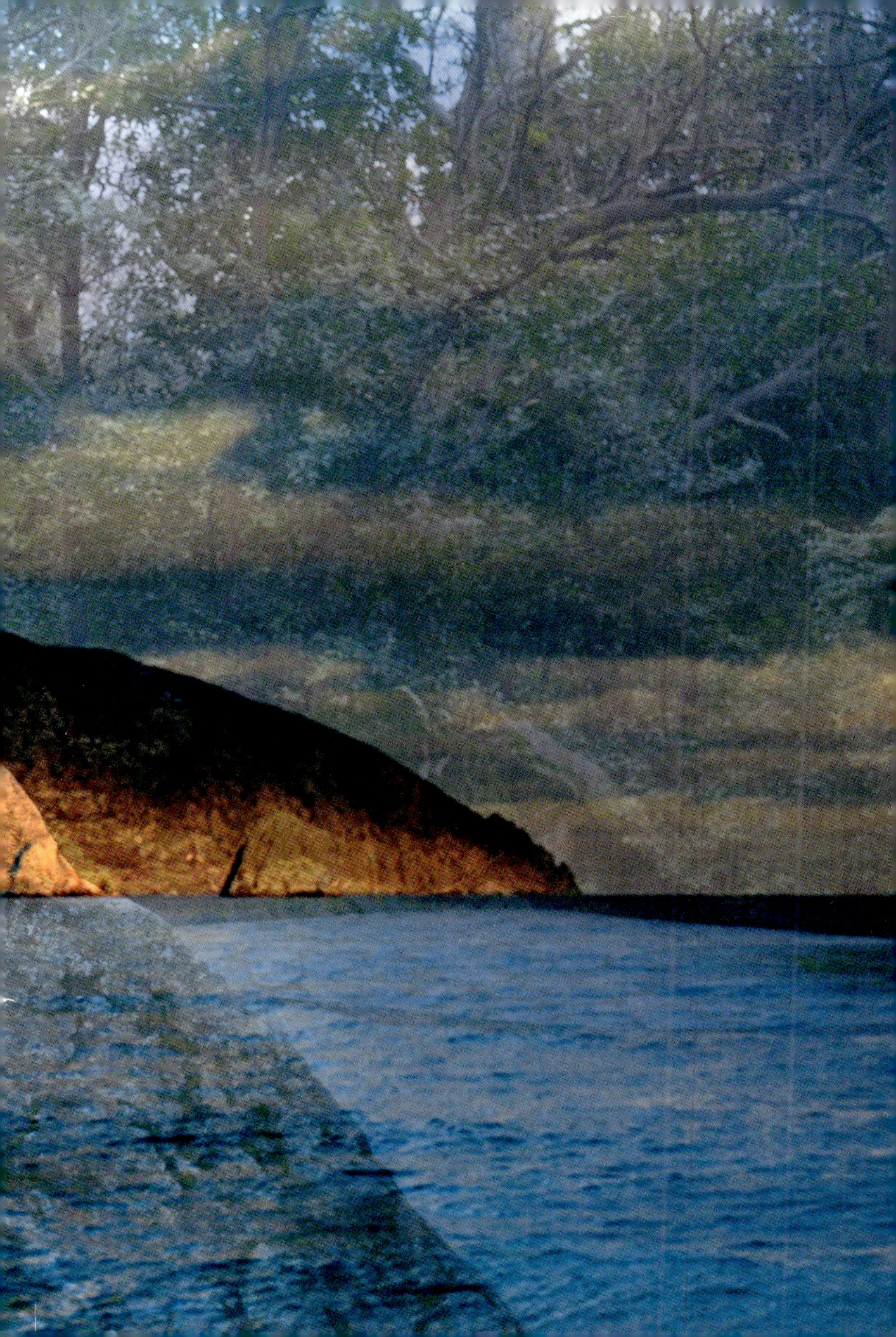